C0-AVX-483
3 3073 00288823 6

Cadmium and Health: A Toxicological and Epidemiological Appraisal

Volume I
Exposure, Dose, and Metabolism

Editors

Lars Friberg, M.D.
Professor
Department of Environmental Hygiene
Karolinska Institute
Stockholm, Sweden

Carl-Gustaf Elinder, M.D.
Associate Professor
Department of Occupational Medicine
National Board of Occupational Safety and Health
Solna, Sweden

Tord Kjellström, Ph.D. (Med. Dr.)
Senior Lecturer
Department of Community Health and General Practice
Univesity of Auckland
Auckland, New Zealand

Gunnar F. Nordberg, M.D.
Professor
Department of Environmental Medicine
Umeå University
Umeå, Sweden

CRC Press, Inc.
Boca Raton, Florida

Library of Congress Cataloging in Publication Data
Main entry under title:

Cadmium and health.

Includes bibliographies and index.
Contents: v. 1. Exposure, dose, and metabolism —
v. 2. Effects and response.
1. Cadmium—Toxicology. 2. Cadmium—Physiological
effect. 3. Poisons—Dose-response relationship.
I. Friberg, Lars. [DNLM: 1. Cadmium—metabolism.
2. Cadmium Poisoning. QV 209 C124]
RA1231.C3C28 1985 615.9′25662 85-5272
ISBN-0-8493-6690-9 (v.1)
ISBN-0-8493-6691-7 (v.2)

This book represents information obtained from authentic and highly regarded sources. Reprinted material is quoted with permission, and sources are indicated. A wide variety of references are listed. Every reasonable effort has been made to give reliable data and information, but the author and the publisher cannot assume responsibility for the validity of all materials or for the consequences of their use.

Direct all inquiries to CRC Press, Inc., 2000 Corporate Blvd., N.W., Boca Raton, Florida, 33431.

International Standard Book Number 0-8493-6690-9 (v.1)
International Standard Book Number 0-8493-6691-7 (v.2)

Library of Congress Card Number 85-5272
Printed in the United States

PREFACE

In 1971 the first edition of the monograph *Cadmium in the Environment* was published by CRC Press. A second revised and updated edition appeared in 1974. These works focused on information essential to the understanding of the potential toxic action of cadmium and the relationship between exposure and effects on human beings and animals.

Since the publication of the second edition of *Cadmium in the Environment,* a vast amount of new information has become available. Research has been stimulated by the great demand for more detailed and accurate information, which in turn is due to the concern about the health hazards for industrially exposed workers and for people living in cadmium-polluted areas. The incentive for research comes from international organizations, government departments, cadmium-using industries, the research community, public health officers, industrial physicians, and other decision-makers concerned with environmental and industrial toxicology.

The aim of *Cadmium and Health* is basically the same as that of the earlier work *Cadmium in the Environment,* i.e., to present a comprehensive and up-to-date treatise containing information relevant to an undestanding of the toxic action of cadmium and the relationship between exposures and effects. There are some passages in *Cadmium and Health* which are identical or similar to parts of *Cadmium in the Environment,* 2nd ed., but on the whole the present text is the result of complete rewriting.

In *Cadmium and Health* we have summarized many studies but do not treat all of them in detail. In order to facilitate access to some of the more significant information some studies are discussed and analyzed at length, including details of study design and results. We have also incorporated a number of references, occasionally in great detail, to studies which lack validity and for which the results and conclusions are highly questionable. The reason for this approach was to draw attention to the fact that these studies are often still cited as valid information.

The extensive data on the subject of cadmium have necessitated the division of the text into two volumes. Volume 1: *Exposure, Dose, and Metabolism,* treats analysis of cadmium, its uses and occurrence in the environment, cadmium and metallothionein, normal values in human tissues and fluids, metabolism, and presents a metabolic model for cadmium. Volume 2: *Effects and Response,* is primarily devoted to the toxicology of cadmium and includes effects on the respiratory system, kidneys, and bone as well as other toxic effects, including those from the hematopoietic and cadiovascular system, the liver, the reproductive organs, and the fetus. Carcinogenic and genetic effects are treated in a separate chapter as are the concepts of critical organ and critical concentration for cadmium. The last chapter in Volume 2 is devoted to some general conclusions and a general discussion of the toxicology of cadmium including diagnosis, treatment, and prevention. In an appendix we present a detailed discussion of the Itai-itai disease. Each chapter is followed by numerous references. In all, *Cadmium and Health* contains more than 1500 references.

As was the case with *Cadmium in the Environment* the preparation of *Cadmium and Health* is the result of teamwork on the part of the editors. All drafts of the chapters have been extensively discussed at numerous meetings and the conclusions drawn have general consensus among the editors. Nevertheless, the responsibility for the preparation of drafts of the different chapters has rested upon dividual editors, sometimes in collaboration with outside contributors. For this reason, names of authors are given for each separate chapter.

It is the editors' intention that these two closely related volumes should be read as a whole and not separately. Only then can one obtain a complete picture of the situation today. To maintain continuity the chapter numbers follow-on over both volumes and references are listed at the end of each chapter in which they are cited.

The authors are indebted to several colleagues in various countries, who critically reviewed the chapter manuscripts and provided both suggestions and corrections. However, sole responsibility for the statements which follow and for any errors which may remain rests with the authors. The names of those who reviewed the different chapters are given at the beginning of the book.

Parallel to the preparation of *Cadmium and Health* the Department of Environmental Hygiene of the Karolinski Institute, and the National Institute of Environmental Medicine in their capacity as a WHO Collaborating Centre for Environmental Health Effects have prepared a draft for a WHO Environmental Health Criteria Document on Cadmium. This draft was discussed at a WHO Working Group meeting in January 1984 with representatives from all WHO regions. To the extent that the discussion at this Working Group meeting motivated changes in *Cadmium and Health*, such changes have been introduced. Collaboration with the World Health Organization is gratefully acknowledged.

ACKNOWLEDGMENTS

The editors would like to thank Elisabeth Kessler who acted as administrative editor during the preparation of this book; Margit Dahlquist and Evi Werendel for valuable editorial assistance, for checking the references and for producing clean and correct manuscripts; Gunnel Gråbergs who drew most of the figures; Ilse Görke and Gun-Inger Loboda for compiling reference lists; and Diana Crowe for typing and checking some of the chapters.

Most of the original figures used in this volume have been slightly modified. In all cases reference is made to the original publications in the figure caption. The full sources can be found in the reference lists at the end of each chapter. The permission for the reproduction of this material is gratefully acknowledged.

THE EDITORS

Lars Friberg, M.D., is Professor and Chairman of the Department of Environmental Hygiene, Karolinska Institute and Director of the National Institute of Environmental Medicine, Stockholm, Sweden. These departments are a World Health Organization (WHO) Collaborating Centre for Environmental Health Effects.

Dr. Friberg graduated as M.D. from Karolinska Institute in 1945 and became Doctor of Medical Sciences in 1950. During 1967 he was Visiting Professor at the University of Cincinnati and during 1978 he worked at the Division of Environmental Health, WHO, Geneva.

Dr. Friberg is a member of the WHO Advisory Board on Occupational Health. In 1975 he served as Chairman of the WHO Task Group for an Environmental Health Criteria Document on Cadmium and in 1984 he served as Temporary Adviser to the WHO Working Group for the revision of the same document.

Dr. Friberg has been a member of the Board of the Permanent Commission and International Association on Occupational Health and is presently Chairman of its Scientific Committee on the Toxicology of Metals.

Since 1981 Dr. Friberg has been a member of the U.N. Joint Group of Experts on the Scientific Aspects of Marine Pollution (GESAMP) and also Chairman of its Working Group on Potentially Harmful Substances.

Dr. Friberg was awarded the prize of Jubilee by the Swedish Society of Medical Sciences for his work on cadmium. In May 1985 Dr. Friberg received the William P. Yant Award from the American Industrial Hygiene Association. He is Honorary Life Member of the New York Academy of Sciences.

Dr. Friberg has published about 200 scientific papers dealing primarily with the toxicology and epidemiology of metals and air pollution. He was Senior Editor of the monograph *Cadmium in the Environment,* 1st ed. 1971, 2nd ed. 1974, and one of the editors of the *Handbook on the Toxicology of Metals,* 1979.

Carl-Gustaf Elinder, M.D., is Associate Professor at the Department of Occupational Medicine at the National Board of Occupational Safety and Health, Solna, Sweden.

Dr. Elinder graduated as M.D. from the Karolinska Institute in 1978 and became Doctor of Medical Sciences at the Karolinska Institute in 1979.

Dr. Elinder has been involved in teaching environmental medicine and engaged in research in relation to metabolism and toxicity of metals, at the Department of Environmental Hygiene, at the Karolinska Institute since 1973. In 1983 he was appointed Associate Professor at the National Board of Occupational Safety and Health. On several occasions Dr. Elinder has acted as Temporary Adviser for the World Health Organization. Dr. Elinder is a member of an expert committee on hazards from environmental contaminants within the Swedish Society for the Conservation of Nature.

Dr. Elinder has published about 50 scientific papers, most of them dealing with metabolism and toxicity of metals, especially with regard to cadmium. He has also published review articles and criteria documents for several other metals.

Tord Kjellström, Ph.D. (Med. Dr.), M.Eng., is Senior Lecturer in occupational and environmental health at the Department of Community Health and General Practice, University of Auckland, New Zealand, since 1976.

Dr. Kjellström became Doctor of Medical Science at the Karolinska Institute, Stockholm, Sweden, in 1977 and has a Master's degree in Mechanical Engineering from the Royal Institute of Technology, Stockholm, since 1967.

Between 1970 and 1975 he was involved in environmental health research and teaching at the Karolinska Institute, Department of Environmental Hygiene. During 1978 he was a research student at Tokyo University. For several years he coordinated international cooperative cadmium research projects involving Sweden, Japan, and the U.S.

In 1975 he served as a member of the WHO Task Group for an Environmental Health Criteria Document on Cadmium and in 1984 he was appointed a member of the WHO Working Group for the revision of the same document. In 1985 he was appointed Medical Officer/Epidemiologist in the Division of Environmental Health, World Health Organization, Geneva.

Since 1979 Dr. Kjellström has been a member of the New Zealand Department of Health Advisory Committees on Occupational Threshold Limit Values. He was also Advisor to the New Zealand Federation of Labour on occupational health matters.

Dr. Kjellström has published about 100 scientific papers dealing mainly with cadmium toxicology and epidemiology. He was co-author of the monograph *Cadmium in the Environment,* 2nd ed., CRC Press, 1974.

Gunnar F. Nordberg, M.D., is Professor and Chairman of the Department of Environmental Medicine, Umeå University, Umeå, Sweden.

Dr. Nordberg graduated from the Karolinska Institute, Stockholm, Sweden, as Doctor of Medical Sciences in 1972 and as M.D. in 1977. He was Professor and Chairman of the Department of Environmental Medicine and Community Health, Odense University, Odense, Denmark in 1977 to 1979. Since 1979 he has been Professor and Chairman of the Department of Environmental Medicine at the University of Umeå.

During 1974 and 1975 Dr. Nordberg spent a year as a Visiting Professor at the Department of Pathology of the University of North Carolina, Chapel Hill, the National Environmental Research Center, and the National Institute of Environmental Health Sciences at Research Triangle Park, North Carolina. In 1978 to 1979 Dr. Nordberg worked as a consultant at the Division of Environmental Health, World Health Organization, Geneva.

Dr. Nordberg has also been engaged as a Temporary Adviser and Consultant for the World Health Organization, the CEC and NATO Scientific Committees, the Swedish National Environment Protection Board, and the National Board of Social Welfare.

In 1975 as well as 1984, Dr. Nordberg served as a member of the WHO Task Group for an Environmental Health Criteria Document on Cadmium representing the Permanent Commission and International Association on Occupational Health. He is also Secretary of the Scientific Committee on the Toxicology of Metals under the Permanent Commission and International Association on Occupational Health and in this capacity he has edited and co-edited a number of documents on metal toxicology (e.g., *Effects and Dose-Response Relationships of Toxic Metals,* Elsevier, Amsterdam, 1976 and *Reproductive and Developmental Toxicity of Metals,* Plenum Press, New York, 1983.

Dr. Nordberg has published over 100 scientific papers dealing mainly with the toxicology of metals. He was co-author of the monograph *Cadmium in the Environment,* 1st ed. 1971, 2nd ed. 1974, and co-editor of the *Handbook on the Toxicology of Metals,* 1979.

CONTRIBUTORS

Birger Lind
Chemical Engineer
National Institute of Environmental Medicine
Stockholm, Sweden

Monica Nordberg, Ph.D.
Associate Professor
Department of Environmental Hygiene
Karolinska Institute
Stockholm, Sweden

REVIEWERS

Volume I

Chapter	Name
1. Introduction	
2. Analyses	**Markus Stoeppler** Institute für Chemie der Kernforschungsanlage Jülich GmbH Jülich, West Germany
3. Uses and occurrence in the environment	**Arne Andersson** Department of Soil Sciences Swedish University of Agricultural Sciences Uppsala, Sweden **Albert L. Page** Department of Soil and Environmental Sciences University of California Riverside, California **Markus Stoeppler** Institut für Chemie der Kernforschungsanlage Jülich Gmbh Jülich, West Germany
4. Metallothionein	**M. George Cherian** Department of Pathology Health Sciences Centre University of Western Ontario London, Ontario, Canada **Chiharu Tohyama** National Institute for Environmental Studies Japan Environment Agency Tsukuba, Ibaraki, Japan

5. Norman values for cadmium in human tissues, blood, and urine in different countries

Reinier L. Zielhuis
Faculteit der Geneeskunde
Coronel Laboratorium
Universiteit van Amsterdam
Amsterdam, The Netherlands

6. Kinetics and metabolism

Thoma W. Clarkson
Division of Toxicology
University of Rochester
School of Medicine
Rochester, New York

Bruce A. Fowler
Laboratory of Pharmacology
National Institutes of Health
National Institute of Environmental Health Sciences
Research Triangle Park, North Carolina

7. Kinetic model of cadmium metabolism

Thomas W. Clarkson
Division of Toxicology
University of Rochester
School of Medicine
Rochester, New York

Bruce A. Fowler
Laboratory of Pharmacology
National Institutes of Health
National Institute of Environmental Health Sciences
Research Triangle Park, North Carolina

Volume II

8. Respiratory effects

Robert Lauwerys
Faculté de Médecine
Université Catholique de Louvain
Brussels, Belgium

David H. Wegman, M.D.
Division of Environmental and Occupational Health Sciences
School of Public Health Center for Health Sciences
University of California
Los Angeles, California

9. Renal effects

Ernest C. Foulkes
Institute of Environmental Health
Kettering Laboratory
Cincinnati, Ohio

Hiroshi Saito
Department of Hygiene and Preventive Medicine
Nagasaki University
School of Medicine
Nagasaki City, Japan

Jaroslav Vostal
Biomedical Science Department
General Motors Research Laboratories
Warren, Michigan

10. Bone effects

Sven Erik Larsson
Department of Orthopedic Surgery
University of Umeå
Umeå, Seden

Seiyo Sano
Department of Public Health
Faculty of Medicine
Kyoto University
Kyoto, Japan

11. Other toxic effects

Bruce A. Fowler
Laboratory of Pharmacology
National Institutes of Health
National Institute of Environmental Health Sciences
Research Triangle Park, North Carolina

12. Carcinogenic and genetic effects

George Kazantzis
TUC Centenary Institute of Occupational Health
London School of Hygiene and Tropical Medicine
London, England

13. Critical organs, critical concentrations, and whole body dose-response relationships

Robert Lauwerys
Faculté de Médecine
Université Catholique de Louvain
Brussels, Belgium

14. General discussion and conclusions

Appendix — Itai-itai disease

TABLE OF CONTENTS

Volume I

Volume II

Chapter 1

INTRODUCTION

Lars Friberg

Already in 1942, Nicaud et al.[7] reported the occurrence of a few cases of osteomalacia with multiple fractures among workers in a French factory producing alkaline batteries. The authors concluded that this osteomalacia was a sign of cadmium intoxication, although no other signs, e.g., proteinuria or emphysema, were reported.

In 1948 and in 1950, Friberg[1,2] reported a high prevalence of lung emphysema, kidney dysfunction, and proteinuria in workers in a similar factory. The protein had different precipitation reactions from other urinary proteins and a lower molecular weight. There was no evidence of osteomalacia. Results from experimental studies on animals strongly incriminated cadmium as the etiological agent, although this was not at that time generally accepted. In a 1950 review of the Swedish data, Sherman Pinto, then medical director of the American Smelting and Refining Corporation, concluded: "The contention that the disabilities observed among this group of workers are due to chronic cadmium poisoning is definitely not proved."[8]

With time more and more evidence appeared showing that long-term exposure to cadmium in different forms can give rise to proteinuria, kidney damage, and chronic obstructive lung disease in humans as well as in several animal species. Of particular importance were reports in the 1950s and 60s by British investigators, showing that the cadmium syndrome could occur not only in workers employed in the production of alkaline batteries, but also in other groups of workers exposed to cadmium. Most interest was focused on the kidney disorders. It should be emphasized, however, that the pulmonary effects seen as a result of the high exposure prevailing during the 1950s and 60s were of even greater importance and the main cause of early death among cadmium-exposed workers.

In the 1950s and 60s, cadmium intoxication was considered to be strictly an occupational disease. It was not until the recognition of the contribution of cadmium exposure to the etiology of the Itai-itai disease (osteomalacia and concurrent kidney disease) among the general population of Japan in the 1960s that the concern about long-term exposure to cadmium in the general environment became more widespread. Itai-itai disease means virtually "ouch-ouch disease" and refers to the severe pain patients suffered from their multiple fractures. The disease occurred among inhabitants in certain areas of Toyama Prefecture, where rice had been heavily contaminated due to irrigation of the soil with water contaminated with cadmium from industrial sources.

A local physician, Dr. Hagino, reported on an unusually high occurrence of osteomalacia in the area in 1957.[4] A few years later it was postulated that cadmium played a role in the etiology of Itai-itai disease.[5] Since then there has been a great deal of debate about the etiology of Itai-itai disease. In 1968, the Japanese Ministry of Health and Welfare concluded that, "Itai-itai disease is caused by chronic cadmium poisoning, on condition of the existence of such inducing factors as pregnancy, lactation, imbalance in internal secretion, aging and deficiency of calcium." In 1984, the Japanese Ministry has not changed this conclusion. However, the causal role of cadmium has not been unequivocally recognized. As late as 1975, Kodama referred to the disease as "a phantom pollution-originated disease".[6] The same year, a WHO task group in the preparation of an environmental health criteria document for cadmium concluded that, "cadmium was a necessary factor in the development of Itai-itai disease." This task group[11] also recognized that other factors, e.g., nutritional factors, had been of importance for the development of the disease.

Up to the present time, approximately 60 cases of osteomalacia due to industrial cadmium exposure have been reported. Some workers have other signs of disturbance of mineral metabolism. In some studies, e.g., a high incidence of kidney stones has been observed.

At least a couple of hundred cases of Itai-itai disease have occurred in Japan. With improved standards of nutrition and decreasing exposure to cadmium, Itai-itai disease should be a thing of the past.

The occurrence of osteomalacia is only the top of an iceberg. The high incidence of proteinuria and kidney dysfunction is the major concern today. In some industries, and in cadmium-polluted areas of Japan, proteinuria has been observed in more than half of the exposed population. More recently, the question of cadmium as a carcinogenic agent has been widely discussed, both as a factor for the development of cancer of the prostate and for pulmonary cancer. Certain experimental data on rats have shown a very high and dose-related incidence of lung cancer after exposure to a cadmium chloride aerosol.

In the early studies of proteinuria, fairly crude methods were used. Some methods in common use at that time, such as the boiling test and precipitation using nitric acid, were very insensitive to the low molecular weight protein which was excreted in the urine. The use of the boiling test was probably the major reason why Nicaud and associates did not recognize proteinuria in their 1942 study of workers in a French battery factory. An examination carried out some years later in the same factory[2] confirmed the occurrence of low molecular weight proteins in some workers.

During the 1960s and 70s, great progress has been made in detection and classification of cadmium-induced low molecular weight proteinuria. The proteinuria in cadmium poisoning was characterized as tubular by Piscator in 1966.[9] Nowadays, this type of proteinuria can be detected using routine analytical methods and it has been shown that renal tubular dysfunction is the cause of the proteinuria. The renal tubular dysfunction may progress and in severe cases glomerular damage and uremia may develop.

One of the major concerns in the field of cadmium toxicology is the very long half-time of cadmium both in the environment and in humans. There is ample evidence that once agricultural soil has been contaminated by cadmium, contamination persists for several years constituting a persistent source of uptake of cadmium by plants and grains, such as wheat and rice. The concept of dose commitment, commonly used when estimating the risks of contamination by radioactive substances, may be appropriate for use in the case of cadmium. In human beings, the biological half-time of cadmium in the kidney is very long — 10 to 30 years.

In some geographical areas there are no or at best only very small margins of safety for the development of kidney dysfunction. In these particular countries concentrations in kidney cortex, which is the critical organ for cadmium intoxication, are close to the recognized critical concentrations. It should be pointed out that the dose-response data for humans are based on measurement of total cadmium levels in tissues, whereas recent animal data indicate that there may be a difference in toxicity of nonmetallothionein-bound and metallothionein-bound cadmium. The cadmium species in the tissues may in turn be a function of differences in exposure routes and exposure forms.

In several countries regulations have been introduced, sometimes far reaching, to prevent cadmium exposure. These measures apply not only to industry, but also to exposure via food and from other sources in the general environment. As cadmium, "the dissipated element", is found in a very large number of commercially available products and also as a contaminant in, e.g., cigarettes and fertilizers, efforts to reduce exposure, or at least to limit any future increase in exposure, may involve considerable problems.

The growing scientific and administrative interest in cadmium toxicology is reflected in the increasing scientific output. In 1950, only a few reports were published annually. Nowadays, approximately 300 to 400 scientific papers are published each year. Cadmium has also been discussed in detail at a number of major international scientific meetings. The conclusions reached at some of these meetings may be of interest to the reader as reference material and as indicators of the international progress of cadmium research. Therefore, a list of important reports from such meetings is given at the end of this chapter.

Despite the intensive research in Japan, which is still very active and important, it is only during recent years that publications from this country have become available in English. Prior to an overview paper by Tsuchiya in 1969,[10] there was no information at all in languages other than Japanese. Already in the first edition of our monograph *Cadmium in the Environment*, 1971, we were able to include much original data from Japan as a result of fruitful collaboration with Japanese colleagues.[3] This tradition has been continued in the present monograph. As one of the editors (Dr. T. Kjellström) speaks and reads Japanese, much of the information is based not only on papers translated into English, but also on original Japanese papers and discussions with Japanese scientists.

We believe that the information presented in the following chapters will provide the reader not only with a comprehensive and up-to-date outline of the present situation in regard to cadmium and health, but also with a retrospective examination of the sources of our present knowledge on this subject.

The following list includes reports pertaining to metabolism and effects of cadmium — proceedings from major international scientific meetings.

Year	Place and date	Published report
1971	Slanchev Briag, Bulgaria September 24	Dukes, K. and Friberg, L., Eds., Absorption and excretion of toxic metals, *Nord. Hyg. Tidskr.* 52, 70—104, 1971.
1972	Buenos Aires, Argentina September	Task Group on Metal Accumulation. Accumulation of toxic metals with special reference to their absorption, excretion and biological half-times, *Environ. Physiol. Biochem.*, 3, 65—107, 1973.
1974	Tokyo, Japan November 18—23	Nordberg, G. F., Ed., *Effects and Dose-Response Relationships of Toxic Metals*, Elsevier, Amsterdam, 1976.
1975	Geneva, Switzerland July 1—12	WHO, Environmental health criteria for cadmium. Summary, *Ambio*, 6, 287—290, 1977.
1977	San Francisco, Calif. January 31—February 2	*Cadmium 77, Edited Proc. 1st Int. Cadmium Conf., San Francisco,* Metal Bulletin Ltd., London, 1978.
1977	Stockholm, Sweden July 17—22	Nordberg, G. F., Ed., Factors influencing metabolism and toxicity of metals, *Environ. Health Perspect.*, 25, 3—41, 1978.
1977	Jena, East Germany August 1—3	*Cd — Kadmium-Symposium,* Friedrich-Schiller-Universität, Jena, 1979.
1978	Luxembourg	*CEC Criteria (Dose/Effect Relationships) for Cadmium,* Pergamon Press, New York, 1978.
1978	Bethesda, Md. June 7—9	Environmental cadmium, *Environ. Health Perspect.*, 28, 1—300, 1979.

Year	Place and date	Published report
1978	Zürich, Switzerland July 17—22	Kägi, H. R. and Nordberg, M., Eds., *Metallothionein*, Birkhäuser Verlag, Basel, 1979.
1979	Geneva, Switzerland February 2—3	WHO environmental health criteria for cadmium, Interim Report No. EHE/EHC/79.20, World Health Organization, Geneva, 1979.
1979	Cannes, France February 6—8	*Cadmium 79, Edited Proc. 2nd Int. Cadmium Conf., Cannes*, Metal Bulletin Ltd., London, 1980.
1979	Geneva, Switzerland June 5—11	WHO, Recommended health-based limits in occupational exposure to heavy metals, Tech. Rep. Ser. 647, World Health Organization, Geneva.
1980	London, England March 20	Occupational exposure to cadmium — Report of a seminar, Cadmium Association, London, 1980.
1980	Atlanta, Ga. March 24—28	Belman, S. and Nordberg, G. F., Eds., Workshop/Conference on the role of metals in carcinogenesis, *Environ. Health Perspect.*, 40, 3—42, 1981.
1981	Miami, Fla. February 3—5	*Cadmium 81, Edited Proc. 3rd Int. Cadmium Conf., Miami*, Cadmium Association, London, 1982.
1981	Cincinnati, Ohio March 22—27	Foulkes, E. C., Ed., *Biological Roles of Metallothionein*, Elsevier, Amsterdam, 1982.
1982	Rochester, N.Y. May 24—26	Clarkson, T. W., Nordberg, G. F., and Sager, P. R., Eds., *Reproductive and Developmental Toxicity of Metals*, Plenum Press, New York, 1983.
1982	Luxembourg July 7—9	Proc. Int. Workshop on Biological Indicators of Cadmium Exposure — Diagnosis and Analytical Reliability, to be published.
1983	München, West Germany March 2—4	*Cadmium 83, Edited Proc. 4th Int. Cadmium Conf., München*, Cadmium Association, London, 1984.
1983	Research Triangle Park, N.C. May 16—18	Goyer, R. A., Fowler, B. A., Nordberg, G. F., Shepherd, G., and Moustafa, L., Eds., Metallothionein and cadmium nephrotoxicity conference, *Environ. Health Perspect.*, 54, 1—295, 1984.

In addition to the above list, a number of relevant reports on cadmium research have been presented at a series of conferences on *Trace Elements Metabolism in Animals and Man*. These TEMA conferences are held every 4 years (see, e.g., *TEMA 3*). For information contact: Institut für Ernährungsphysiologie, Technische Universität, Freising Weihenstephan, Munich, FRG.

Another contributor to this field is a series of conferences on *Heavy Metals in the Environment*, which are held every 2 years. For publications contact: CEP Consultants Ltd., 26, Albany Street, Edinburgh, EH1 3QH, U.K.

In Japan, where cadmium in the environment is a particularly controversial issue, special cadmium research meetings are organized by the Japanese Public Health Association (address below). Detailed proceedings are published in Japanese in the periodical *Kankyo Hoken Report*. For information contact: Japanese Public Health Association, Shinjuku, Shinjuku-ku, Tokyo 160, Japan.

REFERENCES

1. **Friberg, L.,** Proteinuria and kidney injury among workmen exposed to cadmium and nickel dust, *J. Ind. Hyg. Toxicol.*, 30, 32—36, 1948.
2. **Friberg, L.,** Health hazards in the manufacture of alkaline accumulators with special reference to chronic cadmium poisoning (Doctoral thesis), *Acta Med. Scand.*, 138 (Suppl. 240), 1—124, 1950.
3. **Friberg, L., Piscator, M., and Nordberg, G.,** *Cadmium in the Environment*, CRC Press, Boca Raton, Fla., 1971.
4. **Hagino, N.,** About investigations on Itai-itai disease (in Japanese), *J. Toyama Med. Assoc.*, December 21, 1957.
5. **Hagino, N. and Yoshioka, K.,** A study on the cause of Itai-itai disease (in Japanese), *J. Jpn. Orthoped. Assoc.*, 35, 812—815, 1961.
6. **Kodama, T.,** Is Itai-itai disease a phantom pollution-originated disease? (transl.), *Bungei Shunju*, Int. Lead Zinc Research Organization, New York, 1975.
7. **Nicaud, O., Lafitte, A., and Gros, A.,** The problems of chronic cadmium intoxication (in French), *Arch. Mal. Prof. Med. Trav. Secur. Soc.*, 4, 192, 1942.
8. **Pinto, S. S.,** Book review of *Health Hazards in the Manufacture of Alkaline Accumulators with Special Reference to Chronic Cadmium Poisoning* by Lars Friberg, Ivar Häggströms Boktryckeri, Stockholm, 1950, *Arch. Ind. Hyg. Occup. Health*, 2, 120, 1950.
9. **Piscator, M.,** Proteinuria in Chronic Cadmium Poisoning, Doctoral thesis, Karolinska Institute, Stockholm, 1966.
10. **Tsuchiya, K.,** Causation of ouch-ouch disease (Itai-Itai Byō) — an introductory review. I. Nature of the disease, *Keio J. Med.*, 18, 181—194, 1969.
11. WHO environmental health criteria for cadmium, *Ambio*, 6, 287—290, 1977.

Chapter 2

PRINCIPLES AND PROBLEMS OF CADMIUM ANALYSIS

Carl-Gustaf Elinder and Birger Lind

TABLE OF CONTENTS

I. GENERAL ASPECTS

It is not possible to draw conclusions about cadmium levels in the environment, about normal values of cadmium in human tissues and fluids, or to evaluate dose-response relationships for cadmium without seriously considering the analytical methods used. Numerous scientific reports have been published, in which reported levels of cadmium are clearly, or very likely to be, erroneous. These results must be disregarded. In most published reports which give data, or which mention cadmium concentrations in various media, it is difficult to judge whether the analytical results are valid or not. In the absence of better published material, such data will have to be considered when evaluating cadmium and health. Nevertheless, the analytical data referred to in this book will be critically evaluated.

A. Fundamental Criteria for Assessing a Method

An analytical method can be evaluated on the basis of the following criteria: accuracy, precision, detection limit, and sensitivity.

1. Accuracy

In this report the term accuracy will be used to mean that there is no systematic bias involved. An accurate analytical method should provide results which are not, systematically, either too high or too low when compared to the "true" value. In other reports the term accuracy has sometimes been defined differently.[71]

Some authors have defined accuracy as "the percentage of deviation of the method value from the true value".[55] Defining accuracy in this way is a bit confusing, since it would mean that accuracy would actually increase when error increases. This estimate should rather be named "inaccuracy".

The "true" value for cadmium in biological samples is rarely known. If there is a reasonable agreement in the average concentration based on measurements made at several qualified laboratories using different analytical techniques, the overall average can be tentatively regarded as a "true" value.

In recent years, various types of reference materials with a certified cadmium concentration, such as lyophilized orchard leaves and bovine liver, have become available from the National Bureau of Standards[50] and other materials from the International Atomic Energy Agency[56] (Table 1). These materials can be used for testing the accuracy and precision (see below) of a procedure (see Section IV).

2. Precision

Precision denotes the random error of the method. If replicate analysis provides data which are closely clustered, the method is said to have a high precision. A high precision makes it possible to measure small changes in concentrations with confidence but does not imply per se that the results are accurate. If the variation in individual measurements can be assumed to follow a Gaussian distribution, the best description of precision is the standard deviation (Sx) from the mean ($\bar{x}$) of a number of analyses of the same sample. Often precision is given as relative standard deviation, or coefficient of variation = $Sx/\bar{x}$.[55]

The relative standard deviation (RSD), commonly expressed in percent ($Sx \cdot 100/\bar{x}$), usually decreases and precision improves with increasing concentration in the samples. This is natural since the magnitude of the random error (Sx), within optimal limits of the method, is usually the same. An increase in the average concentration ($\bar{x}$) will thus lead to a reduction in the RSD ($Sx/\bar{x}$). As an example, Delves[10] compiled data on blood cadmium levels from different studies and noticed that the RSD decreased from about 20 to 30% at cadmium concentrations in blood below 0.5 μg/ℓ to less than 5% at cadmium concentrations exceeding 5 μg/ℓ.

Table 1
AVAILABLE CERTIFIED REFERENCE MATERIAL FOR CADMIUM ANALYSIS

Type of material	Code	Certified concentration (mg Cd/kg) dry weight
Oyster tissue	NBS-SRM-1566	3.5
Fish powder	IAEA-MA-A-1	0.75
Bovine liver	NBS-SRM-1577	0.27
Orchard leaves	NBS-SRM-1571	0.11
Wheat flour	NBS-SRM-1567	0.032
Rice flour	NBS-SRM-1568	0.029

Note: NBS material is available from U.S. National Bureau of Standards, Department of Commerce, Washington, D.C. 20234; USA and IAEA material from Analytical Quality Control Services, Laboratory Seibersdorf, International Atomic Energy Agency, POB 100, A-1400 Vienna, Austria.

3. Detection Limit and Sensitivity

The detection limit is the lowest concentration, or amount, of metal which can be measured using a given method, i.e., a concentration which gives rise to a signal which with a certain statistical probability is greater than the background "noise" and/or blank value.[55] The term "sensitivity" is sometimes used as having the same meaning as "detection limit". This is not correct.

Sensitivity describes the relationship between a concentration, or amount, of metal and the signal received on the analytical instrument. For atomic absorption spectrophotometry, a special definition of the term "sensitivity" has been applied. In this case it is defined as a concentration, or amount, of metal which will result in 1% light absorption (0.0044 absorbance units).[32] In Figure 1, an example is given of typical signals obtained using different methods having a high and low detection limit, respectively. For accurate and precise measurement of cadmium in low level samples, high sensitivity combined with a low background noise is necessary to achieve a low detection limit.

B. Review of Methods Used

Many different methods have been used to measure cadmium concentrations. Previously, colorimetric methods with dithizone/chloroform extraction, polarography and emission spectroscopy were the most frequently used.

In recent decades, atomic absorption spectrophotometry (AAS) has become the most popular method of analysis. Because of its high capacity (sample through-put) and comparably low costs, AAS is commonly used for cadmium analysis in large series of samples. Different types of electrochemical methods are also used. Other more sophisticated methods are neutron activation analysis (NAA) and spark source mass spectrometry, especially isotope dilution mass spectrometry (IDMS). These latter methods are often used for analytical quality assurance purposes.

More information about methods suitable for cadmium analysis is to be found in a book by O'Laughlin et al.[55] and in recent review articles by Stoeppler.[65,66] Methods for cadmium analysis in blood and urine at low and normal concentrations have been summarized by Delves[10] and Herber.[28]

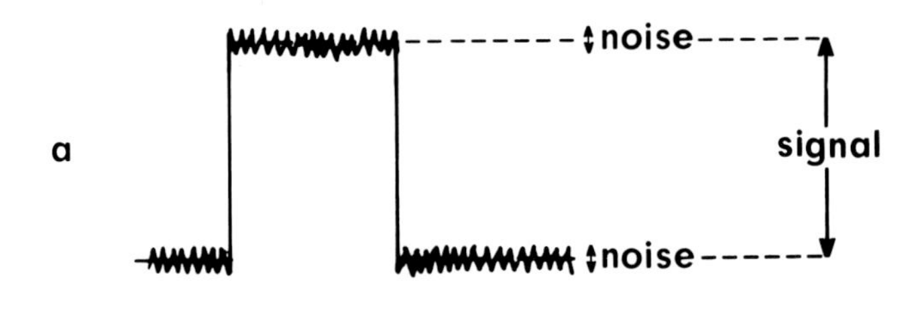

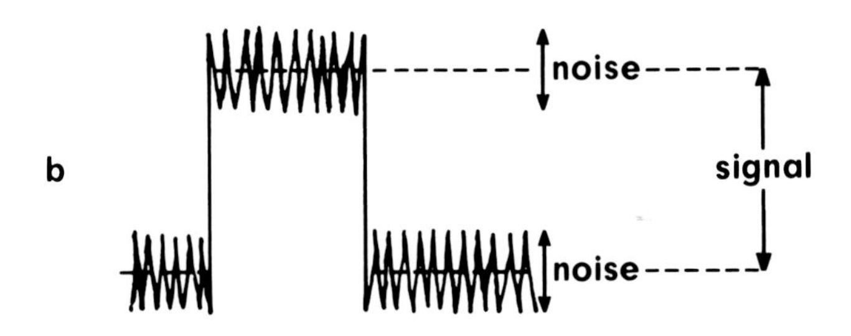

FIGURE 1. Three typical recordings from measurements of cadmium concentration in a solution. The detection limit is considerably lower in recording (a) than in (b) due to less background noise. In recording (c), the detection limit is the same as in (a), but the sensitivity (amplitude) is less than in recording (a) and (b).

To obtain valid data, the following steps should be considered: sample collection, sample preparation and treatment, instrumental analysis, and evaluation.

II. SAMPLE COLLECTION AND PREPARATION

A. Sample Collection

Sample collection is a first and crucial step. The major problem is to obtain a random sample which is large enough to provide information which, with a certain statistical probability, will represent the true value of a specific population. When measuring cadmium in, e.g., food or air, it is often difficult to obtain representative samples. If, e.g., long-term exposure to atmospheric cadmium is to be assessed, it is not sufficient to rely on a small number of measurements over a limited period of time.[27]

The risk of contamination is another serious problem. This may occur in connection with sample collection and during the chemical analysis. Colored materials, especially plastics and rubber, should be avoided. Blood or urine samples which come into contact with, e.g., red plastic caps, are easily spoiled due to contamination.[59] Contamination of blood samples has also been reported to occur when blood is collected in certain types of evacuated blood collection tubes.[49,52]

Utensils used during analysis may also be a source of contamination. For example, Salmela and Vuori[62] noticed that disposable colored micropipette tips could seriously contaminate acid solutions with cadmium. The highest contributions to contamination are normally given by yellow (CdS)- and red (CdSe)-colored tips. It is necessary then that all utensils used for sample collection, storage, and sample treatment be carefully checked before analysis to make sure that they will not contaminate the sample. In general, utensils made from polyethylene or glass are preferable.[80]

There are several good reviews covering the problems encountered in connection with sample collection and handling of samples, e.g., the book entitled *Accuracy in Trace Analysis: Sampling, Sample Handling, Analysis*, edited by LaFleur,[39] and review articles by Piscator and Vouk,[59] Behne,[5] and Stoeppler.[66]

B. Preparation and Sample Treatment

Normally, biological samples have to be digested (destructed) in order to separate cadmium from the organic matrix before analysis. Common methods to destroy and oxidize organic matrixes are wet and dry ashing. In dry ashing, the sample is oxidized by air or oxygen at a relatively high temperature, usually between 400 and 500°C in a muffle furnace, for more than 12 hr. Dry ashing should result in the complete ashing of the samples. The ashes are subsequently dissolved in nitric acid or some other solvent. Low temperature dry ashing at about 100°C can be performed in a plasma asher, containing a high concentration of oxygen. During low temperature ashing, the risk of loss due to volatilization is minimal.[21,48] The problem of loss of cadmium during ashing at temperatures below 500°C is also as a rule minimal.[37,45,61] In a study of the recovery of ^{109}Cd added to wheat samples, no measurable losses occurred when ashing was carried out at 450°C.[35]

In wet ashing, concentrated acids such as nitric acid, sulfuric acid, and perchloric acid are used to digest the sample. The digestion is usually performed on a heating plate or in an aluminum block, and there is no risk of loss through volatilization as long as the samples do not boil dry. Digestion in a closed system with acids in a Teflon® vessel under high pressure, so-called pressurized decomposition, can be useful for samples containing volatile elements or matrixes which are difficult to destroy.[57,66]

Additional information about sample treatment and the destruction of organic matter is to be found in the books by Gorsuch[23] and Bock.[6]

When low cadmium concentrations are involved, it may be necessary to concentrate the sample before analysis. Chelating agents such as dithizone, ammoniumpyrrolidine dithiocarbamate (APDC), or sodium diethyldithiocarbamate (Na-DDC) may be used in order to extract and concentrate cadmium in an organic solvent. Preconcentration can also be achieved by an ion-exchange technique.[55]

III. ANALYTICAL METHODS

A. Spectrophotometric Methods

The dithizone method has been the most commonly used spectrophotometric method. The basic principle is that cadmium forms a stable red-colored complex with dithizone at alkaline pH levels and that this complex can be extracted into chloroform or carbontetrachloride.[63] If proper steps are taken to remove other metals, the method becomes specific for cadmium. The method is suitable mainly for high level samples since the detection limit for cadmium dithizone complex in chloroform or carbontetrachloride is comparably high in the order of 50 μg Cd per liter in a solution.[55] The dithizone method has been used for the determination of cadmium in urine.[31]

There was good agreement for cadmium measurements in human urine performed by laboratories using the dithizone method and those using atomic absorption spectrophotometry. The concentrations reported in this study were high, ranging from 4 to 190 μg Cd per liter, and therefore it cannot be concluded that the dithizone method is suitable for measurements of cadmium in normal urine, where concentrations are in the order of 1 μg Cd per liter or less (see Chapter 5, Section II).

B. Atomic Emission Spectrometry (AES)

Atomic emission is based on the excitation of metal ions by flames or by electric discharges. Excited atoms emit radiation of a specific wavelength which is characteristic for each type of atom. Cadmium can be determined by this method utilizing the emission lines 228.8 or 326.1 nm. Tipton et al.[73] used emission spectroscopy for the determination of cadmium in a large series of human tissue samples and reported a detection limit of 50 mg Cd per kilogram ash weight, corresponding to about 0.5 mg Cd per kilogram wet weight. There have been some reports from studies using emission spectrometry for the measurement of cadmium in food and diets.[4,20] Compared to results obtained using other analytical methods for cadmium measurement in food, data obtained by emission spectrometry appear to be falsely high. Thus, the method at present seems to be suitable mainly for samples with a comparably high cadmium content.

One important advantage of emission spectrometry compared to most other methods is the possibility to measure several metals at the same time.

A new technique, inductively coupled plasma emission (ICP),[53] enables simultaneous determination of a large number of metals. However, the equipment is very expensive and the detection limit for cadmium (2 μg Cd per liter) is about the same as for conventional atomic absorption spectrophotometry using a flame.[65]

C. Atomic Absorption Spectrophotometry (AAS)

The basic principle underlying atomic absorption is the capacity of atoms in their ground energy stage to absorb radiation of specific energies. The specific radiation energies which are absorbed by an atom are emitted from the same element when it is excited (see Section III. B). Atomized cadmium thus absorbs radiation from light emitted by a cadmium lamp. The decrease in energy of the radiation passing through the atomized metal gas is proportional to its concentration. There are two major methods for the atomization of a sample: the flame method and electrothermal atomization (ETA). The latter method is also termed the graphite furnace method, flameless or heated graphite atomization.

Flame methods are generally used for liquid samples which can be aspirated into a flame, usually an air-acetylene flame. The detection limit for cadmium in pure water is in the order of 1 to 5 μg/ℓ.[55] Flame atomic absorption is generally sensitive enough to allow measurement of cadmium in biological materials containing cadmium concentrations exceeding about 0.1 mg/kg.[13] At lower levels it is usually necessary to increase the sensitivity by accessory means or by preconcentration during sample treatment.

One important modification of the flame technique is the use of a microsampling technique utilizing a cup made of nickel.[9,12,15] Sometimes this method is named the "Delves cup technique". From 5 to 50 μℓ of a liquid sample, e.g., blood or urine, are placed in the cup which is inserted after drying just below a hole in a silica tube heated by a flame. Due to more efficient atomization and the fact that the atoms are subject to the light beam that passes through the tube, for a longer period of time, the Delves cup method increases sensitivity considerably. This method has also facilitated the determination of cadmium in much smaller samples than is possible with the ordinary flame method, in which the sample is aspirated into a burner system. Ulander and Axelson[74] using the cup technique were able to measure cadmium in blood below 1 μg/ℓ and reported a coefficient of variation (CV) of

6.3 and 16% in high and low level samples, respectively. Vesterberg and Bergström[77] dry ashed the samples in the cups prior to the AAS analyses. This procedure improved the precision. The CV was reported to be about 6% in blood samples having cadmium concentrations in the range between about 4 to 20 μg Cd per liter.

In recent years, flameless methods have undergone rapid development. The sample, usually in solution (1 to 100 μℓ), is first inserted into the graphite tube. The tube is surrounded with a flow of inert gas, such as argon or nitrogen. The temperature is then increased according to a preset program in order to dry, ash, and atomize the sample. During atomization the specific absorption for cadmium is deduced from the light beam passing through the furnace. The detection limit is extremely low, in the order of a few picograms Cd. Several detailed reports describing cadmium analysis in biological samples such as blood and urine with electrothermal atomization have been published.[7,11,19,42,67,68,70,78] The lowest detectable concentration of cadmium in blood and urine using electrothermal atomization is in the order of 0.1 to 0.3 μg Cd per liter.[10]

Although atomic absorption for cadmium is specific, the method is not entirely free from problems when applied to measurement in biological samples. Several important interferences exist, among which are, e.g., light scattering from particles and nonspecific absorption from the broad molecular absorption band formed by, e.g., sodium and phosphate ions. Piscator[58] showed that 0.5 *M* NaCl emitted a signal corresponding to a concentration of 100 μg Cd per liter when using standard air-acetylene flame atomic absorption equipment without background correction. The actual concentration was less than 0.4 μg Cd per liter. Many problems related to interference from salts may be compensated for by the use of a background correction system. Usually a deuterium or hydrogen lamp is utilized. The nonspecific absorption can thus be measured and the signal proportional to the actual cadmium concentration is measured as the difference between the total and nonspecific absorption.[33] Background correction for fine structure nonspecific absorption can be made more effectively utilizing the Zeeman effect on incoming light when modulated by strong magnetic fields.[3,38,60] When the micro-sampling or electrothermal atomization techniques are used for cadmium analysis, some kind of background correction is necessary since the nonspecific absorption increases due to the atoms being kept in the light beam for a longer period of time.

D. Activation Analysis

Cadmium has a number of stable isotopes. Neutron irradiation yields new radioactive cadmium isotopes. These can be quantitatively measured on the basis of their specific energy and half-life. A procedure for determination of cadmium in human liver samples by neutron activation analysis (NAA) is given by, e.g., Halvorsen and Steinnes.[25] Usually the irradiated sample is digested before radioactivity is measured. Occasionally, it becomes necessary to concentrate cadmium by chemical methods and to separate the cadmium ions from other isotopes, which have an energy spectrum that overlaps the cadmium spectrum, before measurement can be carried out. Nonradioactive cadmium is normally added after irradiation. This procedure enables measurement of the recovery after digestion and the various concentration steps. For neutron activation analysis in most biological materials, sensitivity is high and the detection limit is low, in the order of 1 μg Cd per kilogram or 0.1 to 1 μg Cd per liter in a solution which is very close to the normal values (Chapter 5). However, the method is not normally used for screening programs due to the prohibitive costs entailed in the irradiation of the samples in a reactor. Neutron activation analysis is often used as a reference method to test the accuracy of other methods. Activation analysis is not the ideal method for liquid samples such as blood and urine. Ampoules filled with liquid samples may explode, due to the formation of gases in the sample during irradiation in the reactor.

Irradiation with protons, proton-induced X-ray emission (PIXE),[1] is another method for the activation analysis of cadmium. Using this method several elements can be measured simultaneously. The main advantage of PIXE is its capability to detect and quantify cadmium in very small samples, such as thin slices of tissue which weigh less than 1 mg.[26,43]

E. Electrochemical Methods

Cadmium levels can be determined using various types of electrochemical methods, e.g., classic polarographic methods and the more recently developed differential pulse anodic stripping voltammetry (DPASV), and selective ion electrodes. The basic principle underlying electrochemical methods is to measure the current produced between electrodes at different potential (charge) due to oxidation and/or reduction of ions in a solution. A dropping mercury electrode is placed in the solution in which the metal concentration is to be determined. By varying the potential of the electrode, different metals will be reduced and subsequently form amalgam (a solution of metals in mercury), with the mercury electrode, and polarographic waves can then be recorded. Different metals can be determined simultaneously in a solution since they are reduced at different potentials. Differential pulse anodic stripping voltammetry is based on the reversed process, i.e., the release of metals which have already been reduced and are bound to the mercury electrode. During oxidation and release from the amalgam, a peak current can be recorded at a potential which is characteristic for the particular metal. Differential pulse anodic stripping voltammetry is one of the most sensitive methods available for cadmium determination. Most crucial to this method is the complete destruction of all organic materials and transfer of the cadmium ions from the sample into a noncontaminated electrolyte. This method is particularly suitable for water analysis, where no sample treatment is necessary,[54,55,59] but it has also been used for the measurement of cadmium in various biological materials such as blood,[76] urine,[22,30] foodstuffs,[18] and tissues.[8] In urine a detection limit of about 0.1 $\mu g/\ell$ was found when utilizing a computerized potentiometric stripping analysis.[30]

Selective ion electrodes for cadmium are commercially available, but the sensitivity of these is not adequate for the determination of cadmium in most biological materials.[55] Furthermore, these electrodes are not ion specific and problems may easily arise as a result of the presence of various contaminants in the solution used.[29]

F. Spark Source Mass Spectrometry (SSMS)

The basis for spark source mass spectrometry (SSMS) is the formation of atomized metal ions which occurs when a sample is subjected to high energy discharges in an evacuated chamber. The newly formed ions are accelerated into a mass spectrometer in which the different elements can be separated according to their mass and charge (m/e ratio). Ordinary mass spectrometry has limited application for cadmium analysis since the accuracy and precision are not sufficient at normal cadmium concentrations in biological materials.

In isotope dilution mass spectrometry (IDMS), an important development of the method, a sample with unknown metal content is spiked with a known amount of a stable isotope of the element to be determined.[24] The ratio between the isotopes can be measured and on the basis of the obtained ratio, the metal concentration in the original sample can be calculated with utmost precision. Since the quantification is based on a ratio between isotopes, the method is valid also when the recovery is low. This method is complicated and expensive, and has been used mainly for quality control of other methods and for certifying reference materials.

G. Other Methods

A number of other methods for the determination of cadmium have been reported.[55] It is not possible to review all of them in this chapter.

One interesting method employing atomic fluorescence spectrometry is, however, worth mentioning, although it is not yet commercially available.[41] Michel et al.[46] have described a method where acidified urine and blood samples are aspirated directly into an air-acetylene flame and the atomized sample is then irradiated from a cadmium electrodeless discharge lamp (EDL) and a high pressure xenon arc is used for background correction. The generated atomic fluorescence signals are recorded perpendicular to the light beams. Sensitivity is sufficient to enable measurement of cadmium in normal urine with concentrations of less than 1 μg/ℓ. The agreement was satifactory when compared to AAS in 29 urine samples with cadmium concentrations ranging from 0.3 to 15 μg/ℓ.

Additional information about instruments and methods for cadmium analysis can be obtained from the reviews by Hislop[29] and Stoeppler.[65]

IV. QUALITY ASSURANCE

The importance of analytical quality assurance should be taken into account by any researcher who publishes data on concentrations or amounts of trace substances in various media. Analytical quality assurance deals with precision and accuracy. Precision can easily be controlled within the laboratory by repeated determinations of the same sample. However, accuracy can only be checked by the analysis of cadmium in reference materials or by the analysis of a series of samples using a completely different analytical method. Table 1 presents different types of certified reference materials having cadmium concentrations ranging from 0.029 to 3.5 mg/kg dry weight.[56] Unfortunately, as yet there are no certified standards available for checking cadmium analyses of blood and urine even though Behringswerke, Marburg, FRG, has made available two sets of freeze-dried blood samples (with assigned values) on a commercial basis.[65]

In this context it should be mentioned that the capability of a laboratory or adequacy of a method for measuring cadmium in one material, e.g., lyophilized liver, does not necessarily imply that the method is accurate for the analysis of cadmium in other biological materials. This is due to differences in matrix and concentrations. A specific test for accuracy should be applied to each material or fluid and to a wide range of concentrations. The frequent use of physically independent methods and different analytical procedures within the same laboratory can significantly improve the reliability of data.[69]

In recent years several interlaboratory studies have been performed relating to the quality assessment of cadmium analysis in biological materials.[17,34,36,40,64,71] In high level samples such as kidney and liver, interlaboratory agreement has been acceptable. In the low level range, results from collaborative studies have been rather disappointing. Reported results differed widely. Figure 2 shows the results of cadmium analysis in wheat samples, which contained cadmium concentrations below 0.2 mg/kg. Results from seven different laboratories in Sweden, U.S., and Japan, using atomic absorption spectrophotometry, are compared to results from analyses using neutron activation. Several laboratories using AAS reported results which were about twice too high (Figure 2).

In Figure 3, the results of measurement of cadmium in liver samples with concentrations exceeding 2 mg Cd per kilogram dry weight are shown. It can be seen that in this case there is a good agreement between the results obtained in all the laboratories using AAS, when compared with results from neutron activation analysis.[34]

The need for strict quality assurance of analytical results has been acknowledged by the WHO and UNEP[79] and an extensive quality control program was included in a recently concluded UNEP/WHO international program for the assessment of human exposure to heavy metals through biological monitoring.[16,75]

The UNEP/WHO report[75] proposes that an analytical quality control program should involve analysis of both internal quality control (IQC) samples and external quality control (EQC) samples. The IQC samples should contain a "known" metal concentration. Such samples can, e.g., be certified reference materials obtained from the National Bureau of

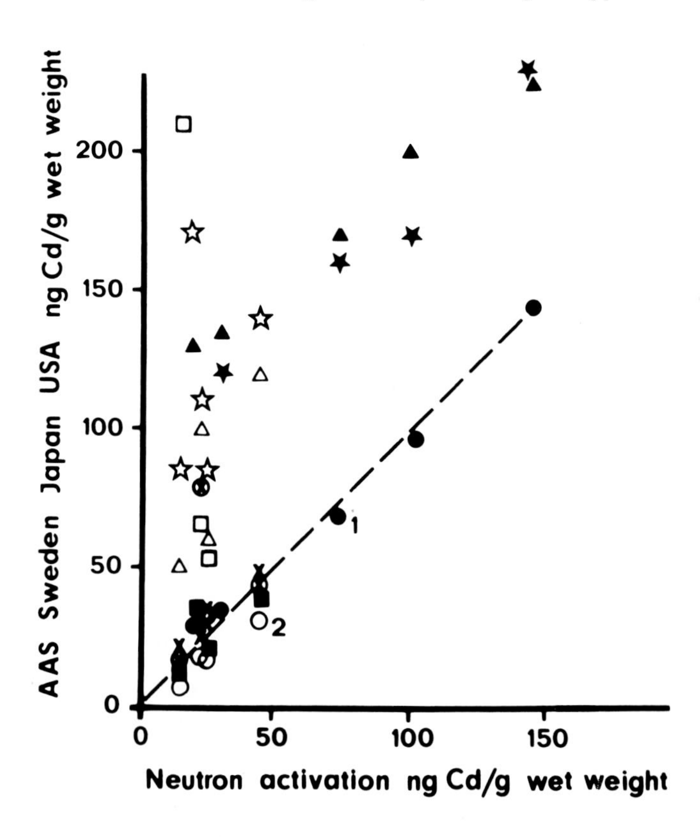

FIGURE 2. Comparison of cadmium analysis of wheat grain performed by seven different laboratories using atomic absorption spectrophotometry (y axis) compared to one laboratory using neutron activation (x axis). Each type of symbol denotes one laboratory. (From Kjellström, T., Tsuchiya, K., Tompkins, E., Takabatake, E., Lind, B., and Linnman, L., *Recent Advances in the Assessment of the Health Effects of Environmental Pollution, II, Proc. Int. Symp., Paris,* Commission of the European Communities, Luxembourg, 1975, 2197—2316. With permission).

Standards (NBS) in the U.S. or from the International Atomic Energy Agency (IAEA) in Europe (Table 1). External quality control samples should contain metal concentrations which are unknown to the laboratory performing the analysis. Preferably, the EQC samples should be prepared by an independent laboratory or agency. Spiking low level samples with small exact amounts of metal in order to obtain a range of concentration in the order of one magnitude is one way that has been used to prepare EQC samples.[75] Results obtained by the laboratory (reported values) can subsequently be plotted on a diagram against the "true" concentration after spiking (reference values), and a regression line can be calculated. Ideally, the regression line should be a line Y = X, passing through origo. This is, however, rarely the case and as a rule the regression line will differ from the equation Y = X. Within the UNEP/WHO Monitoring Programme[16,75] a maximum allowable deviation was decided on and, based on a statistical power analysis, an acceptance criterion was estimated based on how much the observed regression line was allowed to deviate from Y = X. If the regression line fell within the upper and lower acceptance lines (Figure 4), the results were accepted, and if not, the results were rejected. It was rare that a laboratory met the criteria for acceptance throughout the entire project. An accepted performance on a limited number of quality control runs was no guarantee for a continuous good performance. Even well-experienced laboratories had substantial problems at times. One conclusion was that laboratories must

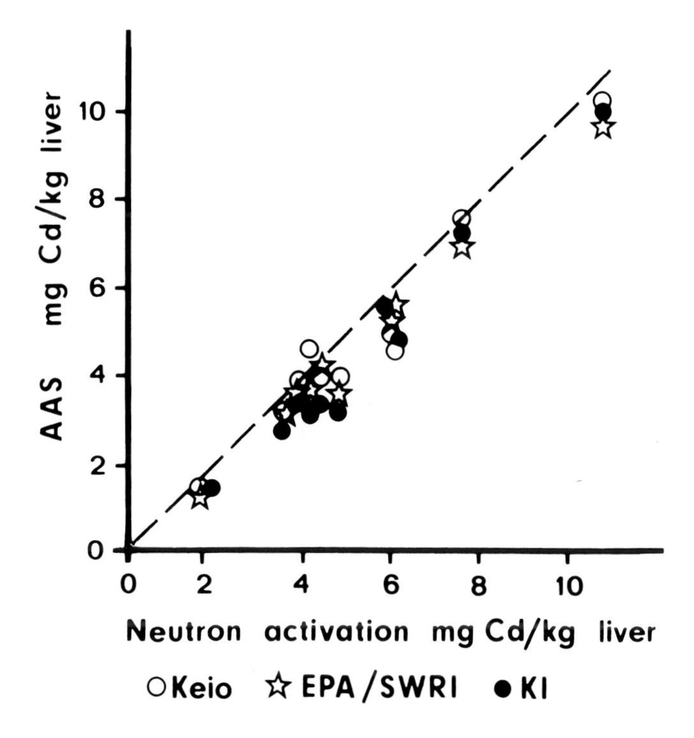

FIGURE 3. Comparison of cadmium analysis in lyophilized liver samples performed by three different laboratories using atomic absorption spectrophotometry (y axis) compared to one laboratory using neutron activation (x axis). Each type of symbol denotes one laboratory. (From Kjellström, T., *Environ. Health Perspect.* 28, 169—197, 1979. With permission.)

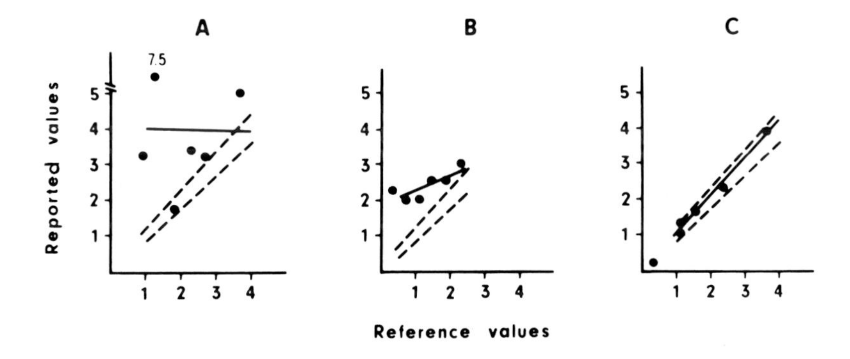

FIGURE 4. Results from three quality control runs, two rejected (a and b) and one (c) accepted. y axis: reported values; x axis: reference values. Solid line is calculated regression line. Dotted lines indicate acceptance interval. (Modified from Vahter, M. Report prepared for United Nations Environment Programme and World Health Organization by National Swedish Institute of Environmental Medicine and Department of Environmental Hygiene, Karolinska Institute, Stockholm, 1982.

continuously check their results by means of quality control analysis, performed *parallel to* the actual monitoring activities. For details reference is made to the reports by Vahter[75] and Friberg and Vahter.[16]

In 1982, the National Food Administration in Sweden summarized data on cadmium in various foodstuffs analyzed by FAO/WHO Collaborating Centres in different countries throughout the world.[51] Samples of freeze-dried vegetables and shellfish in one study by a British laboratory had been sent to 37 of the laboratories participating in the study. This analytical quality assurance exercise revealed that many analytical errors occurred and that the previously reported results could not be regarded as reliable. It is therefore to be recommended that quality assurance exercises should be integrated during the monitoring program itself, and not added merely during the concluding phase.

V. IN VIVO MEASUREMENT OF CADMIUM IN HUMAN LIVER AND KIDNEY

A method for in vivo measurement of cadmium in human liver was presented by McLellan et al.[44] A certain part of a person's body is irradiated by a neutron beam. A naturally occurring isotope, ^{113}Cd, constituting 12.3% of the metal in its native state, arrests neutrons and forms excited ^{114}Cd isotopes. Excited ^{114}Cd decays promptly and emits gamma rays which are detected outside the body. The intensity of the emitted gamma rays is proportional to the cadmium concentration in the liver. Mobile neutron activation equipment is now available,[14,47,72] and by increasing the sensitivity of the method, it has been made possible to measure the cadmium content in the left kidney also. (The left kidney is used in order to diminish interference from the liver on the right side.) In order to obtain a correct measure of the cadmium content, the kidney must first be localized by ultrasonic scan.[47] In kidney it is only possible to measure the amount, or content, of cadmium and not the concentration. Taking the weight of the kidney and the ratio between cadmium concentration in cortex and medulla, it is possible to estimate the approximate concentration of cadmium in kidney cortex, which is considered the critical organ for cadmium (Volume II, Chapters 9 and 13). An alternative method for determination of cadmium concentration in kidney cortex using X-ray-generated atomic fluorescence has also been proposed.[2]

In this context it must be pointed out that the analytical accuracy and precision of in vivo measurements have not been adequately established. They have been tested in models or removed tissue specimens only[47] but not in animal or human tissues *in situ*. There is a great need for studies to confirm that reported cadmium levels in human liver and kidney cortex, measured by in vivo neutron activation analysis, really do agree with the results obtained using the conventional methods described in Section III. A to G of this chapter. One way to examine this, in vivo or at autopsy, would be to take biopsies from the liver and kidney after external measurement and to analyze them using conventional methods.

VI. SUMMARY AND CONCLUSIONS

Accurate determination of cadmium is not a simple task, especially not in low level samples. No ideal, exact, or absolute method exists. Even though the analytical procedure may be very exact, there are, invariably, several problems connected with sampling and preparation of the sample. One analytical technique may be highly accurate for cadmium analysis in one material, but fail in another. It is impossible to conclude that one analytical method is invariably better than another. The choice of method will depend on the cadmium level in the sample, type of sample, type of cadmium bound in matrix, etc.

During the last decades, atomic absorption spectrophotometry (AAS) has become the most commonly used method of analysis. Equipped with facilities for electrothermal atomization (ETA) and automatic compensation for nonspecific atomic absorption, the method enables measurement of cadmium in biological fluids at concentrations as low as 0.1 to 0.3 $\mu g/\ell$. Different electrochemical methods are of particular value for measurement of cadmium in solutions. Activation analysis and spark source mass spectrometry serve as important reference methods which can be used for analytical quality control purposes.

The importance of analytical quality assurance has not always been fully acknowledged. Several interlaboratory studies have pointed to the fact that a great deal of the previously published data is probably in error.

It is recommended that any laboratory performing cadmium analysis should provide evidence of the accuracy of data presented by using internal and external quality control programs as an integrated part of the analytical work. To mention that quality control has been made, or that the laboratory performing the analysis has participated in an interlaboratory comparison program, is not sufficient. Data showing the actual results from quality control runs are desirable.

REFERENCES

1. **Ahlberg, M., Akselsson, R., Brune, D., and Lorenzen, J.,** Proton-induced X-ray analysis of steel surfaces for microprobe purposes, *Nucl. Instrum. Methods,* 123, 385—393, 1975.
2. **Ahlgren, L. and Mattson, S.,** Cadmium in man measured *in vivo* by X-ray fluorescence analysis, *Phys. Med. Biol.,* 26, 19—26, 1981.
3. **Alt, F.,** Comparative determination of cadmium in blood by four different methods (in German) *Fresenius Z. Anal. Chem.,* 308, 137—142, 1981.
4. **Barchet, R. and Wilk, G.,** Studies on the daily cadmium uptake through food by emission spectrography (in German), *Dtsch. Lebensm. Rundsch.,* 10, 348—351, 1980.
5. **Behne, D.,** Problems of sampling and sample preparation for trace element analysis in the health sciences, in *Trace Elements Analytical Chemistry in Medicine and Biology, Proc. 1st Int. Workshop, Neuherberg,* Brätter, P. and Schramel, P., Eds., Walter de Gruyter & Co., New York, 1980, 769—782.
6. **Bock, R.,** *A Handbook of Decomposition Methods in Analytical Chemistry,* (transl., updated and extended by I. L. Marr), Int. Textbook Co., Glasgow, 1978.
7. **Castilho, P. D. and Herber, R. F. M.,** The rapid determination of cadmium, lead, copper and zinc in whole blood by atomic absorption spectrometry with electrothermal atomization. Improvements in precision with a peak-shape monitoring device, *Anal. Chim. Acta,* 94, 269—270, 1977.
8. **Danielsson, L.-G, Jagner, D., Josefson, M., and Westerlund, S.,** Computerized potentiometric stripping analysis for the determination of cadmium, lead, copper and zinc in biological materials, *Anal. Chim. Acta,* 127, 147—156, 1981.
9. **Delves, H. T.,** A micro-sampling method for the rapid determination of lead in blood by atomic absorption spectrophotometry, *Analyst,* 95, 431—438, 1970.
10. **Delves, H. T.,** Analytical techniques for measuring cadmium in blood, in *Proc. Int. Workshop on Biological Indicators on Cadmium Exposure Diagnostic and Analytical Reliability,* Commission of the European Communities, Luxembourg, 1982, 1—15.
11. **Delves, H. T. and Woodward, J.,** Determination of low levels of cadmium in blood by electrothermal atomization and atomic absorption spectrophotometry, *At. Spectrosc.,* 2, 65—67, 1981.
12. **Ediger, R. D. and Coleman, R. L.,** Determination of cadmium in blood by a Delves cup technique, *At. Absorpt. Newsl.,* 12, 3—6, 1973.
13. **Elinder, C.-G., Kjellström, T., Friberg, L., Lind, B., and Linnman, L.,** Cadmium in kidney cortex, liver, and pancreas from Swedish autopsies, *Arch. Environ. Health,* 31, 292—302, 1976.
14. **Ellis, K. J., Morgan, W. D., Zanzi, I., Yasumura, S., Vartsky, D., and Cohn, S. H.,** *In vivo* measurement of critical level of kidney cadmium: dose effect studies in cadmium smelter workers, *Am. J. Ind. Med.,* 1, 339—348, 1980.
15. **Fernandez, F. J. and Kahn, H. L.,** The determination of lead in whole blood by atomic absorption spectrophotometry with the "Delves sampling cup" technique, *At. Absorpt. Newsl.,* 10, 1—5, 1971.
16. **Friberg, L. and Vahter, M.,** Assessment of exposure to lead and cadmium through biological monitoring: results of a UNEP/WHO global study, *Environ. Res.,* 30, 95—128, 1983.
17. **Fukai, R., Oregioni, B., and Vas, D.,** Interlaboratory comparability of measurements of trace elements in marine organisms: results of intercalibration exercise on oyster homogenate, *Oceanol. Acta,* 1(3), 391—396, 1978.
18. **Gajan, R. J., Capar, S. G., Subjoc, C. A., and Sanders, M.,** Determination of lead and cadmium in foods by anodic stripping voltammetry. I. Development of method, *J. Assoc. Off. Anal. Chem.,* 65(4), 970—977, 1982.

19. **Gardiner, P. E., Ottaway, J. M., and Fell, G. S.,** Accuracy of the direct determination of cadmium in urine by carbon-furnace atomic-absorption spectrometry, *Talanta,* 26, 841—847, 1979.
20. **Geldmacher-v. Mallinckrodt, M. and Pooth, M.,** Simultaneous spectrographic analysis of 25 metals and metalloids in biological materials (in German), *Arch. Toxikol.,* 25, 5—18, 1969.
21. **Gleit, Ch. E. and Holland, W. D.,** Use of electrically excited oxygen for the low temperature decomposition of organic substances, *Anal. Chem.,* 34, 1454—1457, 1962.
22. **Golimowski, J., Valenta, P., Stoeppler, M., and Nürnberg, H. W.,** Toxic trace metals in food. II. A comparative study of the levels of toxic trace metals in wine by differential pulse anodic stripping voltammetry and electrothermal atomic absorption spectrometry, *Z. Lebensm. Unters. Forsch.,* 168, 439—443, 1979.
23. **Gorsuch, T. T.,** *The Destruction of Organic Matter,* Vol. 39, Pergamon Press, Oxford, 1970.
24. **Gramlich, J. W., Machlan, L. A., Murphy, T. J., and Moore, L. J.,** The determination of zinc, cadmium and lead in biological and environmental materials by isotope dilution mass spectrometry, in *Trace Substances in Environmental Health — XI,* Hemphill, D. D., Ed., University of Missouri, Columbia, 1977, 376—379.
25. **Halvorsen, C. and Steinnes, E.,** Simple and precise determination of Zn and Cd in human liver by neutron activation analysis, *Z. Anal. Chem.,* 274, 199—202, 1975.
26. **Hasselmann, J., Koenig, W., Richter, F. W., Steiner, U., and Wätjen, U.,** Application of PIXE to trace element analysis in biological tissues, *Nucl. Instrum. Methods,* 142, 163—169, 1977.
27. **Hassler, E.,** Exposure to Cadmium and Nickel in an Alkaline Battery Factory — As Evaluated from Measurements in Air and Biological Material, Doctoral thesis, Karolinska Institute, Stockholm, 1983.
28. **Herber, R. F. M.,** Methods for cadmium analysis in urine: a comparison, in *Proc. Int. Workshop on Biological Indicators on Cadmium Exposure Diagnostic and Analytical Reliability,* Commission of the European Communities, Luxembourg, 1982.
29. **Hislop, J. S.** Choice of the analytical method, in *Trace Element Analytical Chemistry in Medicine and Biology, Proc. 1st Int. Workshop, Neuherberg,* Brätter, P. and Schramel, P., Eds., Walter de Gruyter & Co., New York, 1980, 747—767.
30. **Jagner, D., Josefson, M., and Westerlund, S.,** Simultaneous determination of cadmium and lead in urine by means of computerized potentiometric stripping analysis, *Anal. Chim. Acta,* 128, 155—161, 1981.
31. Japanese Public Health Association, *Studies of Standardization of Analytical Methods for Chronic Cadmium Poisoning,* (in Japanese), Japanese Public Health Association, Toyko, 1970.
32. **Kahn, H. L.,** Principles and practice of atomic absorption, *Adv. Chem. Ser.,* 73, 183—229, 1968.
33. **Kahn, H. L. and Manning, D. C.,** Background correction in atomic absorption spectroscopy, *Am. Lab.,* pp. 51—56, August 1972.
34. **Kjellström, T.,** Exposure and accumulation of cadmium in populations from Japan, the United States and Sweden, *Environ. Health Perspect.,* 28, 169—197, 1979.
35. **Kjellström, T., Lind, B., Linnman, L., and Nordberg, G.,** A comparative study on methods for cadmium analysis of grain with an application to pollution evaluation, *Environ. Res.,* 8, 92—106, 1974.
36. **Kjellström, T., Tsuchiya, K., Tompkins, E., Takabatake, E., Lind, B., and Linnman, L.,** A comparison of methods for analysis of cadmium in food and biological material. A co-operative study between Sweden, Japan and U.S.A., in *Recent Advances in the Assessment of the Health Effects of Environmental Pollution, II, Proc. Int. Symp. Paris,* Commission of the European Communities, Luxembourg, 1975, 2197—2316.
37. **Koirtyohann, S. A. and Hopkins, C. A.,** Losses of trace metals during the ashing of biological materials, *Analyst,* 101, 870—875, 1976.
38. **Koizumi, H., Yasuda, K., and Katayama, M.,** Atomic absorption spectrophotometry based on the polarization characteristics of the Zeeman effects, *Anal. Chem.,* 49(8), 1106—1112, 1977.
39. **LaFleur, P. D.,** Accuracy in trace analysis: sampling, sample handling, analysis, in Proc. 7th Mater. Res. Symp., I and II, U.S. Governmental Printing Office, Washington, D.C., 1976.
40. **Lauwerys, R., Buchet, J. -P., Roels, H., Berlin, A., and Smeets, J.,** Intercomparison program of lead, mercury and cadmium analysis in blood, urine and acqueous solutions, *Clin. Chem.,* 21, 551—557, 1975.
41. **Loon, van, J. D.,** Atomic fluorescence spectrometry. Present status and future prospects, *Anal. Chem.,* 53, 332—361, 1981.
42. **Lundgren, G.,** Direct determination of cadmium in blood with a temperature-controlled heated graphite-tube atomizer, *Talanta,* 23, 309—312, 1976.
43. **Mangelson, N. F., Hill, M. W., Nielson, K. K., Eatough, D. J., Christensen, J. J., Izatt, R. M., and Richards, D. O.,** Proton induced X-ray emission analysis of Pima Indian autopsy tissues, *Anal. Chem.,* 51, 1187—1194, 1979.
44. **McLellan, J. S., Thomas, B. J., Fremlin, H. H., and Harvey, T. C.,** Cadmium — its *in vivo* detection in man, *Phys. Med. Biol.,* 20, 88—95, 1975.
45. **Menden, E. E., Brockman, D., Choudhury, H., and Petering, H. G.,** Dry ashing of animal tissues for atomic absorption spectrometric determination of zinc, copper, cadmium, lead, iron, manganese, magnesium and calcium, *Anal. Chem.,* 49, 1644—1645, 1977.

46. **Michel, R. G., Hall, M. L., and Ottaway, J. M.,** Determination of cadmium in blood and urine by flame atomic-fluorescence spectrometry, *Analyst*, 104, 491—504, 1979.
47. **Morgan, W. D., Vartsky, D., Ellis, K. J., and Cohn, S. H.,** A comparison of ^{252}Cf and ^{238}Pu, Be neutron sources for partial-body *in vivo* activation analysis, *Phys. Med. Biol.*, 26(3), 413—424, 1981.
48. **Mulford, C. E.,** Low-temperature ashing for the determination of volatile metals by atomic absorption spectroscopy, *At. Absorpt. Newsl.*, 5(6), 135—139, 1966.
49. **Nackowski, S. B., Putnam, R. D., Robbins, D. A., Varner, M. O., Whit, L. D., and Nelson, K. W.,** Trace metal contamination of evacuated blood collection tubes, *Am. Ind. Hyg. Assoc. J.*, 38, 503—508, 1977.
50. National Bureau of Standards, Standard Reference Material, General Science Corporation, Washington, D.C., 1980.
51. National Food Administration, Summary and Assessment of Data Received from the FAO/WHO Collaborating Centres for Food Contamination Monitoring, National Food Administration, Uppsala, Sweden, 1982.
52. **Nise, G. and Vesterberg, O.,** Sampling vessels influence on content of heavy metals (Cd, Cr, Hg, Ni and Pb) in blood and urine samples (in Swedish), *Läkartidningen*, 75, 19—20, 1978.
53. **Nixon, D. E., Fassel, V. A., and Kniseley, R. N.,** Inductively coupled plasma-optical emission analytical spectroscopy. Tantalum filament vaporization of microliter samples, *Anal. Chem.*, 46, 210—213, 1974.
54. **Nürnberg, H. W.,** A critical assessment of the voltammetric approach for the study of toxic metals in biological specimens, in *Electroanalysis in Hygiene, Environmental, Clinical and Pharmaceutical Chemistry: Analytical Chemistry Symposia Series II*, Smyth, W. F., Ed., Elsevier, New York, 1980, 351—372.
55. **O'Laughlin, J. W., Hemphill, D. D., and Pierce, J. O.,** *Analytical Methodology for Cadmium in Biological Matter — A Critical Review*, International Lead Zinc Research Organization, New York, 1976.
56. **Parr, R. M.,** The reliability of trace element analysis as revealed by analytical reference materials, in *Trace Element Analytic Chemistry in Medicine and Biology*, Walter de Gruyter & Co., New York, 1980, 631—655.
57. **Paus, P. E.,** Bomb decomposition of biological materials, *At. Absorpt. Newsl.*, 11(6), 129—130, 1972.
58. **Piscator, M.,** in *Cadmium in the Environment*, Friberg, L., Piscator, M., and Nordberg, G. F., Eds., CRC Press, Boca Raton, Fla., 1971.
59. **Piscator, M. and Vouk, V. B.,** Sampling and analytical methods, in *Handbook on the Toxicology of Metals*, Friberg, L., Nordberg, G. F., and Vouk, V. B., Eds., Elsevier, Amsterdam, 1979, 33—45.
60. **Pleban, P. A., Kerkay, I., and Pearson, K. H.,** Polarized Zeeman-effect flameless atomic absorption spectrometry of cadmium, copper, lead and manganese in human kidney cortex, *Clin. Chem.*, 27, 68—72, 1981.
61. **Raaphorst, van, J. G., Weers, van, A. W., and Haremaker, H. M.,** On the loss of cadmium, antimony and silver during dry ashing of biological material, *Fresenius Z. Anal. Chem.*, 293, 401—403, 1978.
62. **Salmela, S. and Vuori, E.,** Contamination with cadmium from micropipette tips, *Talanta*, 26, 175—176, 1979.
63. **Saltzman, B. E.,** Colorimetric microdetermination of cadmium with dithizone, *Anal. Chem.*, 25, 493—496, 1953.
64. **Sperling, K. -R. and Bahr, B.,** Determination of heavy metals in sea water and in marine organisms by flameless atomic absorption spectrophotometry, *Fresenius Z. Anal. Chem.*, 306, 7—12, 1981.
65. **Stoeppler, M.,** Analysis of cadmium in biological materials, in *Cadmium 81, Proc. 3rd Int. Cadmium Conf., Miami*, Cadmium Association, London, 1982, 95—102.
66. **Stoeppler, M.,** Analytical aspects of sample collection, sample storage and sample treatment, in *Trace Element Analytical Chemistry in Medicine and Biology, Proc. 2nd Int. Workshop, Neuherberg, April 1982*, Brätter, P. and Schramel, O., Eds., Walter de Gruyter & Co., New York, 1983.
67. **Stoeppler, M. and Brandt, K.,** Determination of lead and cadmium in whole blood by electrothermal atomic absorption spectroscopy, in *International Symposium — Clinical Biochemistry on Diagnosis and Therapy of Porphyrias and Lead Intoxication*, Doss, M., Ed., Springer-Verlag, Basel, 1978, 185—187.
68. **Stoeppler, M. and Brandt, K.,** Contributions to automated trace analysis. V. Determination of cadmium in whole blood and urine by electrothermal atomic-absorption spectrophotometry, *Fresenius Z. Anal. Chem.*, 300, 372—380, 1980.
69. **Stoeppler, M., Valenta, P., and Nürnberg, H. W.,** Applications of independent methods and standard materials: an effective approach to reliable trace and ultratrace analysis of metals and metalloids in environmental and biological matrices, *Fresenius Z. Anal. Chem.*, 297, 22—34, 1979.
70. **Subramanian, K. S. and Méranger, J. C.,** A rapid electrothermal atomic absorption spectrophotometric method for cadmium and lead in human whole blood, *Clin. Chem.*, 27, 1866—1871, 1981.
71. **Sunderman, F. W., Jr., Brown, S. S., Stoeppler, M., and Tonks, D. B.,** Interlaboratory evaluations of nickel and cadmium analyses in body fluids, in *IUPAC Collaborative Interlaboratory Studies in Chemical Analysis*, Egan, H. and West, T. S., Eds., Pergamon Press, New York, 1982, 25—35.

72. **Thomas, B. J., Harvey, T. C., Chettle, D. R., McLellan, J. S., and Fremlin, J. H.,** A transportable system for the measurement of liver cadmium *in vivo, Phys. Med. Biol.*, 24, 432—437, 1979.
73. **Tipton, I. H., Cook, M. J., Steiner, R. L., Boyes, C. A., Perry, H. M., and Schroder, H. A.,** Trace elements in human tissue. I. Methods, *Health Phys.*, 9, 89—101, 1963.
74. **Ulander, A. and Axelson, O.,** Measurement of blood-cadmium levels, *Lancet*, 1, 682—683, 1974.
75. **Vahter, M., Ed.,** Assessment of human exposure to lead and cadmium through biological monitoring, Report prepared for United Nations Environment Programme and World Health Organization by National Swedish Institute of Environmental Medicine and Department of Environmental Hygiene, Karolinska Institute, Stockholm, 1982.
76. **Valenta, P., Rützel, H., Nürnberg, H. W., and Stoeppler, M.,** Trace chemistry of toxic metals in biomatrices. II. Voltammetric determination of the trace content of cadmium and other toxic metals in human whole blood, *Fresenius Z. Anal. Chem.*, 285, 25—34, 1977.
77. **Vesterberg, O. and Bergström, T.,** Determination of cadmium in blood by use of atomic absorption spectroscopy with crucibles — and a rational procedure for dry-ashing, *Clin. Chem.*, 23(3), 555—559, 1977.
78. **Vesterberg, O. and Wrangskogh, K.,** Determination of cadmium in urine by graphite-furnace atomic absorption spectroscopy, *Clin. Chem.*, 24, 681—685, 1978.
79. WHO/UNEP, Pilot project on assessment of human exposure to pollutants through biological monitoring, Report on a planning meeting on quality control, February 26 to March 2, 1979, HCS/79.4, World Health Organization, Geneva, 1979.
80. **Zief, M. and Mitchell, J. W., Eds.,** *Contamination Control in Trace Element Analysis*, Vol. 47, John Wiley & Sons, New York, 1976.

Chapter 3

CADMIUM: USES, OCCURRENCE, AND INTAKE

Carl-Gustaf Elinder

TABLE OF CONTENTS

I. GENERAL CHEMISTRY

Cadmium is a soft, silvery, and ductile metal. Some of its physical and chemical properties are given in Table 1. Cadmium is readily volatilized at high temperatures, e.g., during welding or smelting, producing a fine dust of cadmium oxide particles with a typical yellow-brownish color. In most of its chemical properties, cadmium resembles zinc. Like zinc, cadmium is always divalent (2+) in stable compounds such as in the acetate, carbonate, chloride, oxide, stearate, sulfate, sulfide, and cadmium sulfoselenide compounds. The water solubility of inorganic cadmium compounds varies from the readily soluble acetate, chloride, and sulfate, to the almost insoluble carbonate, oxide, and sulfide. Certain organic cadmium compounds, in which cadmium is bound directly to a carbon atom, can be synthesized.

The physical and chemical properties of all major cadmium compounds are described in reviews by Aylett[11] and Farnsworth.[58]

II. PRODUCTION AND USES

Cadmium is obtained as a by-product principally in the refining of zinc and to a lesser extent during the refining of copper and lead, from sulfide ores. The ratio between cadmium and zinc in ore ranges from 0.1 to 1%.[60]

Normally, approximately 1 ton of cadmium is obtained from the refining of 300 tons of zinc.

Production and consumption of cadmium have increased considerably since the beginning of this century. Then, less than 10 tons were refined annually, but by the late 1970s, world production was in the order of 17,000 tons/year. Major cadmium-producing countries in 1982 were U.S.S.R., Japan, Canada, U.S., and West Germany.[203a] Only a minor portion of all refined cadmium is recycled, a fact which points to an apparent risk for contamination of the biosphere.[67,180]

Major worldwide uses of cadmium are in electroplating (35%), pigments (25%), plastic stabilizers (15%), and batteries (15%).[181] Substantial variations in regard to the types of use exist between different countries. For example, in order to decrease environmental contamination caused by cadmium emission, Sweden has recently prohibited the use of cadmium in pigments and plastic stabilizers.[104,177]

Cadmium plating protects steel and iron objects from corrosion. Zinc coatings are generally preferable from the environmental and occupational health points of view, but are less effective in certain circumstances, e.g., when plated objects are in close contact with sea water. Cadmium plating is used extensively on, e.g., nuts, bolts, and screws in the automobile and aircraft industries.

Cadmium pigments, mainly cadmium sulfide and cadmium sulfoselenide, with colors from yellow to deep red, are used in various types of products, especially plastics, ceramics, and paints. About 60 to 80% of all cadmium pigments produced today is used in the coloring of plastics. The advantages of cadmium pigments are high temperature tolerance and stability to light.[42] Certain other compounds, mainly cadmium stearate, are used as stabilizers in plastics, especially in PVC. Cadmium stabilizers inhibit the deterioration processes which take place within the plastic, and which may lead to darkening, hardening, and embrittlement. In rechargeable nickel-cadmium batteries, cadmium hydroxide constitutes one of the electrodes, and nickel hydroxide the other. In larger cells, an alkaline electrolyte is usually employed. Formerly, production was concentrated on the larger cells used in, e.g., aircrafts and locomotives and for lighting and telephone systems. Recently, small cadmium batteries have been developed for use in different types of portable electronic equipment, such as calculators, flashlights, etc. At present, there is a rapidly growing market for small rechargeable cadmium batteries.

Miscellaneous uses constitute about 10% of worldwide cadmium consumption. Indeed,

Table 1
PROPERTIES OF ELEMENTAL CADMIUM

Atomic weight	112.41
Atomic number	48
Specific gravity (25°C)	8.65
Melting point (°C)	321
Boiling point (°C)	765
Vapor pressure (mmHg)	
394°C	1.0
578°C	60
711°C	400
767°C	760
Major stable isotopes of cadmium (natural abundance)	%
^{106}Cd	1.2
^{108}Cd	0.9
^{110}Cd	12.4
^{111}Cd	12.8
^{112}Cd	24.1
^{113}Cd	12.3
^{114}Cd	28.9
^{116}Cd	7.6

Note: ^{109}Cd is a gamma-emitting isotope with a half-life of 450 days. This isotope is frequently used in different types of animal experiments. ^{115}Cd is another isotope which emits β-particles and has a half-life of 53.4 hr.

cadmium has a large number of industrial applications and was termed "the dissipated element" by Fulkerson and Goeller.[67] One important use is in certain copper alloys. An addition of 0.3 to 1% cadmium to copper increases its physical resistance especially at high temperatures. Small quantities of cadmium compounds are also used in the chemical, nuclear energy, electrical, and electronic industries. Other applications worth mentioning are in television tubes, radar equipment, and solar and photoelectric cells.

More information about worldwide production and the uses of cadmium and its compounds is given in a review by Nriagu[181] and in the proceedings of the three International Cadmium Conferences — Cadmium 77,[30] Cadmium 79,[31] and Cadmium 81.[32] National data on predominant uses of cadmium are available from, e.g., Japan,[250] Denmark,[173] U.K.,[42] Sweden,[104] and in the annual publications from the U. S. Bureau of Mines.

III. OCCURRENCE

Many publications are available giving data on cadmium in various media, such as air, water, soils, and food. Some of them are obviously in error. Usually these data are several times too high. In most of the publications, information about analytical accuracy is scarce. It is, therefore, difficult to be wholly conclusive about the true concentrations of cadmium in various media. We have evaluated the published analytical results carefully in order to identify likely erroneous data. As quality assurance exercises have rarely been made, in most cases it is not possible to be certain whether the analytical results are reasonably accurate or not. In the absence of better results, analytical data presented without quality assurance will have to be considered. Even if some of the results cited below are in error, it is our belief that such analytical errors will not invalidate the general conclusions drawn in this chapter concerning environmental levels and intake of cadmium.

A. Global Emissions and Changes in Human Exposure

Although a rare element, cadmium is found in small quantities throughout the lithosphere with an average crustal abundance of 0.1 to 0.2 mg Cd per kilogram.[60,85] Cadmium is closely related to zinc and is found wherever zinc is found. Zinc, in contrast to cadmium, is an essential metal for most forms of life.[255] Because of the close relationship of the two metals, it is unlikely that any naturally occurring material will be *completely* free from cadmium. Human beings have always been exposed to low levels of cadmium from the environment. Even though cadmium has been recognized for a relatively short period of time, pollution has taken place for several thousands of years, i.e., ever since the start of metal refining of ores containing cadmium. In this century the widespread industrial use of cadmium and cadmium compounds has caused a sharp increase in emissions to the environment. Cadmium is emitted into the air and water by, e.g., metal smelters and industries utilizing cadmium in accumulators, alloys, paints, and plastics. Combustion of household waste is an important source of emission since this waste frequently includes different types of cadmium-containing products, such as plastics and metal scrap. Agricultural use of fertilizers, either phosphates or sewage sludge, also contributes to contamination. The concentration of cadmium in oil and coal is usually below 0.2 mg/kg.[112,146] In 18 crude oil samples analyzed by Stoeppler,[232] the cadmium concentration was less than 1 μg/kg. According to different estimates (Table 2),[112,180] oil and coal combustion do not contribute very much to the overall worldwide anthropogenic emission of cadmium, but may, via deposition of acidifying substances such as sulfur dioxide and nitrous oxides, contribute to the acidification of soil. The acidification of soil may result in an increased uptake of cadmium in plants and crops (see Section III.C). Within the European Economic Community (EEC), attempts have been made to quantify the major sources of cadmium inputs to the environment of the member states.[93] The total atmospheric emissions were estimated at 132 metric tons/year, major sources being iron and steel industries (34 tons/year) and waste incineration (31 tons/year). The total input of cadmium to soil was estimated to be about 3000 tons/year, with waste disposal as a major source (1434 tons/year). Inputs to soil from sewage sludge and phosphate fertilizers were estimated to be 130 and 346 tons/year, respectively. The latter two sources are of more concern to human health, since cadmium from these two sources will be available to plants, whereas most of the waste is deposited in the form of landfills. The total emission of cadmium into water was estimated to be 273 tons/year.

Table 2 summarizes the worldwide emission of cadmium to the atmosphere from natural and anthropogenic sources as estimated by Nriagu.[180,181] The data are of interest since they indicate that man-made emissions surpass natural emissions by a factor close to 10. Nevertheless, the values given in the table should be regarded with some caution. If one or two of the concentrations used in the calculation of worldwide emission is in error, the estimated total amount could also show a considerable error. The natural emissions from volcanic eruptions especially are extremely difficult to estimate.[182]

Some of the cadmium emitted into the air is inhaled by humans and animals, but most of it is deposited. The annual deposition of cadmium is about ten times higher in industrialized parts of Europe and the U.S., when compared to nonindustrialized areas (see Section III.B.3).

Attempts to use ice cores from glaciers to measure the cadmium deposition which occurred in earlier centuries and to compare this to the present deposition have been made by several groups of investigators.[23,87,103,265] Weiss et al.[265] reported that in Greenland, snow deposited between 1807 and 1917 at one particular site had cadmium concentrations ranging from 0.002 to 0.144 μg Cd per kilogram water, with an average of 0.034 μg/kg. Snow deposited between 1966 and 1971 at the same site contained on an average 0.64 μg/kg.[265] At sampling

Table 2
ESTIMATED WORLDWIDE EMISSIONS OF CADMIUM TO THE ATMOSPHERE FROM NATURAL AND ANTHROPOGENIC SOURCES IN 1975[180]

Source (annual emission, ton/year)	Cd	(range)
Natural sources		
Windblow dust	100	(1—220)
Forest fires	12	(1—70)
Volcanogenic particles	520	(300—7800)
Vegetation	200	(200—2700)
Seasalt sprays	1	
Total	833	
Anthropogenic emissions		
Mining, nonferrous metals	2	
Primary nonferrous metal production		
Cd	110	
Cu	1600	
Pb	200	
Zn	2800	
Secondary nonferrous metal production	600	
Iron and steel production	70	
Industrial applications	50	
Coal combustion	60	
Oil (including gasoline combustion)	3	
Wood combustion	200	
Waste incineration	1400	
Manufacture, phosphate fertilizers	210	
Miscellaneous	—	
Total	7305	

stations still further north on Greenland, no significant increase with time could be revealed.[265] The latter data (0.64 μg/kg) appear unusually high in comparison to other reported data on cadmium concentrations in Arctic or Antarctic ice. Herron et al.[87] were not able to confirm the data by Weiss et al.[265] and found no difference in the cadmium concentration in ice cores from pre-1900 and post-1960. Both had an average cadmium concentration of 0.008 μg/kg. In ice cores obtained from an Antarctic glacier, Boutron and Lorius[23] found a clear tendency for increased cadmium concentration between 1914 and 1971, though this tendency was not significant. Cadmium levels in the Antarctic snow samples ranged from 0.001 to 0.025 μg Cd per kilogram ice. Even lower concentrations, an average of 0.0004 μg/kg, were found in freshly fallen snow collected in the Arctic under ultraclean conditions.[153]

In a Scandinavian glacier, a lower increase than that seen in southern Greenland was found, i.e., about 1 μg Cd per kilogram ice in 1870, lower values before that; and an average of 4 μg Cd per kilogram ice in snow deposited between 1954 and 1972.[103]

The evidence suggesting that the cadmium concentration in Arctic snows has increased during the last century is under dispute,[21,22,137,181] especially in view of the difficulties encountered in obtaining noncontaminated and representative samples of Arctic snow.[183] At a recent meeting on Changing Metal Cycles and Human Health,[182] it was concluded that there is, as yet, no conclusive evidence to prove global contamination by cadmium. On the other hand, there is a large number of reports on local and regional contamination of cadmium (Section III.D.1).

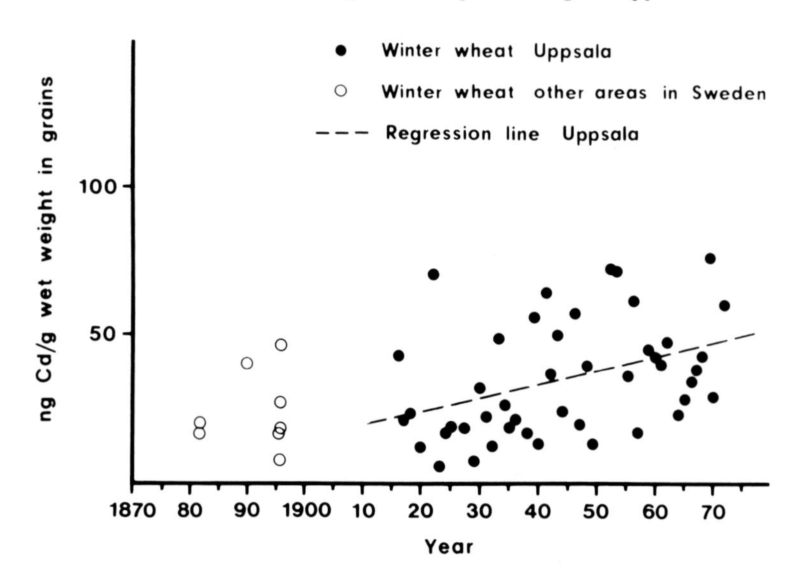

FIGURE 1. Cadmium concentration in winter wheat samples grown close to the city of Uppsala and elsewhere in Sweden. (From Kjellström, T., Lind, B., Linnman, L., and Elinder, C.-G., *Arch. Environ. Health*, 30, 321—328, 1975b. With permission.)

Deposited cadmium can give rise to human exposure via different routes. Cadmium deposited directly on the soil can be absorbed by plants and crops, which may eventually result in increased peroral exposure. Cadmium deposited in water may increase the concentration of cadmium in seafoods, such as oysters and mussels. In the event of flooding or even due to irrigation, cadmium in water might lead to an increase in the cadmium concentration in soil and consequently give rise to an increase in levels of cadmium in the agricultural products consumed by humans or farm animals. Farm animals having long life spans and grazing on cadmium-polluted fields, or given cadmium-polluted feed, will accumulate high concentrations of cadmium especially in liver and kidneys and to a certain extent in muscle. People consuming large amounts of liver and kidney from such animals may risk excessive exposure.

Available data from Sweden indicate that the general population in Sweden is exposed to increasing amounts of cadmium. Kjellström et al.[122] analyzed cadmium in two series of winter and spring wheat samples, one from the end of the 19th century and another from 1916 to 1972. Specimens from the period 1916 to 1972 were of the same genotype and had been harvested from the same fields. Cadmium concentration tended to increase over time. There was, however, considerable scatter in the data, and the increase was statistically significant for winter wheat only (Figure 1). Elinder and Kjellström[51] measured cadmium concentrations in human kidneys from the 19th century, preserved at two anatomical museums in Sweden. The geometric mean cadmium concentration in kidney cortex of 33 old kidneys was 15 mg Cd per kilogram dry weight, with a 95% confidence interval between 11 to 20 mg/kg. This was significantly lower than the corresponding figure for 24 nonsmoking adults who died in 1974. The average cadmium concentration in the latter group was 57 mg Cd per kilogram dry weight with a 95% confidence interval between 46 and 71 mg/kg. The data are, however, not absolutely conclusive since some cadmium may have been lost from the kidney samples while in the storage containers.

Although both of these studies were carried out in Sweden, there is no reason to believe that Sweden is subject to higher cadmium contamination than the rest of the industrialized world. Normal values for cadmium in human tissue in Sweden are no higher than the values obtained from other industrialized countries (see Chapter 5). The daily intake of cadmium via food in Sweden is similar to, or lower than, that in most other countries (see Section IV.).

In another study on historical specimens preserved in West Germany, Drasch[48] found a similar significant tendency for an increase in human kidney samples. The average kidney cortex concentration was less than 1 mg/kg wet weight in preserved kidneys from 1897 to 1914 compared to 41 mg/kg in samples collected during 1980 to 1981. There was, however, no increase in the cadmium concentration in liver samples collected between 1897 to 1914 compared to liver samples from 1980 to 1981. This phenomenon is difficult to explain and makes the interpretation of the results doubtful.

For further information on the behavior of cadmium in the environment, in different ecosystems and speciation in different media, reference is made to *The Chemistry, Biochemistry and Biology of Cadmium*[263] and *Cadmium in the Environment, Part I: Ecological Cycling*.[181]

B. Cadmium in Air

1. Ambient Air

Measurements of cadmium in air have been made in many areas and countries during the last decades. Data from different reports have been compiled by, e.g., CEC[36] and Department of the Environment.[42] In remote uninhabited areas cadmium concentrations in air are generally less than 1 ng/m^3. Heindryckx et al.[84] e.g., reported cadmium concentrations of an average of 0.1 ng/m^3 from northern Norway and 0.3 ng/m^3 from a Swiss mountain area. Air samples collected over the Atlantic Ocean between Iceland and the Bermudas ranged from 0.003 to 0.62 ng/m^3.[49] An average of less than 0.2 ng/m^3 has been reported from the Antarctic.[41]

Similarly, a low monthly average (0.9 ng/m^3) has been reported from sparsely populated rural areas in Sweden,[201] whereas moderately higher averages (1 to 5 ng/m^3) have been found in more populated rural areas in Central Europe, U.S., and Japan.[45,59,93,114,131,194,216]

In urban areas, cadmium concentrations in air are higher and more variable, depending on degree and type of industrialization. Monthly averages in U.S., Europe, and Japan usually range from about 5 to 40 ng/m^3.[40,43,80,93,107,114,125,131,254]

Cadmium concentrations in ambient air may be considerably higher in the vicinity of cadmium-emitting industries. In Sweden in 1970, weekly means of 300 ng/m^3 were recorded on several occasions, 0.5 km from a factory producing cadmium-copper alloys. At a distance of 100 m from the source, a monthly mean value of 600 ng/m^3 was obtained. The highest concentration found in a 24-hr sample was 5400 ng/m^3.[201] Total emissions from the main chimney were estimated to be in the order of 8 tons Cd per year at the time of measurement. In 1971, the main source of pollution by cadmium was equipped with a textile filter after which emissions decreased by more than 95%.[209a]

Similarly, very high levels of cadmium in ambient air have been reported close to zinc smelters in Japan and the U.S. A 3-day mean of 560 ng/m^3 was recorded at a distance of about 0.5 km from a zinc refinery in the Fukuoka Prefecture, Japan.[163,223] The maximum 24-hr average was 3000 ng/m^3. In the vicinity of another zinc smelter in Japan, Annaka, in the Gunma Prefecture, average 24-hr air samples varied from 14 to 166 ng/m^3.[272] In the city of East Helena, Montana, a maximum 24-hr value of 700 ng/m^3 was reached at a sampling station about 1 km away from the smelter.[91] The highest reported ambient air levels of cadmium have come from the U. K. Muskett et al.[171] found levels reaching 11,000 ng/m^3 in the close vicinity of a small secondary metal refinery in London. The abovementioned results display extreme values. Usually cadmium in air is well below 100 ng/m^3 even in highly industrialized areas.[36]

It is important to recognize the form in which cadmium occurs in air in order to estimate the degree of pulmonary absorption of inhaled cadmium air pollutants. Unfortunately, this is in fact not very well known. Probably cadmium oxides constitute the major part of airborne cadmium, but other forms, e.g., chlorides and sulfates, may occur. Knowledge of particle size is also essential for calculating deposition and absorption in the lungs. Lee et al.[143,144] measured the size distribution of cadmium particles in urban and suburban Cincinnati and in St. Louis. Between 40 and 65% of the particles had an aerodynamic diameter below 2 μm. The mass median diameter (MMD) ranged from 1.5 to 10 μm. Dorn et al.[45] measured cadmium in air close to a lead smelter and in a rural area and found that about 70% of the particles in the smelter area and 40% in the rural area were below 2 μm.

Thus, cadmium levels in rural and remote areas are very low, usually below 5 ng/m^3. In industrialized cities, the levels in air are slightly higher, up to 40 ng/m^3. Close to sources of pollution, e.g., near cadmium smelters with insufficient cleansing of exhaust gases, concentrations in ambient air several hundred times greater than the background level may be found.

2. *Workroom Air*

In factories where cadmium is produced, or utilized in manufacturing processes, cadmium concentrations in air are often several magnitudes higher than those in outdoor air. Nowadays, very excessive exposure to cadmium in air, exceeding 1 mg/m^3, is only likely to occur when cadmium-containing materials are smelted, welded, or soldered. High concentrations of total cadmium dust in workroom air (median of 0.6 mg/m^3) have also been reported from a factory producing different types of cadmium salts.[140]

Table 3 presents reported results from measurements of cadmium in workroom air in different types of industries. It should be pointed out that most of these data are based on the very limited number of estimates and that the methods are usually only mentioned very briefly. This makes the reported results very approximate. Nevertheless, it is evident from the table that the order of occupational exposure is highly variable even in comparable industries. Installation of effective exhaust ventilation systems, enclosure of cadmium dust, and oxide-generating processes have made it technically feasible to maintain cadmium exposure levels relatively low, below 20 μg/m^3. This is illustrated in Figures 2 and 3 in which two Swedish cadmium-using industries — one cadmium battery plant (Figure 2) and one cadmium-copper alloy-producing factory (Figure 3) — are shown. Table 3 and Figures 2 and 3 present data on total dust. However, it is known that particles with a mass median aerodynamic diameter of less than 5 μm are more prone to deposit in the alveoli than larger particles.[240] Therefore, "respirable" particles having a mass median aerodynamic diameter less than 5 μm should be measured at the same time as the total dust if possible. The proportion of respirable cadmium to the total amount of cadmium dust in workroom air varies from one type of industry to another. In factories where cadmium oxide fume is generated, e.g., during smelting, most of the total cadmium content in air is respirable, whereas in other industries sometimes only a fraction of the total cadmium dust is "respirable".[156]

3. *Dustfall*

Background data on deposition of cadmium in rural subarctic areas in Finland are available.[136] The deposition varied from less than 1 μg/m^2/month in winter to 6 μg/m^2/month in summer, giving an annual deposition of about 40 to 50 μg/m^2. In the arctic (Greenland-Spitsbergen), the yearly fallout has been estimated to be as low as 0.15 μg/m^2.[153] In rural areas of Sweden, annual (12-month) deposition of cadmium has been estimated to be about 90 μg/m^2.[7] In Denmark, which is closer to the highly industrialized areas of Central Europe, the annual deposition of cadmium has been estimated to be about 200 μg/m^2.[75] The combined wet and dry deposition of cadmium for 1 year (1979 to 1980) in the Federal Republic of Germany ranged from 360 to 1350 μg/m^2.[210] Wet deposition at different locations in West Germany

Table 3
CADMIUM CONCENTRATIONS IN WORKROOM AIR (TOTAL DUST) IN DIFFERENT TYPES OF CADMIUM-USING INDUSTRIES

Type of industry	Year	Country	Concentration (μg/m³)	Ref.
Cadmium smelting or production	1947	U.S.	40—31,000	205
	1974	Belgium	4—27,000[a]	139
	1976	U.S.	>200	226
	1975—78	Finland	6.3—11.0	4
Cadmium-copper alloys	1951—53	U.K.	100—400	138
	1954	U.K.	1—270	113
	1959	U.K.	10—80	19
	1961—71	Sweden	100—200	209a
	1971—75	Sweden	40	209a
	1975—	Sweden	20	209a
Cadmium-silver alloys	1967	Japan	68—241	248
Cadmium battery plants	1946	Sweden	600—15,000	61
	1947—61	Sweden	300—800	119
	1962—74	Sweden	50	119
	1974—	Sweden	5—20	3,81
	1969	U.K.	100	2
	1974	Belgium	1—465[b]	139
Cadmium pigment	1973	Japan	20—7000	78
	1981	Australia	200—1600	44
Cadmium salt production	1979	Belgium	100—2125	140
Cadmium PVC stabilizers	1965	Japan	20—700	234
Cadmium electronic workshop	1974	Belgium	7—19[c]	139

[a] The highest concentration of respirable cadmium being 65 μg/m³.
[b] The highest concentration of respirable cadmium being 4 μg/m³.
[c] The highest concentration of respirable cadmium being 88 μg/m³.

during the same year (1980) ranged from 140 to 2500 μg/m² with a median of 350 μg/m².[184] These data from West Germany indicate that the wet deposition is predominant. Hutton[93] considered 300 μg/m²/year as the most likely estimate for atmospheric input of cadmium to agricultural soils in the rural areas of the European Economic Community.

In residential, commercial, and industrial areas of 77 cities in the U.S., monthly average cadmium dustfall was 40, 63, and 75 μg/m², respectively,[92] corresponding to about 600 μg/m² annually. These estimates appear somewhat higher than those for western Europe.

Deposition of cadmium dust is considerably higher close to emission sources. Within 100 m of the Swedish cadmium-emitting industry, mentioned in Section III.B.1, monthly dustfall, between 1968 and 1970, ranged from 4,000 to 40,000 μg/m². Deposition decreased rapidly with increasing distance from the factory.[190] From the German Democratic Republic, Börlitz[20] reported on deposition of cadmium at different distances from two smelters. Monthly depositions ranged from 130 to 8600 μg/m² at distances ranging from 1.5 to 14 km from the emission sources. In the city of East Helena, Montana, values of 1,000 to 40,000 μg/m²/month were found within a distance of 1 km from a cadmium-emitting smelter.[91] Monthly cadmium deposition within 2 km from a large U.S. lead smelter in Boss, Missouri, ranged

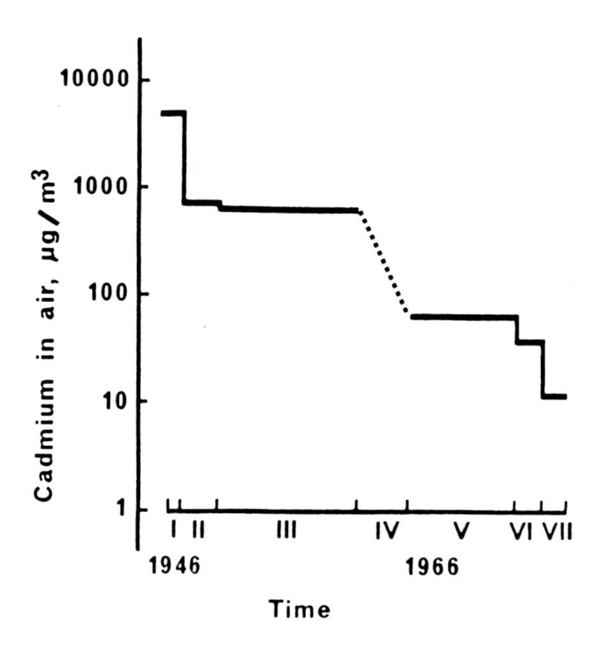

FIGURE 2. Changes in average cadmium concentration in workroom air over time in a cadmium battery plant.[81]

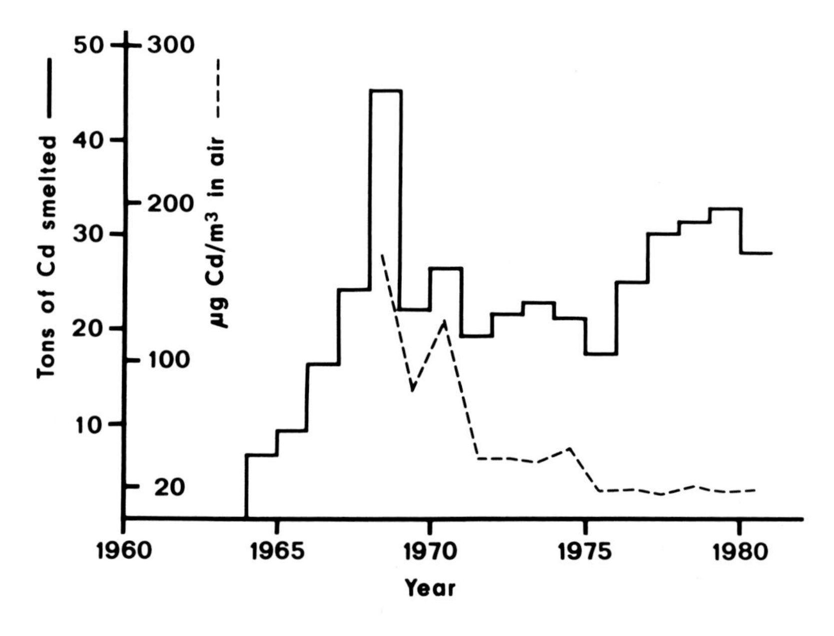

FIGURE 3. Changes in average cadmium concentration in workroom air over time in a cadmium-copper alloy-producing plant.[209a]

from 500 to 20,000 $\mu g/m^2$.[69] Hirata[89] has summarized studies on heavy metal pollution, especially cadmium, in the vicinity of three major zinc smelters — Bandai, Kurobe, and Annaka — in Japan. The monthly cadmium dustfall at 2 km from the Annaka smelter was 4900 $\mu g/m^2$ in 1968, i.e., before the discharge facilities were improved in 1970 and 1971. After reconstruction of the smelters, atmospheric emissions dropped from 1200 to about 50 kg/month (4% of the earlier emission). In spite of this marked decrease in the amount of

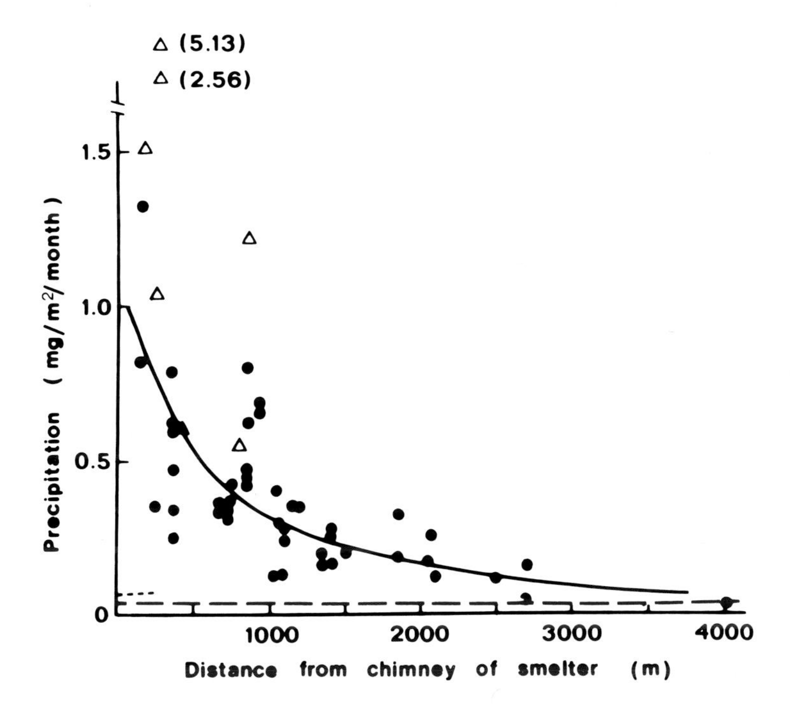

FIGURE 4. Cd precipitation related to distance from Annaka smelter. Collected in dust jars at a height of 2 m from soil surface. Sampling period: (△) October 1971 to October 1973; (●) August 1974 to April 1978; (———) average of six places in Tokyo, July 1976 to June 1978; (- - -) presumed natural abundance at Annaka City. (From Hirata, H., *Heavy Metal Pollution in Soils of Japan*, Kitagishi, K. and Yamane, I., Eds., Japan Scientific Societies Press, Tokyo, 1981, 149—163. With permission.)

cadmium emitted, there was still substantial cadmium dustfall in the vicinity of the smelter between 1973 and 1978 (Figure 4). At 2 km from the smelter, the average was about 200 μg/m²/month, which is about 4% of the dustfall in 1968. This agrees well with the decrease in emission.

4. Indirect Methods for Measurement of Cadmium Deposition and Cadmium in Air

Mosses can be used as a convenient and readily available material for the assessment of air pollution and deposition of cadmium.[213,253] The basis for this phenomenon is that the mineral and cation metabolism of mosses are dependent on the retention of metals taken up from the air and from precipitation. Uptake from the ground is virtually nonexistent and, therefore, the concentration of several metals, e.g., cadmium, copper, lead, mercury, nickel, and zinc, in mosses is proportional to the combined contribution from particles in air and deposited particles.

Background levels of cadmium in mosses from nonpolluted areas are about 0.2 to 0.8 mg/kg Cd dry weight.[212] Increased levels are found in industrial areas and in the vicinity of cadmium-emitting industries. Concentrations in moss exceeding 10 mg/kg may be found close to pollution sources.[29,190] Figure 5 shows regional variations in cadmium concentrations in moss in Sweden, around 1970 and in 1980. At the end of the 1960s, the major cadmium-emitting industries in Sweden were located at the center of the three areas along the east coast of Sweden where cadmium concentrations in moss samples exceeded 1 mg/kg. During the 1970s, emission from several of the industries decreased and the results of this decrease are evident in the form of decreased cadmium deposition in the vicinity of the factories (Figure 5). Measurement of cadmium in moss appears to be a very sensitive indicator of

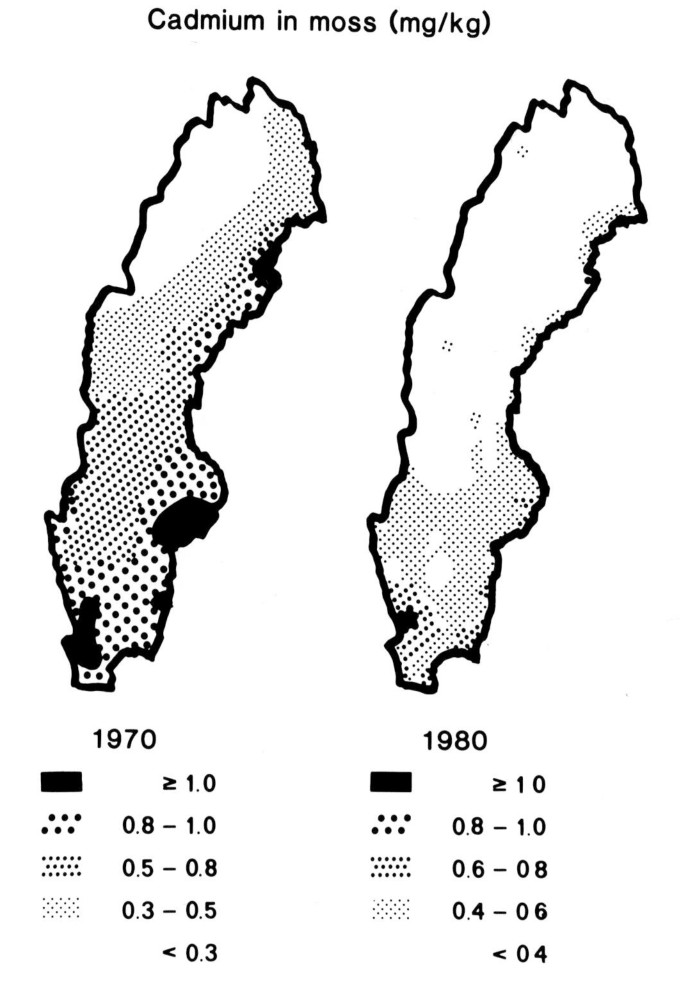

FIGURE 5. Concentration of cadmium in carpets of moss (hylocomium splendens) in Sweden in 1970 and 1980.[76,213]

cadmium pollution. Increased levels of cadmium in moss can be detected at distances exceeding 100 km from the sources of emission.[212]

There are other biological materials which may be used for the screening of cadmium pollution: e.g., Kobayashi[126] measured cadmium in mulberry leaves and noticed a decrease in cadmium concentration with increasing distance from a cadmium-emitting factory in Annaka City, Japan. Similarly, Gale and Wixson[69] and Hemphill et al.[86] reported increased levels of cadmium in white oak and blueberry leaves in an area close to pollution sources of cadmium in Missouri. Elevated levels of cadmium were also found in decomposing leaf litter on the forest floor at distances 10 to 20 km from the smelter. Organs obtained from wild or domestic animals may also be used in order to identify environmental pollution with cadmium.[6,50,90]

C. Cadmium in Water and Fresh Water Sediments

The normal cadmium concentration in sea water is about 0.01 to 0.10 $\mu g/\ell$.[33,150,154,204,225] Values approximately 5 to 10 times higher, i.e., about 0.2 to 0.4 $\mu g/\ell$, have been reported from coastal areas such as the Oslofjord, Norway, and Liverpool Bay, U. K.[204,211]

In rain water collected at areas without point sources of cadmium pollution, the cadmium concentration has been found to range from 0.01 to 0.07 μg/ℓ.[79,147]

In fresh surface and ground waters, the cadmium concentration is usually less than 1 μg/ℓ. In 49 water samples collected from different local wells throughout Sweden, the cadmium concentration was below the detection limit of 1 μg/ℓ in all samples except one which had a cadmium concentration of 3 μg/ℓ.[52] Méranger et al.[160] reported on cadmium concentrations in raw, treated, and distributed water from the main water supply of 71 municipalities in Canada. The mean cadmium concentration was <0.02 μg/ℓ with a range between <0.02 to 0.9 μg/ℓ. Increased concentration of cadmium in drinking water may occur as a result of contamination either from industrial discharges or from the use of metal pipes in the distribution of the drinking water.[218] Sharrett et al.[221] measured cadmium in standing and running tap water from 130 houses in Seattle, Washington, which had copper or galvanized pipes. Median levels of cadmium in water were more than ten times higher in water obtained from galvanized pipes. The median cadmium concentration in standing and running water from galvanized pipes was 0.63 and 0.25 μg/ℓ and from copper pipes 0.06 and 0.01 μg/ℓ, respectively.

Natural waters occasionally contain cadmium concentrations higher than 1 μg/ℓ. In areas where there are zinc-bearing mineral formations, concentrations may reach 10 μg/ℓ.[1]

It is important to recognize that in natural, noncontaminated water, as well as in contaminated water systems, cadmium is mainly found in the bottom sediment and in suspended particles. Yamagata and Shigematsu[274] pointed out that in rivers, polluted by cadmium, the metal is often undetectable in the water phase, while large concentrations are found in suspended particles and in bottom sediments. This is especially true in water of a neutral or alkaline pH. A similar finding was obtained in Sweden, where 0.5 km downstream from a cadmium-emitting factory, 4 μg/ℓ of cadmium was found in water, while 80 mg/kg (dry weight) was found in mud.[201] To avoid errors when determining the degree of contamination in water, cadmium in suspended particles or in sediments must be determined. The contamination of rice fields surrounding the Jintsu River, the area in Japan where the Itai-itai disease occurred, was probably due to the transportation to the paddy soil of cadmium-containing suspended particles when river water was used for irrigation.[274]

D. Cadmium in Soil and Uptake by Plants

Soil is a heterogeneous material. Cadmium concentration in soil is, therefore, highly variable, depending on the material from which it is derived as well as the types of secondary material and organic substances present in the sample. Generally, the concentration of cadmium in soil, with no known cadmium pollution, is less than 1 mg/kg (dry weight).[67,192] However, as much as 30 mg/kg has been observed in nonpolluted soil samples derived from shale (see Lund, 1981, cited in Page).[191] Figure 6 presents the log normal frequency distribution of cadmium concentration in 361 Swedish soil samples analyzed by Andersson.[7] The median cadmium concentration in soil was 0.22 mg/kg. Similar results have been obtained elsewhere. A large national survey in Japan has shown that the median cadmium concentration in nonpolluted soils is about 0.3 to 0.4 mg/kg.[250]

1. Contamination of Soil Caused by Emission from Industries

The contamination of soil by cadmium can take place in different ways. Deposition of cadmium from air and water used for irrigation has already been mentioned. Concentrations in soil near cadmium-emitting industries have also been reported to be highly elevated, ranging from a few to more than 50 mg/kg.[34,67,70,152,185,208] The highest values are normally seen very close to the pollution source. Figure 7 shows cadmium concentrations in surface soil with increasing distance downwind from a cadmium production smelter which has been operating since 1929 in Avonmouth, U.K.[152] Thus, emission of cadmium causes local

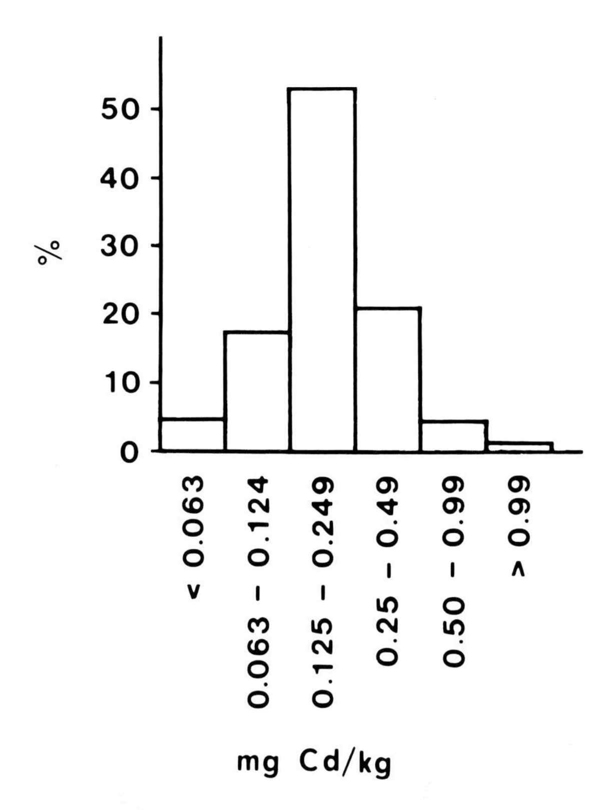

FIGURE 6. Frequency distribution of cadmium in Swedish soils. (From Andersson, A., *Swed. J. Agric. Res.*, 7, 7—20, 1977. With permission.)

problems in many cases. However, long distance transport of air-polluting cadmium may also occur, although it may be difficult to pinpoint any specific source of cadmium pollution in areas far removed from the cadmium-emitting industries.

Mining and refining of metals such as zinc, copper, lead, and cadmium from ore have given rise to substantial cadmium pollution in several areas of the world. In Japan, cadmium emission into water from mines situated upstream of farming areas and atmospheric emissions from zinc smelters have resulted in cadmium pollution of soils in at least 13 different areas.[250] A good review on cadmium pollution of soils in Japan has been provided by Kitagishi and Yamane.[117] Another heavily contaminated area has been recognized in the village of Shipham, U.K. Surface soil samples taken from gardens were found to have cadmium concentrations ranging up to 800 mg/kg, with a median concentration of about 80 mg/kg.[243] In addition to cadmium, the soil was also rich in zinc, median of about 8000 mg/kg, and lead, median of about 2000 mg/kg. The village was actually built on slag heaps from an old zinc mine which was operating in the 18th and 19th centuries.[243] Considerable pollution with cadmium has also been recognized in certain areas of Central Europe.[134,142,198]

2. Contamination of Soil by Sewage Sludge

Application of municipal sewage sludge to soils may be an important source of cadmium contamination. The cadmium concentration in sewage sludge has been reported to range from 2 to 1500 mg/kg dry material, with medians in the order of 5 to 20 mg/kg.[18,68,187,229] Regular use of sludge as a nitrogen fertilizer can cause substantial cadmium enrichment in surface soil. Yearly application of 5 tons/ha (10,000 m^2) of sludge containing 10 mg Cd per kilogram, which is not an extraordinary amount, would result in an annual deposition

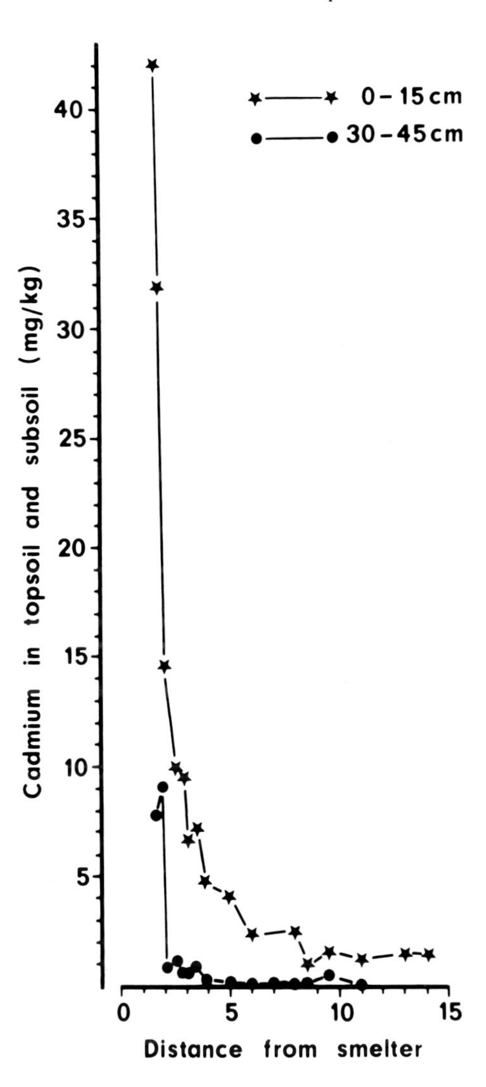

FIGURE 7. Cadmium in surface (0 to 15 cm) and subsoils (30 to 45 cm) with increased distance from a cadmium-producing smelter in Avonmouth, U.K. (From Marples, A. E. and Thornton, I., *Cadmium 74, Proc. 2nd Int. Cadmium Conf., Cannes*, Metal Bulletin Ltd., London, 1980, 74—79. With permision.)

of 50 g Cd per hectare. Andersson[7] has estimated that for a silt loam soil with a bulk density of 1.33 g/cm^{-3}, the amount of cadmium in the surface 20 cm of soil with a Cd concentration of 0.22 mg/kg will be about 0.6 kg/ha. Thus, addition of the above mentioned volume of sewage sludge would give rise to an increase of about 10% in the surface cadmium content. Therefore, if sewage sludge is regularly added to the soil on an annual basis, cadmium content may increase considerably. The amount of cadmium deposited on arable land from dustfall is less, around 300 μg/m^3 or 3 g/ha (See Section III.B.3).

Several investigations have shown that cadmium from sewage sludge is in fact plant available.[37,46,47,148] For example, Gutenmann et al.[71] found 20 times higher cadmium concentrations in tobacco grown on soil amended excessively with sludge compared to tobacco grown on control soil. Figure 8 shows a typical pattern of results obtained when plants (Swiss chard) were grown on sludge-amended soil. The figure also shows that plants grown

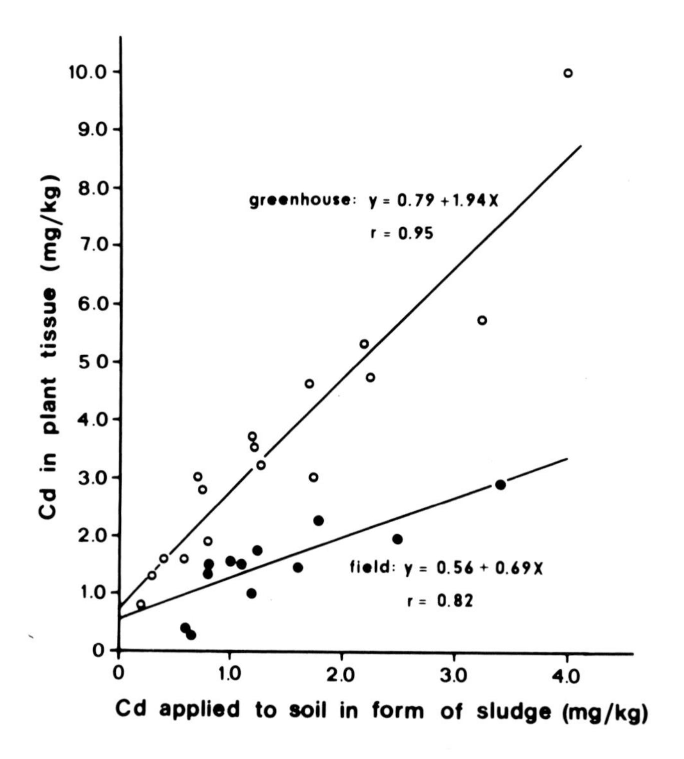

FIGURE 8. Relationship between concentration of cadmium in plants (Swiss Chard) grown on sewage sludge-amended soils. (Redrawn from Page, A. L., Bingham, F. T., and Shang, A. C., *Effect of Heavy Metal Pollution on Plants*, Vol. 1, Lepp, W., Ed., Applied Science Publishers, Barking, Essex, 1981, 77—109. With permission.)

in greenhouses take up cadmium from soil more effectively than plants grown in open fields. It has not yet been ascertained to what extent the cadmium content in crops is related to the annual amounts of cadmium added to soil via sewage sludge or to the accumulated amount added over a period of years. Available experimental results indicate that the annual loading is of greater importance than the cumulative amount, but the cadmium previously applied to soil remains available to plants for extended periods of time.[192]

In Sweden, a maximum average of 1 ton of sludge (dry weight) per hectare (containing at most 15 mg Cd per kilogram is permitted. This corresponds to 15 g Cd per hectare. In Sweden, sludge containing more than 15 mg/kg dry weight should not be applied to soil used for food crops.[235] Similar regulations, although somewhat less strict, on maximum permissible quantity of cadmium addition to soil from sewage sludge have been enforced in several other countries in Western Europe.[206,262,264] In the U.S., an annual application limit of cadmium from sewage sludge to fields used for the production of crops which accumulate cadmium has been set at 500 g/ha.[175] There are considerable differences in the maximal amounts of cadmium added to soil from sludge in different countries.[264]

Comprehensive reviews covering the effects of sewage sludge on the content of cadmium in crops are available, e.g., by Pahren et al.,[193] Sommers,[230] U.S. Council for Agricultural Science and Technology,[257] Ryan et al.[215] and in the proceedings from a recent seminar.[16]

3. Contamination of Soil by Phosphate Fertilizers

Another possible source of contamination is the use of phosphate fertilizers. The concentration of cadmium in phosphate fertilizers varies greatly, depending on the origin of the raw phosphate. Generally, sedimentary rock phosphate has a high cadmium content (Table

Table 4
CADMIUM CONCENTRATIONS IN NATURALLY OCCURRING RAW PHOSPHATE OBTAINED FROM DIFFERENT COUNTRIES[59,93]

Country	Cadmium concentration in phosphate rock (mg/kg)	Cadmium concentration in relation to the phosphate (P_2O_5) content (mg/kg)
U.S. (Florida)	5.5—16	18—52
Morocco	8—30	24—96
Senegal	70—90	225—290
Togo	50	161
U.S.S.R. (Kola)	0.1—0.4	0.3—1.3
Tunisia/Algeria	—	60
Israel/Jordan	—	35

4). The average cadmium concentration in phosphate fertilizers used in the member states of the European Economic Community during 1979/1980 ranged from 48 to 101 mg/kg phosphate (P_2O_5).[93] Based on the average annual use of fertilizers in the member states of the EEC (a range from 30 to 144 kg/ha), Hutton[93] calculated that arable land of the member states received an annual input of 5.1 g of cadmium per hectare, which corresponds to about 1% of the total cadmium content in surface soil (20 cm in depth).

Commercial fertilizers contain from a few to 200 mg/kg,[165] with typical cadmium concentrations ranging from 2 to 20 mg/kg.[7] The corresponding concentrations expressed in milligrams Cd per kilogram phosphate are about 4 to 5 times higher, i.e., 10 to 100 mg/kg phosphate. Long-term use of such fertilizers has been shown to increase the cadmium content in soils.[8,168,268,269]

Williams and David[268,269] reported that topsoil which had been fertilized with phosphates for more than 20 years contained significantly more cadmium, 0.1 mg/kg, compared to similar unfertilized soils, 0.05 mg/kg. Mulla et al.[168] reported that soil, fertilized with approximately 175 kg of phosphate per hectare and year, for 36 years had an average cadmium concentration of 1 mg/kg which was more than 10 times the concentration in unfertilized control soils. Andersson and Hahlin[8] investigated the relationship between variable amounts of cadmium-containing phosphate fertilizers applied between 1963 and 1978, and the cadmium concentration in soil. They were able to show that there was a significant linear increase of cadmium in the soil with increasing amount of applied phosphate fertilizers. From the regression line it was possible to estimate that an annual fertilization with 25 kg of phosphorus (P) per hectare, a normal amount for Swedish conditions, will increase the cadmium content in soil by 0.3 to 1.1%. The average cadmium concentration in the phosphate fertilizer used during the experiment was estimated at about 23 mg/kg (110 mg Cd per kilogram P).[8]

Cadmium from phosphate fertilizers applied to soils is available to plants in a way similar to that of cadmium from sewage sludge. Williams and David[268] showed that cadmium in superphosphate was as available to plants as cadmium chloride. In a later study, the same authors[269] revealed a close relationship between accumulated amount of superphosphate added to soil, and the cadmium concentration in wheat. In agreement with this study, Andersson and Hahlin[8] found a similar increase in the cadmium content of barley grown on experimental fields to which varying amounts of phosphate fertilizer were applied. The average increase in barley fertilized with an annual amount of 25 kg phosphorus (P) per hectare was found to be 0.6 and 1.1% in grain and in straw, respectively, which is close

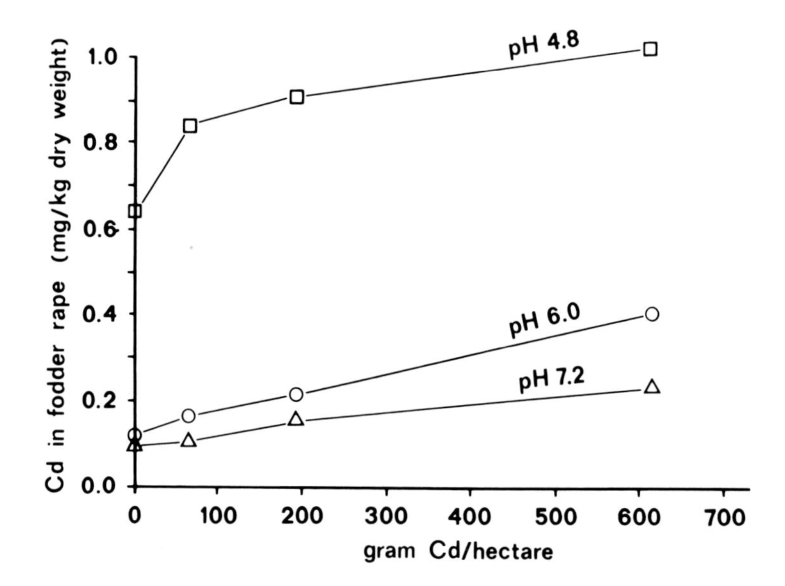

FIGURE 9. The influence of pH on the uptake of cadmium in fodder rape from soils amended with cadmium-containing sewage sludge. (Derived from Andersson, A. and Nilsson, K. O., *Ambio*, 3, 198—210, 1974.)

to the increase in soil. Further, the average annual increase of cadmium in wheat harvested from the same fields in central Sweden from 1916 to 1972[122] was about 1%/year (Figure 1).

4. Uptake of Cadmium in Plants

The uptake of cadmium by plants is not entirely dependent on the concentration of cadmium in soil. Other highly important factors are, e.g., soil type, organic matter content, oxidation reduction potentials, concentration of other trace elements in soil, etc. Plant uptake of cadmium is considerably more efficient in noncalcareous soils when compared to calcareous soils.[247] The most important factor, apart from the concentration of cadmium, appears to be the pH of the soil.[192] A decrease in pH of soil results in a marked increase in plant uptake of cadmium. This was first established in the case of wheat grown on sewage sludge-amended soil,[148] but has subsequently been found to apply generally.[192] There was a marked effect of soil pH on the uptake in fodder rape when cadmium was applied to soil in the form of sewage sludge (Figure 9). Consequently, liming of soil may be an effective method of minimizing the cadmium uptake by plants. Chaney et al.[37] were able to reduce the cadmium content in soybeans from 33 to 5 mg/kg dry weight by increasing the pH in soil from 5.3 to 7. For a discussion of other soil factors which may influence the uptake of cadmium by plants, reference is made to a review by Page et al.[192]

Besides the above-mentioned factors present in the soil and plant environment, there are other important differences in the capability of various plants to take up cadmium. Kobayashi et al.[127] added cadmium oxide to soil pots and noticed that wheat grain accumulated considerably more cadmium than rice. Bingham et al.,[17] in soil tests using sewage sludge as an additive, found high levels of cadmium in spinach and lettuce, median levels in radish and soybeans, and low levels in corn and rice. Similarly, Lönsjö[149] using radioactive cadmium, observed a high uptake of cadmium in spinach and lettuce, median uptake in carrot, tomato, and rapeseed, and low uptake in beans, peas, and wheat. Thus, it may be concluded that leafy vegetables, such as spinach, lettuce, and Swiss chard, may contain the highest concentrations of cadmium. Root crops, such as carrot, potato, and radish, will contain intermediate concentrations, whereas grains and fruits generally contain lower concentrations. In addition to species differences in the uptake of cadmium, it should also be mentioned that significant differences have been found between different cultivars, or genotypes of the same species.[10,88,197] For example, Hinesley and co-workers[88] showed that certain hybrids of corn do not accumulate cadmium even if it is grown on soil heavily amended with cadmium-rich sewage sludge.

5. Measures to Decrease Cadmium Uptake in Plants

As mentioned earlier (Section III.C.4), one way to minimize cadmium uptake in plants is to increase pH of the soil to above 6.5.[192] Another alternative would be to grow grains such as barley and corn, or fruits, which have a comparably low uptake of cadmium from the soil, or to use certain plant genotypes which are known not to accumulate cadmium from soil.[88] Sometimes the pH of the soil is already close to neutral or even alkaline, and is not suitable for growing plants with a low uptake of cadmium. In such cases it may be necessary to remove the surface soil, since measures to remove cadmium from soil by acids or complexing agents have so far been ineffective.[39] In Toyama, as well as in other cadmium-polluted areas of Japan, the local farmers have successfully demanded that the cadmium-polluted soil should be replaced by nonpolluted soil.[164] The surface soil down to a depth of about 30 cm was replaced at considerable expense. Contaminated soil can, without health hazards, be used for growing flowers or trees or be used for housing or nonagricultural industry. This has in fact happened to a great extent in the most polluted areas of Japan, but not all farmers want to stop growing rice and vegetables on their scarce land.

E. Cadmium in Food

1. Nonpolluted Areas

In the last 20 years, especially during the 1970s, a large number of reports dealing with cadmium concentrations in various foodstuffs has been published. There is a substantial variation in the reported results which most likely reflects, not only true differences in concentration, but to a great extent analytical variations. Certain early data on cadmium concentrations in different types of food submitted by, e.g., Schroeder et al.[218] using atomic absorption spectrophotometry without previous extraction or background correction, and by Kropf and Geldmacher-v. Mallinckrodt[135] using emission spectrophotometry, are occasionally several magnitudes higher than those found at a later date. Unfortunately, old and incorrect data are still being referred to as correct in certain scientific publications. For example, Ryan et al.,[215] in their extensive review on cadmium in the human food chain, cited Pinkerton et al.[199,200] who reported unrealistically high levels for cadmium in normal human and bovine milk, 9 to 134 and 20 to 34 $\mu g/\ell$, respectively. The actual concentration of cadmium in milk is most likely below 1 $\mu g/\ell$ (Table 5).[106a,259]

Table 5 presents data on cadmium in a number of selected foodstuffs derived from 40 publications. As discussed previously (Section III.), we have no guarantee that the data given in Table 5 are correct. The table should not be considered to provide true concentrations, but rather to give the reader an idea about the tentative range of cadmium concentrations in various types of foodstuffs. The data have been arranged in two groups according to year of publication. It should be noted that there is a tendency towards obtaining lower values in the more recent studies. In view of the improved analytical facilities developed in later years, it is reasonable to assume that the recently published analytical results are more accurate.

A more extensive compilation of published data on the cadmium content in various foodstuffs is given in a document from the Commission of European Communities and elsewhere.[36,106a] Measurements of cadmium in various types of foodstuffs made by laboratories in different countries have also been compiled by the Food and Agriculture Organization within the United Nations and the World Health Organization.[57] Unfortunately, no

Table 5
CADMIUM IN SELECTED FOODSTUFFS (MG/KG WET WEIGHT) FROM NONPOLLUTED AREAS IN VARIOUS COUNTRIES

	Reports published between 1961 and 1975	Ref.	Reports published in 1976 and later	Ref.
Meat				
Beef	0.002—0.1	56,73,105,115,162,188,217	0.001—0.02	106a,108,128,133,195,207,228
Pork	0.008—0.14	56,73,115,162,188,217,270	0.001—0.01	106a,108,128,195,207,228
Kidney (beef)	0.2—1.6	14,56,73,162,188,217	0.2—1.3	57,63,106a,128,133,195,207,228
Liver (beef)	0.06—0.2	14,56,73,162,188,217	0.05—1.13	108,128,133,195,207,228
Seafood				
Fish (flesh)	0.001—0.01	27,56,83,105,121,122, 162,188,217,270	0.001—0.1	57,106a,108,128,207
Mussels	0.05—7.3	121,122,162,217,270	0.07—0.6	57,106a
Oysters	0.4—4.7	27,105,121,122,176,217	0.1—1.2	57,106a
Grains				
Wheat (grain)	0.03—0.29	73,105,120—122,217	0.005—0.080	10,57,106a,128,161,244,271
Wheat (flour)	0.02—0.24	56,73,98,105,120—122,145,217,270	0.019—0.12	57,106a,128,214
Rice	0.02—0.62	56,73,105,145,188,217	0.008—0.13	57,106a,123,128,155,214
Vegetables				
Carrots	0.03—0.13	56,145,162,188,242	0.005—0.1	5,57,63,106a,128,207,214,222,244
Potatoes	0.001—0.09	56,73,98,145,162,188,217,242	0.01—0.06	5,57,63,106a,108,128, 207,214,222,244,271
Tomatoes	0.002—0.15	56,73,98,162,188,217,242	0.006—0.03	5,57,63,106a,128,207,214,222,244
Cabbage	0.008—0.08	56,73,162,217,242	0.004—0.079	5,57,63,106a,128,207,214,244,
Lettuce	0.01—0.04	56,73,188,217	0.012—0.4	5,57,63,106a,128,214,222,244,271
Spinach	0.02—0.08	73,162,217	0.045—0.150	5,63,106a,128,214,239
Fruits				
Apples	0.01—0.04	56,73,217	0.001—0.004	5,27,106a,128,277
Pears	0.01—0.03	56,73,162,242	0.003—0.020	5,106a,128,214
Milk	0.007—0.01	56,73,98,145,162,188,217	0.00017—0.002	35,106a,108,119,128,207,244,259

analytical quality assurance was made by the participating laboratories in connection with the study[174] and, therefore, all results are not necessarily correct.

It may be concluded from available data that cadmium is present in all foodstuffs, although concentrations are often very low and difficult to measure accurately. The lowest concentration, in the order of a few micrograms per kilogram or less, is found in milk and dairy products. Cadmium concentrations in the order of 1 to 50 μg/kg wet weight are usually found in meat, fish, and fruits. Somewhat higher concentrations are found in vegetables and cereal crops, such as wheat and rice, which in nonpolluted areas usually contain concentrations ranging from 0.01 to 0.15 mg/kg. In this chapter, we use wet weight concentration of cadmium. The dry weight concentration of cadmium in leafy vegetables is usually considerably higher than in, e.g., grains, since leafy vegetables may contain more than 95% water, while grains contain less than 5%. Higher concentrations of cadmium are found in liver and kidneys from adult animals, with cadmium concentrations being in the order of 0.1 to 1 mg/kg. Certain species of mussels, scallops, and oysters frequently have cadmium concentrations exceeding 1 mg/kg. In New Zealand, cadmium concentrations of up to 8 mg/kg wet weight have been found in oysters.[176] A high cadmium content, about 1 to 30 mg/kg, has also been found in the brown meat of crabs[162] and in certain species of wild-growing white mushrooms.[167,220] Liver and kidney obtained from adult horses also contained very high concentrations of cadmium, in the order of 0.5 to 10 mg/kg in liver and 10 to 150 mg/kg in kidney cortex.[14,202]

There is a lack of data describing the form, or chemical species, in which cadmium occurs in different types of foodstuffs. It is not yet known whether or not cadmium in certain foodstuffs is more readily absorbed from the gastrointestinal tract when compared to cadmium from other sources. If differences exist in availability, this will obviously have a great impact on the risk evaluations and on the estimations for highest recommended daily intake. Recent data from a study of oyster fishermen with a very high total cadmium intake (up to 500 μg/day) indicate that cadmium in certain oysters is not absorbed or distributed in the body in the same way as cadmium in other foodstuffs.[157] Experimental data suggest that cadmium in these oysters is bound to a metallothionein-like protein.[179]

2. Polluted Areas

Elevated levels of cadmium in foodstuffs as a result of cadmium pollution have been observed in various areas throughout the world. Most well known is the contamination of rice grown in cadmium-polluted areas of Japan.[55,117,250] In 1971, the Environmental Agency of Japan[55] carried out a survey in order to identify areas which contained elevated cadmium concentrations in soil and rice. Samples were taken from areas where cadmium pollution was suspected, i.e., areas close to mines, smelting plants, and other factories using cadmium. The total number of rice fields tested was 117, covering an area of 11,700 ha. Cadmium concentrations in rice exceeding 1 mg/kg were found in 6.5% of the 4477 rice samples analyzed. One or more rice samples with a cadmium concentration exceeding 1 mg/kg were found in 28 (24%) of the 117 examined fields.

So far, published reports have emerged from at least 13 cadmium-polluted areas in Japan, where average cadmium concentrations in rice ranged from 0.3 to 2.0 mg/kg (Table 6 and Figure 10). Reviews on cadmium pollution in Japan have been provided by Tsuchiya[250] and Kitagishi and Yamane.[117]

Increased levels of cadmium in grains and vegetables as a result of cadmium pollution have also been reported from countries other than Japan (Table 6), e.g., from Belgium,[141,142] Federal Republic of Germany,[198,254] German Democratic Republic,[134] Sweden,[118,166,196,197] U.K.,[162,222,243] and Zambia.[185]

The relative increase of the cadmium concentration in food in these areas cannot be ascertained accurately, but in the areas of Japan where renal effects were recorded (Appendix,

Table 6
CADMIUM IN FOODSTUFFS COLLECTED IN DIFFERENT POLLUTED AREAS THROUGHOUT THE WORLD

Country	District	Type of food analyzed	Cadmium concentration (mg/kg fresh weight)	Increase ratio[a]	Source of cadmium pollution	Number of people living in exposed area[b]	Ref.
Japan	Fuchu, Toyama Pref.	Rice	0.5—2.0	5—2.0[c]	Zinc, lead, and cadmium mines	7,650	97, 109
	Ikuno, Hyogo Pref.	Rice	0.6—1.0	6—10	Silver, copper, and zinc mines	13,000	94, 252
	Tsushima, Nagasaki Pref.	Rice	0.75	7	Lead and zinc mines	2,400	224, 238
	Kakehashigawa, Ishikawa Pref.	Rice	0.2—0.8	2—8	Copper mines	2,800	96, 110
	Kosaka, Akita Pref.	Rice	0.5—0.6	5—6	Silver and copper mines	800	129, 223
	Yoshinogawa, Yamagata Pref.	Rice	0.58	6	Gold, silver, copper, and zinc mines	8,000	256, 223
	Annaka and Takasaki, Gunma Pref.	Rice	0.35—0.5	3—5	Zinc refinery	4,400	224, 64
	Uguisuzawa, Miyagi Pref.	Rice	0.4	4	Lead and zinc mines	800	224, 238
	Watarase, Gunma Pref.	Rice	0.32	3	Copper mine	4,700	65, 64
	Shimoda, Shizuoka Pref.	Rice	Up to 1.1	1—11	Gold and copper mines	1,100	250
	Bandai, Fukushima Pref.	Rice	0.3—0.9	3—9	Zinc refinery	1,800	223
	Kiyokawa, Oita Pref.	Rice	0.23—0.32	2—3	Tin, copper, lead, zinc, and arsenic mines	700	238, 276
	Ohmuta, Fukuoka Pref.	Rice	0.72	7	Zinc refinery	2,540	223
Belgium	Liège	Daily diet	—	No increase	Nonferrous smelters	600,000	141, 142, 26

F.R.G.	Lower Saxony	Vegetables	0.04—2	5—20	Zinc and lead smelters	N.R.[d]	260
	Dortmund	Vegetables, fruit	0.01—0.15	N.R.[d]	Different industries	N.R.[d]	198
G.D.R.	Freiberg	Wheat	0.47	8	Nonferrous smelters	5,000	134
		Potatoes	0.31	8			
Sweden	Skellefteå	Barley	0.05	3	Metal refinery	15,000	118,166, 196,197
		Potatoes	0.20	3			
U.K.	N.R.[d]	Leafy vegetables	0.11—0.76	1—7	Metal refinery	N.R.[d]	162
	Shipham	Winter and summer vegetables	0.28—0.52	19	Old mine	1,100	243
							12
							222
Zambia	Kabwe	Maize	0.064	2.5	Zinc and lead mine smelters	10,000	185

[a] Increase ratio in specified food items. Comparison made between foodstuffs of the same type in polluted and reference area. Please note that these ratios do not correspond to an identical increase in the daily intake of cadmium.

[b] Indicates the approximate order of people living in exposed area. In Japanese areas, the figure usually includes only persons aged over 30 years, who are considered to consume rice containing more than 0.4 mg Cd/kg.

[c] Assuming a background, or normal, level of cadmium in rice of 0.1 mg/kg.[250]

[d] N.R., not reported.

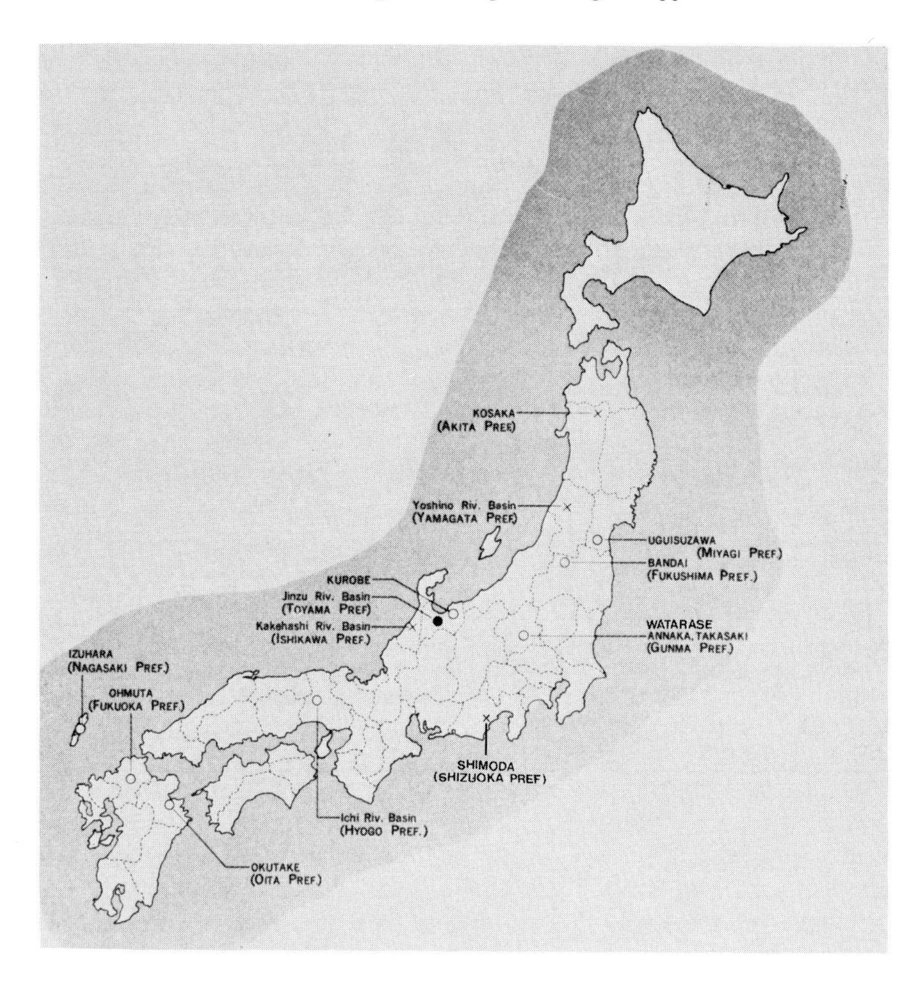

FIGURE 10. Cadmium-polluted areas of Japan. (●) Itai-itai disease area. (○, x) Other cadmium-polluted areas. (Modified from Tsuchiya, K., Ed., *Cadmium Studies in Japan — A Review*, Elsevier, Amsterdam, 1978. With permission.)

Volume II), the estimates vary between 2 and 20 times the "background" level. The increased cadmium concentrations in foodstuffs only reflect an increased daily intake to the extent that the foodstuff represents what is actually consumed. Rice contributes about 80% of the daily energy intake in Japanese farming populations (see Appendix, Volume II), so in those areas the increased concentrations in rice possibly reflect an increased daily intake (see Section IV.A.2). The increased cadmium concentration in foodstuffs other than rice, recorded in areas outside Japan (Table 6), is likely to be an inferior indicator of the daily intake.

F. Cadmium in Cigarettes

Smoking is an important source of cadmium exposure since cigarettes contain comparably high concentrations of cadmium. Furthermore, inhaled cadmium is more readily taken up than ingested cadmium. Szadkowski et al.[237] analyzed 8 brands of cigarettes in West Germany and found a mean cadmium content of 1.4 μg per cigarette. Nandi et al.[172] the same year (1969) found 1.2 μg per cigarette as a mean for 6 different brands of cigarettes produced in Virginia. These early data on cadmium content in cigarettes were later confirmed by several investigators, e.g., Menden et al.,[158] Oelschläger,[188] Wescott and Spincer,[266] Müller,[169] and Elinder et al.[54]

For example, Wescott and Spincer[266] found cadmium concentrations in cigarette tobacco ranging from 0.7 to 3.2 mg/kg. Somewhat lower levels (0.23 to 0.93 mg/kg) were reported by Brooks and Trow[24] who analyzed cadmium in 62 brands of cigarettes from the U.S., Europe, and New Zealand. One explanation for the lower values could be that Brooks and Trow[24] measured cadmium in cigarettes which prior to weighing were equilibrated in air at 60% humidity, whereas most other researchers determined cadmium in dry cigarette tobacco.[54]

Lower concentrations have also been reported in cigarettes produced in Africa. Nwankwo et al.[186] found 0.12 to 0.38 μg Cd per gram dry weight in Zambian and Tanzanian cigarettes compared to 1.3 μg/g in American brands. Likewise, Elinder et al.[54] noted a tendency towards lower cadmium concentrations in cigarettes produced in several developing countries (less than 0.9 μg per cigarette) when compared to cigarettes purchased in industrialized countries, which usually had cadmium contents ranging from 1 to 2 μg per cigarette. A very high concentration of cadmium in cigarette tobacco, 67 mg/kg, was obtained when tobacco plants were grown on heavily sludge-amended experimental soil.[71]

It may be concluded that, in general, cigarettes produced in Europe, U.S., New Zealand, and Japan contain 0.5 to 2 μg Cd per cigarette, whereas cigarettes produced in certain less industrialized countries contain lower amounts of cadmium.

The amount of cadmium actually inhaled when smoking cigarettes has also been studied. Szadkowski et al.[237] measured the total amount of cadmium occurring in the mainstream when cigarettes were smoked in a smoking apparatus under the following experimental conditions: two puffs a minute, each of 2-sec duration, suction volume 30 mℓ with a pressure of 80 to 120 mm H_2O. The mainstream smoke particles were collected on a Cambridge filter and the gases in nitric acid. They found a mean value of 0.15 μg Cd per cigarette in the particulate phase of the 8 brands tested and 0.03 μg in the gaseous phase. It can also be seen from the data that about 0.5 μg went to the sidestream smoke, i.e., about one third of a cigarette containing 1.4 μg Cd. In American cigarettes, Menden et al.[158] found that the particulate phase of the mainstream contained 0.10 to 0.12 μg Cd per cigarette. Similar data, 0.21 μg Cd per cigarette in the particulate phase and 0.03 μg Cd per cigarette in the gaseous phases, were reported by Gutenmann et al.[71] Elinder et al.[54] studied the influence of different numbers of puffs per cigarette and different numbers of puffs per minute for one brand of cigarettes. One puff per minute and 10 puffs altogether resulted in a collection of 0.14 μg Cd per cigarette on a Cambridge filter; 2 and 3 puffs per minute and 15 puffs altogether resulted in 0.16 and 0.19 μg Cd per cigarette on the filter. Wescott and Spincer[266] studied the transfer of cadmium from cigarettes with and without filter tip. In plain cigarettes, the transfer was about 12 to 25% of the total cadmium content in the cigarette, corresponding to a total amount of cadmium transferred from one cigarette in the range of 0.09 to 0.54 μg, depending on the type of cigarette smoked. In filter-tipped cigarettes, the transfer of cadmium to the particulate phase was less, ranging from 4 to 8% of the total cadmium amount in the cigarette.

In conclusion, all data show positive agreement indicating that 0.1 to 0.2 μg Cd might be inhaled by smoking one cigarette with a cadmium content of 1 to 2 μg Cd. Thus, on average, about 10% of the cadmium in a cigarette would be inhaled.

Cadmium content in snuff has also been determined.[13,111] American snuff contained 0.7 to 0.9 mg Cd per kilogram, while snuff used in South Africa contained 1.1 to 1.5 mg/kg. This latter type of snuff is a mixture of powdered tobacco and ash of incinerated plants and herbs.

IV. DAILY INTAKE OF CADMIUM

A. Via Food and Water

1. Nonpolluted Areas

Different approaches have been used to estimate the daily intake of cadmium from food.

One method is based on the average consumption of different foodstuffs and the average cadmium content in these products (national nutrition census method or composite sample method) . One problem with this method is that the cadmium concentration in certain basic foodstuffs, such as dairy products, is usually so low that it is difficult to measure accurately. If detection limit values are used as tentative actual concentration data, the estimated daily intake will be too high. Any erroneous data will be multiplied by the average daily consumption of these products and may significantly affect the estimation of the daily intake.

Another method used to estimate daily intake is based on analyses of cadmium in total diet samples collected by volunteers over a certain period of time. Each volunteer collects duplicate portions of everything he or she consumes over a 24-hr period. Homogenized 24-hr diet samples are subsequently analyzed for cadmium (total diet collection method).

A third approach is to analyze the amount of cadmium in feces. Since ingested cadmium is taken up from the gastrointestinal tract to a limited extent only, 90 to 99% of the daily intake of cadmium can be recovered from feces and thus may be used as an indirect measure of the daily intake (feces method). An advantage with the feces method is that only one material needs to be analyzed. Furthermore, the concentration of cadmium in feces is generally higher than in most of the foodstuffs. One drawback is the difficulty involved in obtaining homogeneous and representative 24-hr samples.

In Table 7, estimates of the daily intake of cadmium in different countries are presented. The problems connected with the accuracy of these estimates are similar to those for foodstuffs, discussed in Section III.E.1. Data by Schroeder et al.,[218] Kropf and Geldmacher-v. Mallinckrodt,[135] Tipton and Stewart,[245] Murthy et al.,[170] and Méranger and Smith[159] have been exluded from the table, since these results were obtained by atomic absorption without extraction or background correction, or by emission spectrophotometry. Data are arranged according to country and year of publication. There is a substantial variation between estimates also within countries, which possibly, at least to some extent, is related to analytical problems. Again there is a tendency for lower estimates in more recent studies.

In *Cadmium in the Environment*, Friberg et al.[62] wrote that "available data suggest that, on average, 50 μg Cd per day may be ingested in most countries, with probable variations from 25 to 75 μg/day." Today, these figures appear to be an overestimation of the situation in most countries. When more recent and reliable estimates are considered, it is clear that the daily intake of cadmium in the U.S., in most European countries, and in New Zealand is 10 to 25 μg/day. These are mean values, and it should be kept in mind that large individual variations do occur due to different dietary habits and age-dependent changes in energy intake. The highest daily intake of cadmium is likely to occur among teenagers since they have the highest energy intake.[124] Estimates of the daily intake of cadmium from areas regarded as nonpolluted in Japan are usually higher than those reported elsewhere in the world. Some of these areas were rural control areas in epidemiological studies[102] and values obtained may not be typical for the average Japanese person, whose cadmium intake is probably around 35 to 50 μg/day (Table 7).

Normally, cadmium in drinking water does not contribute much to total daily intake. A concentration in drinking water of 1 μg/ℓ (Section III.C) would increase the daily intake by only 1 to 2 μg cadmium with a water consumption of 1 to 2 ℓ/day, which is considered normal for an adult.[221] If the concentration in water is as high as 10 μg/ℓ, the intake of cadmium from water would be in the same order as that from food in the U.S. and Europe.

2. *Polluted Areas*

In any area where cadmium pollution has given rise to elevated levels in foodstuffs, the daily intake of cadmium from food is likely to be increased (Table 6). The extent to which locally grown and possibly contaminated food is consumed varies markedly. In the vicinity of a cadmium-emitting smelter in Sweden, it was found that locally grown foodstuffs con-

Table 7
ESTIMATES OF THE DAILY INTAKE OF CADMIUM VIA FOOD IN DIFFERENT COUNTRIES

Country	Year of publication	Sampling method[a]	Analyzing method	Cd (μg/day)	Ref.
Belgium	1983	B	AAS	15	26
Canada	1977	A	AAS	52[b]	116
Czechoslovakia	1970	A	Dithizone	60[b]	145
Denmark	1980	A	AAS	30	173
Finland	1980	A	AAS	13	128
F.R.G.	1969	A	AAS	48[b]	56
	1969	C	AAS	31	56
	1972	C	AAS	31	236
Japan	1969	C	Dithizone	57	249
	1970	A	Dithizone	59	274
	1970	A	AAS	59—113	102
	1970	B	AAS	99—108	102
	1972	A	—	47	66
	1974	C	AAS	56—81	77
	1975	B	AAS	31	273
	1976	C	AAS	56	241
	1976	B	AAS	48	233
	1976	C	AAS	36	
	1977	B	AAS	49	258
	1977	B	AAS	35	99
	1977	C	AAS	24—36	99
	1977	C	AAS	41	130
	1978	C	AAS	49	251
	1981	A	—	49	189
	1981	C	AAS	70	100
N.Z.	1977	B	AAS	21	74
Roumania	1970	A	Dithizone	36—64[b]	209
Sweden	1974	B	NA	10	267
	1974	C	NA	8	267
	1977	A	AAS	17	119
	1978	C	AAS	18	124
	1979	B	NA	11	219
	1983	B	AAS	10	227
U.K.	1973	A	AAS	15—30	162
	1981	A	—	10—20	261
	1981	B	—	20	261
U.S.	1961	A	Dithizone	28	217
	1969	C	AAS	42	246
	1975	A	AAS	41	151
	1977	C	AAS	19[c]	106
	1979	C	AAS	10—15[c]	132
	1979	B	AAS	33	231
	1979	C	AAS	20—30	231
	1979	B	AAS	26—51	28

[a] A, national nutrition census method; B, total diet collection method; C, feces method.

[b] Probably erroneous data.

[c] The report by Kowal et al.[132] partly presents the same data as those presented by Johnson et al.[106] The daily intake of cadmium has, however, been estimated according to slightly different principles which explain the lower figures obtained by Kowal et al.[132]

stituted only a minor portion of the total energy intake. Furthermore, locally grown vegetables and berries were only consumed during a short period of the year. Thus, probably less than 1% of the food intake actually came from foodstuffs containing increased levels of cadmium.[53] Likewise, diaries kept by people living in the heavily cadmium-contaminated village of Shipham, U.K., revealed that vegetables comprised about only 20% of the diets, and that a substantial proportion of these were brought from noncontaminated sites.[222] In farming areas of Japan, however, locally grown rice has normally constituted a large part of the daily energy intake. The Japanese Association of Health Statistics[101] has estimated that on average 300 g of rice is consumed daily. Thus, people living in areas where cadmium concentration in rice averages 0.5 mg/kg will have an intake of 150 μg/day from rice alone, other sources of cadmium not being taken into account. Table 8 gives data on the daily intake of cadmium in six different cadmium-polluted areas in Japan, estimated according to different methods, e.g., the national nutrition census method (A), total diet collection method (B), and feces method (C). It can be seen from the table that in different polluted areas, the average daily intake has been about 150 to 250 μg, in one case as high as 391 μg/day. It is obvious that there are large individual variations among people living in the same area. For example, Yamagata et al.[275] using the total diet collection method (B), studied the average daily intake of cadmium among adults in 9 farming families living close to the same river and found a range from 39 to 257 μg/day.

Contaminated drinking water can be an additional source of cadmium exposure. If the concentration in drinking water is as high as 10 μg/ℓ (Section III.C), this may contribute an additional amount of 10 to 20 μg Cd per day. If bottom sediments and suspended particles also contaminate the drinking water, this route of cadmium exposure may be of even greater importance.[274]

B. Via Ambient Air and Smoking

Cadmium in air will contribute very little to total exposure in nonpolluted areas. Assuming that around 10 to 15 m^3 are normally inhaled every day, concentrations of 1 to 10 ng/m^3 would contribute not more than 0.01 to 0.15 μg Cd per day via the respiratory route. The amount deposited in the lung would be even less. It has been mentioned that concentrations of 0.1 to 0.5 $\mu g/m^3$ have been found in the vicinity of cadmium-emitting factories. These concentrations would cause inhalation of 1 to 7.5 μg/day, which is still less than the oral intake, but as will be discussed in Chapter 6, the absorption rate from the respiratory route is considerably higher than that from the gastrointestinal route. Thus, exposure via air can be of importance when cadmium concentrations are very high.

An additional source of cadmium intake is smoking. Twenty cigarettes per day would probably cause an inhalation of 2 to 4 μg Cd per day, which constitutes a significant amount. A 25 to 50% absorption of inhaled cadmium (Chapter 6) from cigarettes would correspond to a daily retention of 0.5 to 2 μg Cd. This figure should be compared with about 0.5 to 1.5 μg Cd absorbed daily from a typical western diet (5% gastrointestinal absorption, Chapter 6) and an average daily intake of 10 to 25 μg/day. Thus, smokers generally have a higher exposure to cadmium compared to nonsmokers and this is also reflected in higher tissue levels of cadmium (see Chapters 5 and 6).

C. Via Occupational Exposure Routes

During occupational exposure to cadmium, the inhalation of cadmium will be the dominating exposure route. Eight hours of moderately heavy work at a cadmium concentration of 20 $\mu g/m^3$ will result in an inhalation of about 160 μg Cd (20 $\mu g/m^3 \times 1\ m^3/hr \times 8\ hr$). Depending on particle size, up to 50% of the inhaled amount will be absorbed (see Chapter 6). Assuming a 25 to 50% absorption, this inhaled amount would result in a daily accumulation of about 40 to 80 μg of cadmium, which should be compared to about 0.5 to 1.5

Table 8
DAILY INTAKE OF CADMIUM IN POLLUTED AREAS OF JAPAN

Area	Sampling method[a]	Estimated daily intake (μg/day)	Ref.
Annaka	A	211	250
	B	281	250
Kakehashizawa	C	149	100
Kiyokawa	A	222	250
	B	391	250
Kosaka	C	149—177	77
	C	139	130
	C	158	100
Tsushima	A	215	250
	B	213	250
	C	255	100
Uguisuzawa	A	245	250
	B	180	250

[a] All analyses made by AAS. A, national nutrition census method; B, total diet collection method; C, feces method.

in a typical western diet (See Section IV.A). Smoking habits and personal hygiene are also of great importance as a source of indirect exposure in a cadmium-contaminated working environment. Cigarettes and pipe tobacco may become heavily contaminated by cadmium when carried in the pockets of workers in cadmium-producing plants.[203] Therefore, the combination of occupational exposure to cadmium and smoking may markedly increase an individual's exposure. Hassler et al.[82] found that smokers in the same cadmium battery-producing plant on average have twice the cadmium concentration in blood and urine when compared to nonsmokers.

D. Via Other Sources of Exposure

Other possibilities for oral and respiratory exposure to cadmium may occur in areas contaminated by cadmium when dust deposited during dry seasons is again transferred to the air by whirlwinds, etc., or when children lick their contaminated fingers. Schoolyards, playgrounds, and sports areas for children could also constitute sources of such exposure.[25]

It is important to remember that contamination of food and beverages may occur due to leakage in acid media from cadmium glazed pottery and enameled steel.[72] High leakage has been found in certain glazed and enameled kitchenware with red, orange, or yellow coloring.[15] Standard methods for testing the release of cadmium from glazed goods have been set up in several countries. Usually the amount of cadmium which is released in 4% acetic acid over a 24-hr period is measured at room temperature, occasionally the ceramics are also heated for a period. The maximum acceptable concentration following extraction varies from one country to another, between 0.02 to 1 mg/ℓ.[36] The European Economic Community recommends that at most 0.1 mg/ℓ should be released.[36] The International Organization for Standardization[95] has given 0.25 and 0.5 mg/ℓ as a maximum cadmium release from large hollow and small hollow glassware and ceramics.

Another source of cadmium exposure may be cooling devices in soft drink vending machines. Nordberg et al.[178] found that contamination in water (16 mg/ℓ in solution and 40 mg/ℓ in precipitate) was such that acute gastrointestinal symptoms were induced. Later investigations of two similar vending machines revealed cadmium concentrations of 0.004 and 0.5 mg/ℓ.

V. SUMMARY AND CONCLUSIONS

Cadmium occurs in nature in conjunction with zinc and is similar to zinc in its physical and chemical properties. Cadmium is obtained as a by-product during the refining of zinc-bearing ores, usually at the ratio of 3 tons of cadmium per 1000 tons of zinc. Cadmium is mainly used in the plating industry, but also in the preparation of various alloys. It is used in one type of alkaline batteries. Some cadmium compounds, primarily cadmium stearate, are used as stabilizers in plastics, particularly in PVC. Cadmium sulfide and cadmium sulfoselenide are used as yellow and red pigments in paints and plastics.

Consumption and production of cadmium have increased considerably during the 20th century. At the beginning of the century, only quantities of the order of tens of tons were produced annually; by 1980 world production had reached approximately 17,000 tons/year.

Only a minor proportion of all refined cadmium is recycled. There is a risk that the cadmium which is not recycled is distributed, giving rise to pollution of the general environment. Certain data indicate that the environmental levels of cadmium, and human exposure, have increased during the last century.

Cadmium-contaminated soil and water may occur naturally or as the result of emissions from industries, the use of cadmium-rich fertilizers and/or sewage sludge, or in the form of deposited air pollution. Crops grown on contaminated soils and/or irrigated by cadmium-contaminated water take up cadmium efficiently. Uptake is strongly enhanced by a low pH level.

Humans are exposed to cadmium via food, water, air, and dust. For people not occupationally exposed to cadmium, food is the most important source of exposure. In noncontaminated areas, most foodstuffs will contain less than 0.05 mg Cd per kilogram wet weight. Cadmium concentrations are generally low in milk, meat, fish, and fruits. Intermediate concentrations are found in leafy vegetables and in grains such as rice and wheat. High concentrations of cadmium are found in kidney and liver from adult animals, certain seafoods such as mussels, oysters, and crabs, and in certain species of wild-growing white mushrooms. The average daily intake of cadmium in the U.S. and Europe is about 10 to 25 μg with large individual variations. In Japan, the cadmium intake in nonpolluted areas is 35 to 50 μg/day.

There is a lack of knowledge about the chemical species and bioavailability of cadmium in different types of foodstuffs. If cadmium in certain foodstuffs is less available to gastrointestinal absorption than in others, or vice versa, this would obviously have important practical consequences for the evaluation of the possible harmful effects. More data on the binding of cadmium in different types of foodstuffs are urgently needed.

The normal concentration of cadmium in water is less than 1 μg/ℓ. At cadmium concentrations exceeding 5 to 10 μg/ℓ, drinking water may contribute significantly to the daily intake of cadmium.

"Normal" concentrations of cadmium in air, about 5 ng/m^3 or less, do not contribute much to the daily intake of cadmium, but concentrations of 0.1 to 0.5 μg/m^3 (weekly or monthly means) have been recorded in areas close to cadmium-emitting factories. Such high levels may result in inhalation of 1 to 7.5 μg Cd per day, which is a significant amount since the absorption rate in the respiratory tract is greater than in the gastrointestinal tract. During occupational exposure, inhalation as a rule is the dominating exposure route. Workroom concentrations of cadmium in the order of 20 μg/m^3 will result in an inhalation of more than 150 μg/8 working hours.

Smoking also contributes to daily intake. Smoking 20 cigarettes per day will probably cause inhalation of 2 to 4 μg Cd per day. In cadmium industries, cigarettes often become contaminated by cadmium-containing dust and this will further increase the importance of this exposure route.

REFERENCES

1. **Abdullah, M. I. and Royle, L. G.,** Cadmium in some British coastal and fresh water environments, in *Problems of the Contamination of Man and His Environment by Mercury and Cadmium,* CEC European Colloquium, Commission of European Communities, Luxembourg, 1974, 69—81.
2. **Adams, R. G., Harrison, J. F., and Scott, P.,** The development of cadmium-induced proteinuria, impaired renal function, and osteomalacia in alkaline battery workers, *Q. J. Med.,* 38, 425—443, 1969.
3. **Adamsson, E.,** Long-term sampling of air-borne cadmium dust in an alkaline battery factory, *Scand. J. Work Environ. Health,* 5, 178—187, 1979.
4. **Ahlman, K. and Koponen, M.,** Cadmium production with low exposure level, in *Cadmium 79, Proc. 2nd Int. Cadmium Conf. Cannes,* Metal Bulletin Ltd., London, 1980, 203—205.
5. **Andersen, A.,** Lead, cadmium, copper and zinc in Danish fruit and vegetables, 1977 to 1978 (in Danish with English summary), Publ. No. 40, National Danish Food Institute, Division of Pesticides and Contaminants, Copenhagen, 1979.
6. **Andersen, A. and Hovgaard Hansen, H.,** Cadmium and zinc in kidneys from Danish cattle, *Nord. Vet. Med.,* 34, 340-349, 1982.
7. **Andersson, A.,** Heavy metals in Swedish soils: on their retention, distribution and amounts, *Swed. J. Agric. Res.,* 7, 7—20, 1977.
8. **Andersson, A. and Hahlin, M.,** Cadmium effects from phosphorus fertilization in field experiments, *Swed. J. Agric. Res.,* 11, 3—10, 1981.
9. **Andersson, A. and Nilsson, K. O.,** Influence of lime and soil pH and Cd availability to plants, *Ambio,* 3, 198—210, 1974.
10. **Andersson, A. and Pettersson, O.,** Cadmium in Swedish winter wheat, *Swed. J. Agric. Res.,* 11, 49—55, 1981.
11. **Aylett, B. J.,** The chemistry and bioinorganic chemistry of cadmium, in *The Chemistry, Biochemistry and Biology of Cadmium,* Webb, M., Ed., Elsevier, Amsterdam, 1979, 1—43.
12. **Barltrop, D. and Strehlow, C. D.,** Clinical and biochemical indices of cadmium exposure in the population of Shipham, in *Cadmium 81, Proc. 3rd Int. Cadmium Conf. Miami,* Cadmium Association, London, 1982, 112—114.
13. **Baumslag, N., Keen, P., and Petering, H.,** Carcinoma of the maxillary antrum and its relationship to trace metal content of snuff, *Arch. Environ. Health,* 23, 1—5, 1971.
14. **Beckman, I., Hägglund, J., Lundström, H., Sark, M., and Slorach, S.,** Cadmium and lead contents in samples of kidney and liver from Swedish cattle, pigs, sheep and horses (in Swedish with English summary), *Var Foda,* 26(4), 70—75, 1974.
15. **Beckman, I., Movitz, I., Nygren, M., and Slorach, S.,** Release of lead and cadmium from glazed and enamelled foodware (in Swedish with English summary, *Var Foda,* 31(3), 193—197, 1979.
16. **Berglund, S., Davis, R. D., and L'Hermite, P.,** *Utilisation of Sewage Sludge on Land: Rates of Application and Long-Term Effects of Metals,* D. Reidel Publ., Dordrecht, Holland, 1984.
17. **Bingham, F. T., Page, A. L., Mahler, R. J., and Ganje, T. J.,** Growth and cadmium accumulation of plants grown on a soil treated with a cadmium-enriched sewage sludge, *J. Environ. Qual.,* 4(2), 207—210, 1975.
18. **Blakeslee, P. A.,** Monitoring considerations for municipal wastewater effluent and sludge application to the land, in *Recycling Municipal Sludges and Effluents on Land,* Natl. Assoc. State Univ. Land, Grant Colleges, Washington, D.C., 1974, 183—198.
19. **Bonnell, J. A., Kazantzis, G., and King, E.,** A follow-up study of men exposed to cadmium oxide fume, *Br. J. Ind. Med.,* 16, 135—145, 1959.
20. **Börlitz, S.,** Cadmium content of vegetation polluted by emissions from smelters (in German), in *Kadmium-Symposium, August 1977,* Jena, Friedrich-Schiller Universität, Jena, 1979, 234—237.
21. **Boutron, C.,** Trace metals in remote artic snows: natural or anthropogenic?, *Nature (London),* 284, 575—576, 1980.
22. **Boutron, C. F.,** Atmospheric heavy metals in the snow and ice layers deposited in Antarctica and Greenland from prehistoric times to present, in Int. Conf. Heavy Metals in the Environment, Heidelberg, September 1983, 174—177.
23. **Boutron, C. and Lorius, C.,** Trace metals in Antarctic snows since 1914, *Nature (London),* 277, 551—554, 1979.
24. **Brooks, R. R. and Trow, J. M.,** Lead and cadmium content of some New Zealand and overseas cigarettes, *J. Sci. N.Z.,* 22, 289—291, 1979.
25. **Buchet, J. P., Roels, H., Lauwerys, R., Bruaux, P., Clays-Thoreau, F., Lafontaine, A., and Verduyn, G.,** Repeated surveillance of exposure to cadmium, manganese and arsenic in school-age children living in rural, urban and nonferrous smelter areas in Belgium, *Environ. Res.,* 22, 95—108, 1980.

26. **Buchet, J. P., Lauwerys, R., Vandevoorde, A., and Pycke, J. M.,** Oral daily intake of cadmium, lead, manganese, copper, chromium, mercury, calcium, zinc and arsenic in Belgium: a duplicate meal study, *Food Chem. Toxicol.*, 21(1), 19—24, 1983.
27. **Bugdahl, V. and von Jan, E.,** Quantitative determinations of trace metals in frozen fish, fish oil and fish meal (in German), *Z. Lebensm. Unters. Forsch.*, 157, 133—140, 1975.
28. U.S. Bureau of Foods, Public Health Service, Food and Drug Administration, Department of Health and Human Services, Total Diet Studies — Adult, F475, Washington, D.C., 1979.
29. **Burkitt, A., Lester, P., and Nickless, G.,** Distribution of heavy metals in the vicinity of an industrial complex, *Nature (London)*, 238, 327—328, 1972.
30. *Cadmium 77, Edited Proc. 1st Int. Cadmium Conf., San Francisco,* January 31 to February 2, 1977, Metal Bulletin Ltd., London, 1978.
31. *Cadmium 79, Edited Proc. 2nd Int. Cadmium Conf. Cannes,* February 6 to 8, 1979, Metal Bulletin Ltd., London, 1980.
32. *Cadmium 81, Edited Proc. 3rd Int. Cadmium Conf. Miami,* February 3 to 5, 1981, Cadmium Association, London, 1982.
33. **Campbell, J. A. and Loring, D. H.,** Baseline levels of heavy metals in the waters and sediments of Baffin Bay, *Mar. Pollut. Bull.*, 11, 257—261, 1980.
34. **Cartwright, B., Merry, R. H., and Tiller, K. G.,** Heavy metal contamination of soils around a lead smelter at Port Pirie, South Australia, *Aust. J. Soil Res.*, 15, 69—81, 1976.
35. **Casey, C. E.,** The content of some trace elements in infant milk foods and supplements available in New Zealand, *N.Z. Med. J.*, 85, 275—278, 1977.
36. Commission of the European Communities (CEC), *Criteria (Dose/Effect Relationships) for Cadmium,* Pergamon Press, New York, 1978.
37. **Chaney, R. L., White, M. C., and Simon, P. W.,** Plant uptake of heavy metals from sewage sludge applied to land, in *Proc. 1975 Natl. Conf. Municipal Sludge Management and Disposal,* Information TransferInc., Rockville, Md., 1975, 169—178.
38. **Chaney, R. L., Hornick, S. B., and Simon, P. W.,** Heavy metal relationships during land utilization of sewage sludge in the Northeast, in *Land as a Waste Management Alternative,* Loehr, R. C., Ed., Ann Arbor Science Publ., Ann Arbor, Mich., 1977, 283—314.
39. **Chino, M.,** The assessment of various countermeasures against Cd pollution in rice grains (Appendix-2), in *Heavy Metal Pollution in Soils of Japan,* Kitagishi, K. and Yamane, J., Eds., Japan Scientific Societies Press, Tokyo, 1981, 281—285.
40. **Creason, J. P., McNuty, O., Heiderscheit, L. T., Swanson, D. H., and Buechley, R. W.,** Roadside gradients in atmospheric concentrations of cadmium, lead and zinc, in *Trace Substances in Environmental Health — VI,* Hemphill, D. D., Ed., University of Missouri, Columbia, 1972, 129—142.
41. **Cunningham, W. C. and Zoller, W. H.,** The chemical composition of remote area aerosols, *J. Aerosol Sci.*, 12(4), 367—384, 1981.
42. Department of the Environment, *Cadmium in the Environment and its Significance to Man. An Inter-Departmental Report. Pollution Pap. No. 17,* Her Majesty's Stationary Office, London, 1980.
43. **Demuynck, M., Rahn, K. A., Janssens, M., and Dams, R.,** Chemical analysis of airborne particulate matter during a period of unusually high pollution, *Atmos. Environ.*, 10, 21—26, 1976.
44. **de Silva, P. E. and Donnan, M. B.,** Chronic cadmium poisoning in a pigment manufacturing plant, *Br. J. Ind. Med.*, 38, 77—86, 1981.
45. **Dorn, C. R., Pierce, J. O., Phillips, P. E., and Chase, G. R.,** Airborne Pb, Cd, Zn and Cu concentration by particle size near a Pb smelter, *Atmos. Environ.*, 10, 443—446, 1976.
46. **Dowdy, R. H., Larsson, R. E., and Epstein, E.,** Sewage sludge and effluent use in agriculture, in *Land Application of Waste Materials,* Soil Conservation Society of America, Akeny, Iowa, 1976, 138—153.
47. **Dowdy, R. H., Larsson, W. E., Titrud, J. M., and Latterell J. J.,** Growth and metal uptake of snap beans grown on sewage sludge-amended soil: a four-year field study, *J. Environ. Qual.*, 7(2), 252—257, 1978.
48. **Drasch, G.,** An increase of cadmium body burden for this century — an investigation on human tissues, *Sci. Total Environ.*, 26, 111—119, 1983.
49. **Duce, R. A., Hoffman, G. L., and Zoller, W. H.,** Atmospheric trace metals at remote northern and southern hemisphere sites: pollution or natural?, *Science,* 187, 59—61, 1975.
50. **Elinder, C. -G.,** Early effects of cadmium accumulation in horse kidney cortex, in *Heavy Metals in the Environment, International Conference, Amsterdam, September 1981,* CEP Consultants Ltd., Edinburgh, 1981, 530—533.
51. **Elinder, C.-G. and Kjellström, T.,** Cadmium concentration in samples of human kidney cortex from the 19th century, *Ambio,* 6, 270—272, 1977.
52. **Elinder, C.-G., Stenström, T., Piscator, M., and Linnman, L.,** Water hardness in relation to cadmium accumulation and microscopic signs of cardiovascular disease in horses, *Arch. Environ. Health,* 35, 81—84, 1980.

53. **Elinder, C. -G., Millqvist, K., Andersson, H., Lind, B., Nilsson, B., and Pettersson, B.,** Cadmium and Lead in Blood and Urine Samples from Persons with Heavy Consumption of Locally Cultivated Food in the Vicinity of the Rönnskär Smeltery (in Swedish), Department of Environmental Hygiene, Karolinska Institute, Stockholm, 1982.
54. **Elinder, C. -G., Kjellström, T., Lind, B., Linnman, L., Piscator, M., and Sundstedt, K.,** Cadmium exposure from smoking cigarettes. Variations with time and country where purchased, *Environ. Res.*, 32, 220—227, 1983.
55. Environment Agency (Kankyo cho), Pollution Hygiene Section, *Countermeasures Against Environmental Pollution by Cadmium (Kadmium Kankyo Osen Taisaku),* Mimeographed document including a number of earlier reports from the Ministry of Health and Welfare, March 1972 (in Japanese, translated by Seizaburo Aoki, Japanese Language Translation Service, Fujisawa, Japan).
56. **Essing, H. G., Schaller, K. H., Szadkowski, D., and Lehnert, G.,** Normal cadmium load on man by food and beverages (in German), *Arch. Hyg. Bakteriol.*, 153(6), 490—494, 1969.
57. FAO/WHO Food and Animal Feed Contamination Monitoring Programme, Summary of Data Received from Collaborating Centres 1977 to 1980, Part B, Contaminants, World Health Organization, Geneva, 1981.
58. **Farnsworth, M., Ed.,** *Cadmium Chemicals,* International Lead Zinc Research Organization, New York, 1980.
59. **Feenstra, J. F.,** Control of water pollution due to cadmium discharged in the process of producing and using phosphate fertilizers, Contract No. ENV/223/74-E REV 3, Instituut voor Milieuvraagstukken, Vrije Universitet, Amsterdam, 1978.
60. **Fleischer, M., Sarofin, A. F., Fassett, D. W., Hammond, P., Shacklette, H. T., Nisbet, I. C. T., and Epstein, S.,** Environmental impact of cadmium: a review by the panel on hazardous trace substances, *Environ. Health Perspect.*, 7, 253—323, 1974.
61. **Friberg, L.,** Health hazards in the manufacture of alkaline accumulators with special reference to chronic cadmium poisoning (Doctoral thesis), *Acta Med. Scand.*, 138 (Suppl. 240), 1—124, 1950.
62. **Friberg, L., Piscator, M., Nordberg, G. F., and Kjellström, T.,** *Cadmium in the Environment,* 2nd ed., CRC Press, Boca Raton, Fla., 1974.
63. **Fuchs, G., Haegglund, J., and Jorhem, L.,** The levels of lead, cadmium and zinc in vegetables, *Var Foda,* 128, 160—167, 1976.
64. **Fukushima, I.,** Environmental pollution and health effects — Gunma Prefecture, in *Cadmium Studies in Japan — A Review,* Tsuchiya, K., Ed., Elsevier, Amsterdam, 1978, 170—192.
65. **Fukushima, I., Nakajima, T., Higuchi, Y., Tsuchiya, T., Nakajma, K., Tanaka, T., Fukamachi, M., Arai, K., and Schiono, T.,** Results of screening examination on tubular dysfunction in aged persons confined to bed (in Japanese), in *Kankyo Hoken Rep. No. 36,* Japanese Public Health Association, Tokyo, 1975, 138—142.
66. **Fukushima, M.,** Cadmium content in various foodstuffs (in Japanese), in *Kankyo Hoken Rep. No. 11,* Japanese Public Health Association, Tokyo, 1972, 22.
67. **Fulkerson, W. and Goeller, H. E.,** *Cadmium, the Dissipated Element,* ORNL/NSF/EP-21, Oak Ridge National Laboratory, Oak Ridge, Tenn., 1973, 473.
68. **Furr, A. K., Lawrence, A. W., Tong, S. S. C., Grandolfo, M. C., Hofstader, R. A., Bache, C. A., Gutenmann, W. H., and Lisk, D. J.,** Multielement and chlorinated hydrocarbon analysis of municipal sewage sludges of American cities, *Environ. Sci. Technol.*, 10, 683—687, 1976.
69. **Gale, N. L. and Wixson, B. G.,** Cadmium in forest ecosystems around lead smelters in Missouri, *Environ. Health Perspect.,* 28, 23—37, 1979.
70. **Gough, L. P. and Severson, R. C.,** Impact of point source emissions from phosphate processing on the element content of plants and soils, Soda Springs, Idaho, in *Trace Substances in Environmental Health — X,* Hemphill, D. D., Ed., University of Missouri, Columbia, 1976, 225—233.
71. **Gutenmann, W. H., Bache, C. A., Lisk, D. J., Hoffmann, D., Adams, J. D., and Elfving, D. C.,** Cadmium and nickel in smoke of cigarettes prepared from tobacco cultured on municipal sludge-amended soil, *J. Toxicol. Environ. Health,* 10, 423—431, 1982.
72. **Guthenberg, H. and Beckman, I.,** Investigation of leakage of lead from glazed pottery and enamelled steel in household items designed for direct contact with food elements (in Swedish), *Var Foda,* 5, 41—56, 1970.
73. **Guthrie, B. E.,** Chromium, manganese, copper, zinc and cadmium content of New Zealand foods, *N.Z. Med. J.*, 82, 418—424, 1975.
74. **Guthrie, B. E. and Robinson, M. F.,** Daily intakes of manganese, copper, zinc and cadmium by New Zealand women, *Br. J. Nutr.*, 38, 55—63, 1977.
75. **Gydesen, H. and Rasmussen, L.,** Differences in the regional deposition of cadmium, copper, lead and zinc in Denmark as reflected in bulk precipitation, epiphytic cryptogams and animal kidneys, *Ambio,* 10(5), 229—230, 1981.

76. **Gydesen, H., Pilegaard, K., Rasmussen, L., and Rühling, Å.,** Moss analyses used as a means of surveying the atmospheric heavy-metal deposition in Sweden, Denmark and Greenland in 1980, SNV PM 1670, National Swedish Environment Protection Board, Solna, Sweden, 1983.
77. **Haga, Y. and Yamawaki, T.,** Research report on environmental health studies of heavy metal pollution (in Japanese), Rep. No. 18, Akita Prefecture Institute of Public Health, Akita, Japan, 1974.
78. **Harada, A. and Shibutanni, E.,** Medical examination of workers in a cadmium pigment factory (in Japanese), in *Kankyo Hoken Rep. No. 24,* Japanese Public Health Association, Tokyo, 1973, 16—22.
79. **Haraldsson, C. and Magnusson, B.,** Heavy metals in rainwater collection, storage and analysis of samples, in Int. Conf. Heavy Metals in the Environment, Heidelberg, September 1983, 82—86.
80. **Harrison, P. R. and Winchester, J. W.,** Area-wide distribution of lead, copper, and cadmium in air particulates from Chicago and northwest Indiana, *Atmos. Environ.*, 5, 863—880, 1971.
81. **Hassler, E.,** Exposure to Cadmium and Nickel in an Alkaline Battery Factory — As Evaluated from Measurements in Air and Biological Material, Doctoral thesis, Karolinska Institute, Stockholm, 1983.
82. **Hassler, E., Lind, B., and Piscator, M.,** Cadmium in blood and urine related to present and past exposure. A study of workers in an alkaline battery factory, *Br. J. Ind. Med.*, 40, 420—425, 1983.
83. **Havre, G. N., Underdal, B., and Christiansen, C.,** The content of lead and some other heavy elements in different fish species from a fjord in western Norway, in Int. Symp. Environ. Health Aspects of Lead, Amsterdam, October 1972, Commission of the European Communities, Luxembourg, 1973.
84. **Heindryckx, R., Demuynck, M., Dams, R., Janssens, M., and Rahn, K. A.,** Mercury and cadmium in Belgian aerosols, in Problems of the Contamination of Man and his Environment by Mercury and Cadmium, CEC European Colloquium, July 1973, Commission of the European Communities, Luxembourg, 1974, 135—148.
85. **Heinrichs, H., Schulz-Dobrick, B., and Wedepohl, K. H.,** Terrestrial geochemistry of Cd, Bi, Tl, Pb, Zn and Rb, *Geochim. Cosmochim. Acta,* 44, 1519—1553, 1980.
86. **Hemphill, D. D., Wixson, B. G., Gale, N. L., and Clevenger, T. E.,** Dispersal of heavy metals into the environment as a result of mining activities, in Int. Conf. Heavy Metals in the Environment, Heidelberg, September, 1983, 917—924.
87. **Herron, M. M., Langway, C. C., Jr., Weiss, H. V., and Cragin, J. H.,** Atmospheric trace metals and sulfate in the Greenland Ice Sheet, *Geochim. Cosmochim. Acta.*, 41, 915—920, 1977.
88. **Hinesley, T. D., Alexander, D. E., Redborg, K. E., and Ziegler, E. L.,** Differential accumulations of cadmium and zinc by corn hybrids grown on soil amended with sewage sludge, *Agron. J.*, 74, 469—474, 1982.
89. **Hirata, H.,** Annaka: land polluted mainly by fumes and dust from zinc smelter, in *Heavy Metal Pollution in Soils of Japan,* Kitagishi, K. and Yamane, I., Eds., Japan Scientific Societies Press, Tokyo, 1981, 149—163.
90. **Holm, J.,** Lead, cadmium and arsenic contents in meat and organ samples from game caught in areas with different levels of heavy metal contamination (in German), *Fleischwirtschaft,* 9, 1345—1349, 1979.
91. **Huey, N. A.,** Survey of airborne pollutants, in Helena Valley, Montana. Area Environmental Pollution Study, U. S. Environmental Protection Agency, Research Triangle Park, N. C., 1972, 25—60.
92. **Hunt, W. F., Pinkerton, C., McNulty, O., and Creason, J. A.,** A study in trace element pollution of air in 77 midwestern cities, in *Trace Substances in Environmental Health — IV,* Hemphill, D. D., Ed., University of Missouri, Columbia, 1971, 56—68.
93. **Hutton, M., Ed.,** Cadmium in the European community: a prospective assessment of sources, human exposure and environmental impact, MARC Rep. No. 26, Chelsea College, University of London, 1982.
94. Hyogo Prefectural Government, Results of a comprehensive study of cadmium pollution in the area around Ikuno Mine and an outline of counter-measures (in Japanese), 1972.
95. International Standard Organization, International Standards 7086/2, Glassware and glass ceramic ware in contact with food, release of lead and cadmium. II. Permissible limits, Ref. No. ISO 7086/2—1982(E), Geneva.
96. **Ishizaki, A.,** About results of a health examination (in Japanese), in *Kankyo Hoken Rep. No. 11,* Japanese Public Health Association, Tokyo, April 1972.
97. **Ishizaki, A., Fukushima, M., Sakamoto, M., Kurachi, T., and Hayashi, E.,** Cadmium content of rice eaten in the Itai-itai disease area (in Japanese), in *Proc. 27th Annu. Meet. Jpn. Public Health Assoc.*, Japanese Public Health Association, Tokyo, 1969, 111.
98. **Ishizaki, A., Fukushima, M., and Sakamoto, M.,** Distribution of Cd in biological materials. II. Cadmium and zinc contents of foodstuffs (in Japanese), *Jpn. J. Hyg.*, 25, 207—222, 1970.
99. **Iwao, S.,** Cadmium, lead, copper and zinc in food, feces and organs of humans, *Keio J. Med.*, 26, 63—78, 1977.
100. **Iwao, S., Sugita, M., and Tsuchiya, K.,** Some metabolic interrelationships among cadmium, lead, copper and zinc: results from a field survey in Cd-polluted areas in Japan. I. Dietary intake of the heavy metals, *Keio J. Med.*, 30, 17—36, 1981.

101. Japanese Association of Health Statistics (Koosei Tookei Kyokai), *Trends in National Health (Kokumin Eisei no Dookoo)*, Tokyo, 1972 (in Japanese).
102. Japanese Public Health Association, *Research about Intake and Accumulation of Cadmium in Areas "Requiring Observation"*, Japanese Public Health Association, Tokyo, March 30, 1970 (in Japanese).
103. **Jaworowski, Z., Bilkiewicz, J., Dobosz, E., and Wodkiewicz, L.,** Stable and radioactive pollutants in a Scandinavian glacier, *Environ. Pollut.*, 9, 305—315, 1975.
104. **Jedvall, J.,** The Swedish ban on cadmium, in *Utilisation of Sewage Sludge on Land: Rates of Application and Long-Term Effects of Metals*, D. Reidel Publ. Co., 1984, 198—211
105. **Jervis, R. E., Tiefenbach, B., and Chattopadhyay, A.,** Determination of trace cadmium in biological materials by neutron and pluton activation analyses, *Can. J. Chem.*, 52, 3008—3020, 1974.
106. **Johnson, D. E., Prevost, R. J., Tillery, J. B., and Thomas, R. E.,** The distribution of cadmium and other metals in human tissue, Environ. Prot. Tech. Ser. 232, U.S. Environmental Protection Agency, Washington, D.C., 1977.

106a. **Jorhem, L., Mattson, P., and Slorach, S.,** Lead, cadmium, zinc and certain other metals in foods on the Swedish market, *Var Foda*, 36, Suppl. 3, 1984.

107. **Just, J. and Kelus, J.,** Cadmium in the air atmosphere of ten selected cities in Poland, *Rocz. Panstw. Zakl. Hig.*, 22, 249—256, 1971 (translation obtainable from U.S. Environmental Protection Agency, Office of Air Programs, Translation Section, Research Triangle Park, N.C. 27711).
108. **Käferstein, F. D.,** Toxic heavy metals in foodstuffs (in German), *Zentralbl. Bakteriol. Hyg.(B)*, 171, 352—358, 1980.
109. **Kato, T. and Abe, K.,** Environmental pollution and health effects — Toyama Prefecture, in *Cadmium Studies in Japan — A Review*, Tsuchiya, K., Ed., Elsevier, Amsterdam, 1978, 225—249.
110. **Kawano, S. and Kato, T.,** Environmental pollution and health effects — Ishikawa Prefecture in *Cadmium Studies in Japan — A Review*, Tsuchiya, K., Ed., Elsevier, Amsterdam, 1978, 220—225.
111. **Keen, P.,** Trace elements in plants and soil in relation to cancer, *S. Afr. Med. J.*, 48, 2363—2364, 1974.
112. Kol-Hälsa-Miljö (KHM), Final report ISBN 91-7186-189-0, Coal-Health-Environment, Committee under the Swedish Government, The Swedish State Power Board, Vällingby, Sweden, 1983 (in Swedish).
113. **King, E.,** An environmental study of casting copper-cadmium alloys, *Br. J. Ind. Med.*, 12, 198—205, 1955.
114. **King, R. B., Fordyce, J. S., Antonie, A. C., Leibecki, H. F., Neustadter, H. E., and Sidik, S. M.,** Elemental composition of airborne particulates and source identification: an extensive one-year survey, *J. Air Pollut. Control Assoc.*, 26, 1073—1078, 1976.
115. **Kirkpatrick, D. C. and Coffin, D. E.,** Cadmium, lead and mercury content of various cured meats, *J. Sci. Food Agric.*, 24, 1595—1598, 1973.
116. **Kirkpatrick, D. C. and Coffin, D. E.,** The trace metal content of a representative Candian diet in 1972, *Can. J. Public Health*, 68, 162—164, 1977.
117. **Kitagishi, K. and Yamane, I., Eds.,** *Heavy Metal Pollution in Soils of Japan*, Japan Scientific Societies Press, Tokyo, 1981.
118. **Kjellström, T.,** An epidemiological exposure and effect study on cadmium. An investigation of the general and industrial environment around a Swedish copper and lead refinery (in Swedish), Report to the Swedish Environment Protection Board, Solna, Sweden, August 30, 1973.
119. **Kjellström, T.,** Accumulation and Renal Effects of Cadmium in Man. A Dose-Response Study, Doctoral thesis, Karolinska Institute, Stockholm, 1977.
120. **Kjellström, T., Lind, B., Linnman, L., and Nordberg, G.,** A comparative study on methods for cadmium analysis of grain with an application to pollution evaluation, *Environ. Res.*, 8, 92—106, 1974.
121. **Kjellström, T., Tsuchiya, K., Tompkins, E., Takabatake, E., Lind, B., and Linnman, L.,** A comparison of methods for analysis of cadmium in food and biological material. A co-operative study between Sweden, Japan and U.S.A., in *Recent Advances in the Assessment of the Health Effects of Environmental Pollution. II. Proc. Int. Symp., Paris, June 1974*, Commission of the European Communities, Luxembourg, 1975a, 2197—2316.
122. **Kjellström, T., Lind, B., Linnman, L., and Elinder, C. -G.,** Variation of cadmium concentration in Swedish wheat and barley. An indicator of changes in daily cadmium intake during the 20th century, *Arch. Environ. Health*, 30, 321—328, 1975b.
123. **Kjellström, T., Shiroishi, K., and Evrin, P. -E.,** Urinary β_2-microglobulin excretion among people exposed to cadmium in the general environment. An epidemiological study in cooperation between Japan and Sweden, *Envir. Res.*, 13, 318—344, 1977.
124. **Kjellström, T., Borg, K., and Lind, B.,** Cadmium in feces as an estimator of daily cadmium intake in Sweden, *Environ. Res.*, 15, 242—251, 1978.
125. **Kneip, T. J., Eisenbud, M., Strehlow, C. D., and Freudenthal, P. C.,** Airborne particulates in New York City, *J. Air Pollut. Control Assoc.*, 20, 144—149, 1970.

126. **Kobayashi, J.,** Air and water pollution by cadmium, lead and zinc attributed to the largest zinc refinery in Japan, in *Trace Substances in Environmental Health — V*, Hemphill, D. D., Ed., University of Missouri, Columbia, 1972, 117—128.
127. **Kobayashi, J., Morii, F., Muramoto, S., and Nakashima, S.,** Effects of air and water pollution on agricultural products by Cd, Pd and Zn attributed to a mine and refinery in Annaka City, Gunma Prefecture (in Japanese), *Jpn. J. Hyg.*, 25, 364—375, 1970.
128. **Koivistoinen, P.,** Mineral element composition of Finnish foods: N, K, Ca, Mg, P, S, Fe, Cu, Mn, Zn, Mo, Co, Ni, Cr, F, Se, Si, Rb, Al, B, Br, Hg, As, Cd, Pb and ash, *Acta Agric. Scand.*, Suppl. 22, 1980.
129. **Kojima, S., Haga, Y., Kurihara, T., and Yamawaki, T.,** Report from Akita Prefecture (in Japanese), in *Kankyo Hoken Rep. No. 36*, Japanese Public Health Association, Tokyo, 1975, 114—123.
130. **Kojima, S., Haga, Y., Kurihara, T., Yamawaki, T., and Kjellström, T.,** A comparison beween fecal cadmium and urinary β_2-microglobulin, total protein, and cadmium among Japanese farmers. An epidemiological study in cooperation between Japan and Sweden, *Environ. Res.*, 14, 436—451, 1977.
131. **Komai, Y.,** Heavy metal pollution in urban soils, in *Heavy Metal Pollution in Soils of Japan*, Kitagishi, K. and Yamane, I., Eds., Japan Scientific Societies Press, Tokyo, 1981, 193—217.
132. **Kowal, N. E., Johnson, D. E., Kraemer, D. F., and Pahren, H. R.,** Normal levels of cadmium in diet, urine, blood and tissues of inhibitants of the United States, *J. Toxicol. Environ. Health*, 5, 995—1014, 1979.
133. **Kreuzer, W., Sansoni, B., Kracke, W., and Wiszmath, P.,** Cadmium background content in meat, liver and kidney from cattle and its consequences to cadmium tolerance levels, *Chemosphere*, 4, 231—240, 1976.
134. **Kronemann, H., Anke, M., Grün, M., and Partschefeld, M.,** The cadmium concentration of feedstuffs, foodstuffs and of water in the GDR and in an area with non-ferrous metal industry, in *Kadmium-Symposium, August 1977, Jena*, Friedrich-Schiller Universität, Jena, 1979, 230—234.
135. **Kropf, R. and Geldmacher-v. Mallinckrodt, M.,** Cadmium content in foodstuffs and daily intake of cadmium (in German), *Arch. Hyg. Bakteriol.*, 152, 218—224, 1968.
136. **Laamanen, A.,** Functions, progress and prospects for an environmental subarctic base level station, *Work Environ. Health*, 9, 17—25, 1972.
137. **Landy, M. P.,** Trace metals in remote arctic snows: natural or anthropogenic?, *Nature (London)*, 284, 574—576, 1980.
138. **Lane, R. E. and Campbell, A. C. P.,** Fatal emphysema in two men making a copper cadmium alloy, *Br. J. Ind. Med.*, 11, 118—122, 1954.
139. **Lauwerys, R., Roels, H., Buchet, J. -P., Bernard, A., and Stanescu, D.,** Investigations on the lung and kidney functions in workers exposed to cadmium, *Environ. Health Perspect.*, 28, 137—145, 1979a.
140. **Lauwerys, R., Roels, H., Regniers, M., Buchet, J. -P., Bernard, A., and Goret, A.,** Significance of cadmium concentration in blood and in urine in workers exposed to cadmium, *Environ. Res.*, 20, 375—391, 1979b.
141. **Lauwerys, R., Roels, H., Bernard, A., and Buchet, J. -P.,** Renal response to cadmium in a population living in a nonferrous smelter area in Belgium, *Int. Arch. Occup. Environ. Health*, 45, 271—274, 1980.
142. **Lauwerys, R., Roels, H., Buchet, J. -P., Bernard, A., and de Wals, Ph.,** Environmental pollution by cadmium in Belgium and health damage, in *Cadmium 81, Proc. 3rd Int. Cadmium Conf. Miami, 1981*, Cadmium Association, London, 1982, 123—126.
143. **Lee, R. E., Patterson, R. K., and Wagman, J.,** Particle-size distribution of metal components in urban air, *Environ. Sci. Technol.*, 2, 288—290, 1968.
144. **Lee, R. E., Göranson, S. S., Enrione, R. E., and Morgan, G. B.,** National air surveillance cascade impactor network. II. Size distribution measurements of trace metal components, *Environ. Sci. Technol.*, 6, 1025—1030, 1972.
145. **Lener, J. and Bibr, B.,** Cadmium content in some foodstuffs in respect of its biological effects, *Vitalst. Zivilisationskr.*, 15, 139—141, 1970.
146. **Lim, M. Y.,** Trace elements from coal combustion atmospheric emissions, Report ICTIS/TRO5, IEA Coal Research, London, 1979.
147. **Lindberg, S. E. and Turner, R. R.,** Trace metals in rain at forested sites in the Eastern United States, in Int. Conf. Heavy Metals in the Environment, Heidelberg, September, 1983, 107—114.
148. **Linnman, L., Andersson, A., Nilsson, K. O., Lind, B., Kjellström, T., and Friberg, L.,** Cadmium uptake by wheat from sewage sludge used as a plant nutrient source, a comparative study using flameless atomic absorption and neutron activation analysis, *Arch. Environ. Health*, 27, 45—47, 1973.
149. **Lönsjö, H.,** Studies using isotope technique to test for crop absorption of cadmium from sewage sludge used on various types of soil (in Norwegian), NJR: Seminary on the Circulation of Heavy Metals in Agriculture, Norske Jordbruksrådet, Oslo, Norway, 1975, 101—113.
150. **Magnusson, B. and Westerlund, S.,** The determination of Cd, Cu, Fe, Ni, Pb and Zn in Baltic Sea water, *Mar. Chem.*, 8, 231—244, 1980.

151. **Mahaffey, K. R., Corneliussen, P. E., Jelinek, C. F., and Fiorino, J. A.,** Heavy metal exposure from foods, *Environ. Health Perspect.*, 12, 63—69, 1975.
152. **Marples, A. E. and Thornton, I.,** The distribution of cadmium derived from geochemical and industrial sources in agricultural soils and pasture herbage in parts of Britain, in *Cadmium 79, Proc. 2nd Int. Cadmium Conf. Cannes*, Metal Bulletin Ltd., London, 1980, 74—79.
153. **Mart, L.,** Seasonal variations of Cd, Pb, Cu and Ni levels in snow from eastern Artic Oceans, *Tellus*, 35B, 131—141, 1983.
154. **Mart, L., Rützel, H., Klahre, P., Sipos, L., Platzek, U., Valenta, P., and Nürnberg, H. W.,** Comparative studies on the distribution of heavy metals in the oceans and coastal waters, *Sci. Total Environ.*, 26, 1—17, 1982.
155. **Masironi, R., Koirtyohann, S. R., and Pierce, J. O.,** Zinc, copper, cadmium and chromium in polished and unpolished rice, *Sci. Total Environ.*, 7, 27—43, 1977.
156. **Materne, D., Lauwerys, R., Buchet, J.-P., Roels, H., Brouwers, J., and Stanescu, D.,** Investigations on the risks as a result of cadmium exposure in two cadmium using factories (in French), *Cah. Med. Trav.*, 12(4), 5—73, 1975.
157. **McKenzie, J. M., Kjellström, T., and Sharma, R.,** *Cadmium Intake, Metabolism and Effects in People with a High Intake of Oysters in New Zealand*, U.S. Environmental Protection Agency, Health Effects Research Laboratory, Cincinnati, Ohio, 1982.
158. **Menden, E. E., Elia, V. J., Michael, L. W., and Petering, H. G.,** Distribution of cadmium and nickel of tobacco during cigarette smoking, *Environ. Sci. Technol.*, 6, 830—832, 1972.
159. **Méranger, J. C. and Smith, D. C.,** The heavy metal content of a typical Canadian diet, *Can. J. Public Health*, 63, 53—57, 1972.
160. **Méranger, J. C., Subramanian, K. S., and Chalifoux, C.,** Metals and other elements: survey for cadmium, cobalt, chromium, copper, nickel, lead, zinc, calcium and magnesium in Canadian drinking water supplies, *J. Assoc. Off. Anal. Chem.*, 64, 44—53, 1981.
161. **Miller, G. J., Wylie, M. J., and McKeown, D.,** Cadmium exposure and renal accumulation in an Australian urban population, *Med. J. Aust.*, 10, 20—23, 1976.
162. Ministry of Agriculture, Fisheries and Food, Survey of cadmium in food. Working Party on the Monitoring of Foodstuffs for Heavy Metals, Fourth Report, Her Majesty's Stationery Office, London, 1973.
163. Ministry of Health and Welfare, Results of studies on cadmium pollution around mines and refineries (in Japanese), April 5, 1971.
164. **Morishita, T.,** The Jinzu River basin: contamination of soil and paddy rice with cadmium discharged from Kamioka Mine, in *Heavy Metal Pollution in Soils of Japan*, Kitagishi, K. and Yamane, I., Eds., Japan Scientific Societies Press, Tokyo, 1981, 107—124.
165. **Mortvedt, J. J. and Giordano, P. M.,** Crop uptake of heavy metal contaminants in fertilizers, in *Biological Implications of Heavy Metals in the Environment*, ERDA Rep. Conf. 750929, Wildung, R. A. and Drucker, H., Eds., Oak Ridge, Tenn., 1977, 402—416.
166. **Movitz, J.,** Investigation of metals in vegetables cultivated in the area around the Rönnskär smelter (in Swedish), The National Food Administration, Uppsala, Sweden, 1979.
167. **Movitz, J.,** High levels of cadmium in Swedish wild mushrooms (in Swedish with summary in English), *Var Foda*, 32, 270—277, 1980.
168. **Mulla, D. J., Page, A. L., and Ganje, T. J.,** Cadmium accumulation and bioavailability in soils from long-term phosphorus fertilization, *J. Environ. Qual.*, 913, 408—412, 1980.
169. **Müller, G.,** Heavy metal content (Cd, Zn, Pb, Cu, Cr) in tobacco from cigarettes generally smoked in West Germany (in German), *Chem. Z.*, 103, 133—137, 1979.
170. **Murthy, G. K., Rhea, U., and Peeler, J. T.,** Levels of antimony, cadmium, chromium, cobalt, manganese, and zinc in institutional total diets, *Environ. Sci. Technol.*, 5, 436—442, 1971.
171. **Muskett, C. J., Roberts, L. H., and Page, B. J.,** Cadmium and lead pollution from secondary metal refinery operations, *Sci. Total Environ.*, 11, 73—87, 1979.
172. **Nandi, M., Slone, D., Jick, H., Shapiro, S., and Lewis, G. P.,** Cadmium content of cigarettes, *Lancet*, 2, 1329—1330, 1969.
173. National Danish Agency of Environmental Protection, A report on use, occurrence and hazardous effects of cadmium in Denmark (in Danish with summary in English), Miljöministeriet, Copenhagen, 1980.
174. National Food Administration, Summary and Assessment of Data Received from the FAO/WHO Collaborating Centres for Food Contamination Monitoring, National Food Administration, Uppsala, Sweden, 1982.
175. **Naylor, L. M. and Loehr, R. C.,** Increase in dietary cadmium as a result of application of sewage sludge to agricultural land, *Am. Chem. Soc.*, 15, 881—886, 1981.
176. **Nielsen, S. A.,** Cadmium in New Zealand dredge oysters, geographic distribution, *Int. J. Environ. Anal. Chem.*, 4, 1—7, 1975.
177. **Nilsson, R.,** Sweden bans major uses of cadmium, *Ambio*, 8, 275—277, 1979.

178. **Nordberg, G. F., Slorach, S., and Stenström, T.,** Cadmium poisoning caused by a cooled soft-drink machine (in Swedish), *Lakartidningen*, 70, 601—604, 1973.
179. **Nordberg, M., Cherian, M. G., and Kjellström, T.,** Defence mechanisms against metal toxicity and their potential importance for risk assessments with particular reference to the importance of various binding forms in food stuff, Report 55 (KHM), The Swedish State Power Board, Vällingby, Sweden, 1983.
180. **Nriagu, J. O.,** Global inventory of natural and anthropogenic emissions of trace metals to the atmosphere, *Nature (London)*, 279, 409—411, 1979.
181. **Nriagu, J. O., Ed.,** *Cadmium in the Environment, Part I: Ecological Cycling*, John Wiley & Sons, New York, 1980.
182. **Nriagu, J. O., Ed.,** *Changing Metal Cycles and Human Health*, Dahlem Workshop Report, Springer-Verlag, Basel, 1983.
183. **Nürnberg, H. W.,** Voltammetric trace analysis in ecological chemistry of toxic metals, *Pure Appl. Chem.*, 54(4), 853—878, 1982.
184. **Nürnberg, H. W., Valenta, P., and Nguyen, V. D.,** Wet deposition of toxic metals from the atmosphere in the Federal Republic of Germany, in *Deposition of Atmospheric Pollutants*, Georgü, H. -W. and Pankrath, J., Eds., D. Reidel Publ., Dordrecht, Holland, 1982, 143—157.
185. **Nwankwo, J. N. and Elinder, C. -G.,** Cadmium, lead and zinc concentrations in soils and in food grown near a zinc and lead smelter in Zambia, *Bull. Environ. Contam. Toxicol.*, 22, 625—631, 1979.
186. **Nwankwo, J. N., Elinder, C. -G., Piscator, M., and Lind, B.,** Cadmium in Zambian cigarettes: an interlaboratory comparison in analysis, *Zambia J. Sci. Technol.*, 2, 1—4, 1977.
187. **Odén, S., Berggren, B., and Engwall, A. G.,** Heavy metals and chlorinated hydrocarbons in sludge (in Swedish), *Grundforbattring*, 23, 55, 1970.
188. **Oelschläger, W.,** On the contamination of feedstuffs and foodstuffs with cadmium (in German), *Landwirtsch. Forsch.*, 27, 247, 1974.
189. **Ohmomo, Y. and Sumiya, M.,** Estimation on heavy metal intake from agricultural products, in *Heavy Metal Pollution in Soils of Japan*, Kitagishi, K. and Yamane, J., Eds., Japan Scientific Societies Press, Tokyo, 1981, 235—244.
190. **Olofsson, A.,** Report on investigations of atmospheric emissions from the Finspong plant (in Swedish), *Sven. Metallverken*, No. HL757, 12, 1970.
191. **Page, A. L.,** Cadmium in soils and its accumulation by food crops, in *Int. Conf. Heavy Metals in the Environment, Amsterdam, September 1981*, CEP Consultants Ltd., Edinburgh, 1981, 206—213.
192. **Page, A. L., Bingham, F. T., and Shang, A. C.,** Cadmium, in *Effect of Heavy Metal Pollution on Plants, Vol. 1*, Lepp, N. W., Ed., Applied Science Publ., Barking, Essex, England, 1981, 77—109.
193. **Pahren, H. R., Lucas, J. B., Ryan, J. A., and Dotson, K. K.,** Health risks associated with land application of municipal sludge, *J. Water Pollut. Control Fed.*, 51, 2588—2601, 1979.
194. **Peirson, D. H., Cawse, P. A., Salmon, L., and Cambray, R. S.,** Trace elements in the atmospheric environment, *Nature (London)*, 241, 252—256, 1973.
195. **Penumarthy, L., Oehme, F. W., and Hayes, R. H.,** Lead, cadmium and mercury tissue residues in healthy swine, cattle, dogs and horses from the Midwestern United States, *Arch. Environ. Contam. Toxicol.*, 9, 193—206, 1980.
196. **Pettersson, O.,** The pollution of agricultural land and areal crops by heavy metals from Rönnskärsveerken (in Swedish), Research Report from School of Agriculture, Uppsala, Sweden, 1976.
197. **Pettersson, O.,** Differences in cadmium uptake between plant species and cultivars, *Swed. J. Agric. Res.*, 7, 21—24, 1977.
198. **Pfeilsticker, K. and Markard, C.,** Cadmium content in vegetables from gardens in an industrial region (in German), in *Kadmium-Symposium, August 1977, Jena*, Friedrich-Schiller-Universität, Jena, 1979, 224—230.
199. **Pinkerton, C., Creason, J. P., Shy, C. M., Hammer, D. I., Buechley, R. W., and Murthy, G. K.,** Cadmium content of milk and cardiovascular disease mortality, in *Trace Substances in Environmental Health — V*, Hemphill, D. D., Ed., University of Missouri, Columbia, 1971, 285—292.
200. **Pinkerton, C., Hammer, D. I., Bridbord, K., Creason, J. P., Kent, J. L., and Murthy, G. K.,** Human milk as a dietary source of cadmium and lead, in *Trace Substances in Environmental Health — VI*, Hemphill, D. D., Ed., University of Missouri, Columbia, 1972, 39—43.
201. **Piscator, M.,** in *Cadmium in the Environment*, Friberg, L., Piscator, M., and Nordberg, G., Eds., CRC Press, Boca Raton, Fla., 1971.
202. **Piscator, M.,** Cadmium-zinc interactions, in *Proc. CEC-EPA-WHO Int. Symp. Recent Advances in the Assessment of Health Effects of Environmental Pollution, Paris, June 1974*, Commission of the European Communities, Luxembourg, 1975, 951—958.
203. **Piscator, M., Kjellström, T., and Lind, B.,** Contamination of cigarettes and pipe tobacco by cadmium-oxide dust, *Lancet*, II, 587, 1976.

203a. **Plunkert, P. A.,** Minerals Yearbook, Vol. 1, U.S. Department of the Interior, Washington, D.C., 1982.

204. **Preston, A., Jefferies, D. F., Dutton, J. W. R., Harvey, B. R., and Steele, A. K.,** British Isles coastal waters: the concentrations of selected heavy metals in sea water, suspended matter and biological indicators — a pilot survey, *Environ. Pollut.*, 3, 69—82, 1972.
205. **Princi, F.,** A study of industrial exposures to cadmium, *J. Ind. Hyg. Toxicol.*, 29, 315—324, 1947.
206. **Purves, D.,** National standards for metal additions to soil in sewage sludge, in *Int. Conf. Heavy Metals in the Environment, Amsterdam, September 1981*, CEP Consultants Ltd., Edinburgh, 1981, 176—179.
207. **Raffke, W., Cumbrowski, J., and Jacobi, J.,** The cadmium contents in foods and human organs in a large town (in German), *Nahrung*, 24, 797—802, 1980.
208. **Ragaini, R. C., Ralston, H. R., and Roberts, N.,** Environmental trace metal contamination in Kellogg, Idaho, near a lead smelting complex, *Environ. Sci. Technol.*, 11, 773—781, 1977.
209. **Rautu, R. and Sporn, A.,** Contributions to the determination of cadmium supply by foods (in German), *Nahrung*, 14, 25—31, 1970.

209a. **Rogenfelt, A.,** Unpublished data, 1981.

210. **Rohbock, E., Georgii, H. -W., Perseke, C., and Kins, L.,** Wet and dry deposition of heavy metal-aerosols in the Federal Republic of Germany, in *Int. Conf. Heavy Metals in the Environment, Amsterdam, September 1981*, CEP Consultants Ltd., Edinburgh, 1981, 310—313.
211. **Rojahn, T.,** Determination of copper, lead, cadmium and zinc in estuarine water by anodic-stripping alternating-current voltammetry on the hanging mercury drop electrode, *Anal. Chim. Acta*, 62, 438—441, 1972.
212. **Rühling, Å and Skärby, L.,** *National Survey of Regional Heavy Metal Concentrations in Moss*, Liber Tryck, Stockholm, 1979 (in Swedish).
213. **Rühling, Å and Tyler, G.,** Deposition of heavy metals over Scandinavia (in Swedish), Internal Report, Department of Plant Biology, Lund University, Lund, Sweden, 1972.
214. **Ruick, G. and Schmidt, M.,** Square-wave-polarographic determination of heavy metals in vegetable foods with special regard to cadmium (in German), *Nahrung*, 23, 39—48, 1979.
215. **Ryan, J. A., Pahren, H. R., and Lucas, J. B.,** Controlling cadmium in the human food chain: a review and rationale based on health effects, *Environ. Res.*, 28, 251—302, 1982.
216. **Schroeder, H. A.,** A sensible look at air pollution by metals, *Arch. Environ. Health*, 21, 798—806, 1970.
217. **Schroeder, H. A. and Balassa, J. J.,** Abnormal trace metals in man: cadmium, *J. Chronic Dis.*, 14, 236—258, 1961.
218. **Schroeder, H. A., Nason, A. P., Tipton, I. H., and Balassa, J. J.,** Essential trace metals in man: zinc. Relation to environmental cadmium, *J. Chronic Dis.*, 20, 179—210, 1967.
219. **Schütz, A.,** Cadmium and lead, *Scand. J. Gastroenterol.*, 14, Suppl. 52, 223—231, 1979.
220. **Seeger, R.,** Cadmium in mushrooms (in German), *Z. Lebensm. Unters. Forsch.*, 166, 23—34, 1978.
221. **Sharrett, A. R., Carter, A. P., Orheim, R. M., and Feinleib, M.,** Daily intake of lead, cadmium, copper, and zinc from drinking water: the Seattle study of trace metal exposure, *Environ. Res.*, 28, 456—475, 1982.
222. **Sherlock, J. C., Smart, G. A., and Walters, B.,** Dietary surveys on a population at Shipham, Somerset, United Kingdom, *Sci. Total Environ.*, 29, 121—142, 1983.
223. **Shigematsu, I. and Kawaguchi, T.,** Environmental pollution and health effects — Akita, Yamagata, Fukushima and Fukuoka, Prefectures, in *Cadmium Studies in Japan — A Review*, Tsuchiya, K., Ed., Elsevier, Amsterdam, 1978, 192—199, 199—211, 211—215, 249—253.
224. **Shigematsu, I., Yanagawa, H., and Kawaguchi, T.,** Record filling of health examinations for inhabitants in the Cd polluted areas (2nd and 3rd reports) (in Japanese), *Kankyo Hoken Rep. No. 36*, Japanese Public Health Association, Tokyo, 1975, 72—101.
225. **Simpson, W. R.,** A critical review of cadmium in the marine environment, *Prog. Oceanogr.*, 10, 1—70, 1981.
226. **Smith, T. J., Petty, T. L., Reading, J. C., and Lakshminarayan, S.,** Pulmonary effects of chronic exposure to airborne cadmium, *Am. Rev. Respir. Dis.*, 114, 161—169, 1976.
227. **Slorach, S., Gustafsson, I. B., Jorhem, L., and Mattsson, P.,** Intake of lead, cadmium and certain other metals via a typical Swedish weekly diet (in Swedish), *Var Foda*, 35(Suppl. 1), 1983.
228. **Solly, S. R. B., Revfeim, K. J. A., and Finch, G. D.,** Concentrations of cadmium, copper, selenium, zinc and lead in tissues of New Zealand cattle, pigs and sheep, *N.Z. J. Sci.*, 24, 81—87, 1981.
229. **Sommers, L. E.,** Chemical composition of sewage sludges and analysis of their potential use as fertilizers, *J. Environ. Qual.*, 6, 225—231, 1977.
230. **Sommers, L. E.,** Toxic metals in agricultural crops, in *Sludge — Health Risk of Land Application*, Bitton, G., Damron, B. L., Eds, G. T., and Davidson, J. M., Eds., Ann Arbor Science Publ., Ann Arbor, Mich., 1980, 105—140.
231. **Spencer, H., Asmussen, C. R., Holtzman, R. B., and Kramer, L.,** Metabolic balances of cadmium, copper, manganese and zinc in man, *Am. J. Clin. Nutr.*, 32, 1867—1875, 1979.
232. **Stoeppler, M.,** Data on cadmium in oil, in preparation.

233. **Suzuki, S. and Lu, C. -G.,** A balance study of cadmium — an estimation of daily input, output and retained amount in two subjects, *Ind. Health*, 14, 53—65, 1976.
234. **Suzuki, S., Suzuki, T., and Ashizawa, M.,** Proteinuria due to inhalation of cadmium stearate dust, *Ind. Health*, 3, 73—85, 1965.
235. Swedish National Board of Health and Welfare, *Advice and Instructions for the Use of Sewage Sludge in Soil Improvement*, Liber, Stockholm, Sweden, 1973, 8 (in Swedish).
236. **Szadkowski, D.,** Cadmium — an ecological disturbance at the working place (in German), *Med. Monatsschr.*, 26, 553—556, 1972.
237. **Szadkowski, D., Schultze, H., Schaller, K. H., and Lehnert, G.,** On the ecological consequence of the heavy metal content in cigarettes (in German), *Arch. Hyg. Bakteriol.*, 153, 1—8, 1969.
238. **Takabatake, E.,** Environmental pollution and health effects — Nagasaki, Oita, and Miyagi Prefectures, in *Cadmium Studies in Japan — A Review*, Tsuchiya, K., Ed., Elsevier, Amsterdam, 1978, 181—192, 215—218, 218—220.
239. **Tanaka, Y., Ikebe, K., Tanaka, R., and Kunita, N.,** Extents and average contents of heavy metals in vegetable foods, *Shokuhin Diseigaku Zasshi*, 18, 75—85, 1977.
240. Task Group on Lung Dynamics, Deposition and retention models for internal dosimetry of the human respiratory tract, *Health Phys.*, 12, 173—208, 1966.
241. **Tati, M., Katagiri, Y., and Kawai, M.,** Urinary and fetal excretion of cadmium in normal Japanese: an approach to non-toxic levels of cadmium, in *Effects and Dose-Response Relationships of Toxic Metals*, Nordberg, G. F., Ed., Elsevier, Amsterdam, 1976, 331—342.
242. **Thomas, B., Roughan, J. A., and Watters, E. D.,** Lead and cadmium content of some vegetable foodstuffs, *J. Sci. Food Agric.*, 23, 1493—1498, 1972.
243. **Thornton, I., John, S., Moorcroft, S., and Watt, J.,** Cadmium at Shipham — an unique example of environmental geochemistry and health, in *Trace Substances in Environmental Health — XIV*, Hemphill, D. D., Ed., University of Missouri Press, Columbia, 1981, 27—32.
244. **Tiller, K. G., De Vries, M. P. C., and Spouncer, L. P.,** Some metals in selected vegetables purchased in the Adelaide metropolitan area, Division of Soil Divisional Rep. No. 13, CSIRO, Australia, 1976; cited by **Page et al.,** 1981.
245. **Tipton, I. H. and Stewart, P. L.,** Pattern of elemental excretion in long-term balance studies. II, in *Internal Dosimetry*, Synder, W. S., Ed., Annu. Prog. Rep. ORNL-4446, Health Physics Division, Oak Ridge, Tenn., 1970, 303-304.
246. **Tipton, I. H., Stewart, P. L., and Dickson, J.,** Patterns of elemental excretion in long-term balance studies, *Health Phys.*, 16, 455—462, 1969.
247. **Tjell, J.,** Long-term effects of metals, in *Utilisation of Sewage Sludge on Land: Rates of Application and Long-Term Effects of Metals*, Berglund, S., Davis, R. D., and L'Hermite, P., Eds., D. Reidel Publ. Co., Dordrecht, Holland, 1984, 218—219.
248. **Tsuchiya, K.,** Proteinuria of workers exposed to cadmium fume. The relationship to concentration in the working environment, *Arch. Environ. Health*, 14, 875—880, 1967.
249. **Tsuchiya, K.,** Causation of ouch-ouch disease, an introductory review. I. Nature of the disease, *Keio J. Med.*, 18, 181—194, 1969.
250. **Tsuchiya, K., Ed.**, *Cadmium Studies in Japan — A Review*, Elsevier, Amsterdam, 1978.
251. **Tsuchiya, K. and Iwao, S.,** Interrelationships among zinc, copper, lead and cadmium in food, feces and organs of humans, *Environ. Health Perspect.*, 25, 119—124, 1978.
252. **Tsuchiya, K. and Nakamura, K. -I.,** Environmental pollution and health effects — Hyogo Prefecture, in *Cadmium Studies in Japan — A Review*, Tsuchiya, K., Ed., Elsevier, Amsterdam, 1978, 146—167.
253. **Tyler, G.,** Moss analysis — a method for surveying heavy metal deposition, in *Proc. 2nd Int. Clean Air Congr.*, Englund, H. M. and Berry, W. T., Eds., Academic Press, New York, 1970.
254. Umweltbundesamt, Air quality criteria for cadmium (in German), *Berichte 4/77*, Umweltbundesamt, Berlin 1977.
255. **Underwood, E. J., Ed.,** *Trace Elements in Human and Animal Nutrition*, Academic Press, New York, 1977.
256. **Uruno, K., Otoyama, S., Nakano, N., and Itagaki, Z.,** Results of health examination for inhabitants in Yoshinokawa area polluted with cadmium (in Japanese), in *Kankyo Hoken Rep. No. 36*, Japanese Public Health Association, Tokyo, 1975, 109—113.
257. U.S. Council for Agricultural Science and Technology, Effects of sewage sludge on the cadmium and zinc content of crops, Rep. No. 83, CAST, Ames, Iowa, 1980.
258. **Ushio, F. and Doguchi, M.,** Dietary intakes of some chlorinated hydrocarbons and heavy metals estimated on the experimentally prepared diets, *Bull. Environ. Contam. Toxicol.*, 17, 707—711, 1977.
259. **Valenta, P., Ostapczuk, P. H., Pihlar, B., and Nürnberg, H. W.,** New applications of voltammetry in the determination of toxic trace metals in food, in *Int. Conf. Heavy Metals in the Environment, Amsterdam, September 1981*, CEP Consultants Ltd., Edinburgh, 1981, 619—621.

260. **Vetter, H., Mählhop, R., and Früchtenicht, K.,** Substances emitted in neighbourhood of a lead and zinc smelting plant (in German), *Ber. Ludw.*, 52, 327—350, 1974.
261. **Walters, B. and Sherlock, J.,** Studies on the dietary intake of heavy metals, in *Int. Conf. Heavy Metals in the Environment, Amsterdam, September 1981*, CEP Consultants Ltd., Edinburgh, 1981, 506—512.
262. **Wambeke, van, L.,** Sewage sludge in the European community, in *Cadmium 79, Proc. 2nd Int. Cadmium Conf., Cannes*, Metal Bulletin Ltd., London, 1980, 96.
263. **Webb, M., Ed.,** *The Chemistry, Biochemistry and Biology of Cadmium*, Elsevier, Amsterdam, 1979.
264. **Webber, M.,** in *Utilisation of Sewage Sludge on Land: Rates of Application and Long-Term Effects of Metals*, Berglund, S., Davis, R. D., and L'Hermite, P., Eds., D. Reidel Publ., Dortrect, Holland, 1984, 220—224.
265. **Weiss, H., Bertine, K., Kode, M., and Goldberg, E. D.,** The chemical composition of a Greenland glacier, *Geochim. Cosmochim. Acta*, 39, 1—10, 1975.
266. **Wescott, D. T. and Spincer, D.,** The cadmium, nickel and lead content of tobacco and cigarette smoke, *Beitr. Tabakforsch.*, 4(7), 217—221, 1974.
267. **Wester, P. O.,** Trace element balances in relation to variations in calcium intake, *Atherosclerosis*, 20, 207—215, 1974.
268. **Williams, C. H. and David, D. J.,** The effect of superphosphate on the cadmium content of soils and plants, *Aust. J. Soil Res.*, 11, 43—56, 1973.
269. **Williams, C. H. and David, D. J.,** The accumulation in soil of cadmium residues from phosphate fertilizers and their effect on the cadmium content of plants, *Soil Sci.*, 121, 86—93, 1976.
270. **Woidich, H. and Pfannhauser, W.,** Contribution to the analysis of cadmium in foods (in German), *Z. Lebensm. Unters. Forsch.*, 155, 72—76, 1974.
271. **Wolnik, K. A., Fricke, F. L., Caper, S. G., Braude, G. L., Meyer, M. W., Satzger, R. D., and Bonnin, E.,** Elements in major raw agricultural crops in the United States. I. Cadmium and lead in lettuce, peanuts, potatoes, soybeans, sweet corn, and wheat, *J. Agric. Food Chem.*, 31, 1240—1244, 1983.
272. **Yamagata, N.,** Cadmium in the environment and in humans, in *Cadmium Studies in Japan — A Review*, Tsuchiya, K., Ed., Elsevier, Amsterdam, 1978, 19—37.
273. **Yamagata, N. and Iwashima, K.,** Average cadmium intake of the Japanese people, *Bull. Inst. Public Health (Tokyo)*, 24, 18—24, 1975.
274. **Yamagata, N. and Shigematsu, I.,** Cadmium pollution in perspective, *Bull. Inst. Public Health (Tokyo)*, 19, 1—27, 1970.
275. **Yamagata, N., Iwashima, K., Kuzuhara, Y., and Yamagata, T.,** A model surveillance for cadmium pollution, *Bull. Inst. Public Health (Tokyo)*, 20, 170—186, 1971.
276. **Yamamoto, Y.,** Present status of cadmium environmental pollution (in Japanese), in *Kanyko Hoken Rep. No. 11*, Japanese Public Health Association, Tokyo 1972, 7.
277. **Zook, G., Powell, J. J., Hackley, B. M., Emerson, J. A., Brooker, J. R., and Knobl, G. M., Jr.,** National marine fisheries service, preliminary survey of selected seafoods for mercury, lead, cadmium, chromium and arsenic content, *J. Agric. Food Chem.*, 24, 47—53, 1976.

Chapter 4

METALLOTHIONEIN

Carl-Gustaf Elinder and Monica Nordberg

TABLE OF CONTENTS

I. INTRODUCTION

The first studies on metallothionein were carried out during the 1950s and 1960s. In the past 10 years, there has been an increasing interest in metallothionein among scientists in the field of toxicology, environmental hygiene, medicine, pathology, biochemistry, and nutrition. A vast number of publications dealing with different aspects of metallothionein have appeared, far too many to cover within the scope of this chapter. The purpose of this presentation is to provide certain basic information about metallothionein which may be of value for a better understanding of the mechanisms involved in cadmium metabolism and toxicity.

Several review articles on metallothionein have been published by, e.g., Friberg et al.,[36] Webb and Etienne,[116] Bremner,[9] Cherian and Goyer,[19] Kojima and Kägi,[55] Nordberg,[71] Nordberg and Kojima,[72] Webb,[113] Kotsonis and Klaassen,[58] Brady,[7] Webb and Cain,[114] Bakka,[2] and Cherian and Nordberg.[20] Two books based on information gained from recent meetings on metallothionein have been published — *Metallothionein*[46] and *Biological Roles of Metallothionein*.[32] Results of a third conference, "Proceedings of the Metallothionein and Cadmium Nephrotoxicity Conference", held in 1983, were published in 1984.[39]

II. HISTORY OF DISCOVERY AND IDENTIFICATION

In 1957, Margoshes and Vallee[62] reported the isolation of a low molecular weight cadmium-binding protein from horse renal cortex. Subsequent work by Kägi and Vallee[47,48] resulted in the purification of a metalloprotein, with a molecular size of about 10,000 daltons, which they named metallothionein because of its high content of metals (cadmium and zinc) and sulfur. Later studies have shown that the actual molecular weight is about 6,500,[72] based on amino acid residues. This metalloprotein was shown to have many unique physicochemical properties (Table 1).

In contrast to most other proteins, metallothionein did not show an absorption maximum at 280 nm, indicative of absence of aromatic amino acids. This was later verified by amino acid analysis which also showed high cysteine content (about 30%). Metallothionein showed an absorption maximum at 250 nm when cadmium was the predominant ion. This absorption was shown to be due to charge transfer from sulfur ligands to cadmium in metallothionein. The metal-free apoprotein — thionein — did not show this absorption which reappeared when reconstituted in the presence of cadmium salts.

In 1964, Piscator[82] was able to show that cadmium exposure of rabbits induced the synthesis of metallothionein in rabbit liver. He postulated that metallothionein may be involved in the intracellular sequestering of cadmium and may serve as a hepatic metal storage protein. Piscator[82] also suggested that metallothionein could be a carrier of cadmium from liver to kidney. Two years later, Pulido et al.[85] isolated metallothionein with a high cadmium content from human liver.

III. ISOLATION AND PURIFICATION

The most common techniques used in the preparation and isolation of metallothionein are gel filtration, ion-exchange chromatography, and isoelectric focusing. Salt and/or organic solvent fractionation and heat precipitation procedures have also been used. Isolation procedures should be carried out preferably in alkaline pH to prevent dissociation of metals from the thionein. When preparing copper-containing metallothionein, special measures have to be taken in order to prevent autooxidation of the copper-thionein.[11,41,90] Mercaptoethanol is sometimes added during the preparation of metallothionein from tissue homogenates to stabilize the SH-groups of the cysteine. However, addition of mercaptoethanol to crude

Table 1
PROPERTIES OF METALLOTHIONEIN[19,46]

Induced synthesis of metallothionein by certain metals
Low molecular weight (6000—7000)
High content of cysteine (30%)
High metal content (5—10% w/w)
Absence of disulfide bonds, aromatic amino acids, and histidine
Absorption maximum at 250 nm for cadmium-thionein
Heat stability
Cytoplasmic subcellular localization
Unique amino acid sequence (fixed distribution of cysteinyl residues)

homogenates can result in extraction of copper from other proteins and the binding of this soluble copper to metallothionein in vitro.[89] An alternative method to prevent oxidation of metallothionein during isolation is to use anaerobic conditions by saturation of buffers with nitrogen.[100] As the gel filtration system only separates the different proteins according to their three-dimensional size, separation of the protein from tissue homogenates by gel filtration reveals only one form of metallothionein since the different forms have the same Stokes radius.[72] Figure 1 presents a typical elution profile using gel filteration on supernatant from horse kidney cortex homogenate.

Separation techniques based on the charge properties of metallothionein, such as ion-exchange chromatography and isoelectric focusing, have shown that different forms of metallothionein often exist in the same organ, human and horse kidney,[85] rabbit liver,[68] human liver,[14] and rat liver.[17,51] Usually two main forms of metallothionein are found: MT-1 and MT-2. Figure 2 presents a typical elution pattern where metallothionein-containing fractions obtained from a gel filtration step have been purified further by diethylaminoethyl (DEAE) ion-exchange chromatography. As a rule, the total amount of metal ions bound to each metallothionein molecule is constant, but the types of metal ions might differ. The metal ions found bound to metallothionein in vivo are usually cadmium, zinc, copper, and mercury. Apart from having a different molar ratio of metals, the two different forms of metallothionein (MT-1 and MT-2) from the same species and tissues have also been shown to have slightly different amino acid composition.[68,119]

Indirect methods, using in vitro binding of radioactive mercury[81] or cadmium[17,78] to metallothionein, have been used to estimate metallothionein concentration in various tissues. In certain methods, excess of nonmetallothionien-bound mercury or cadmium is removed from the assay by repeated additions of other, less stable, metal binding proteins, e.g., hemoglobin, which is easily separated from the metallothionein.

Other methods which have been used for the determination of metallothionein are an electrochemical method described by Olafson and Sim[77] and high-performance liquid chromatography coupled with atomic absorption spectrophotometry (HPLC-AAS).[98]

In recent years, radioimmunoassay (RIA) techniques for the determination of metallothionein have become available. This has increased sensitivity in the determination of small amounts of metallothionein by several magnitudes and it is, nowadays, possible to measure very low concentrations of metallothionein in plasma and urine.[8,103,104,109] A radioimmunoassay for human metallothionein has been reported to have a detection limit of 0.1 to 0.2 ng.[16]

Metallothionein concentrations in urine and plasma from cadmium-exposed workers[16,29,37,70,105] and in urine samples from cadmium-exposed Japanese farmers, including Itai-itai disease patients, have been reported.[106,107] In humans, the plasma or serum concentration of metallothionein has been reported to vary between <1 and 16 μg/ℓ and in urine between 5 and about 3000 μg/g creatinine. Urinary excretion of metallothionein is increased among persons with cadmium-induced tubular dysfunction,[29,70,105—107] but the increase is

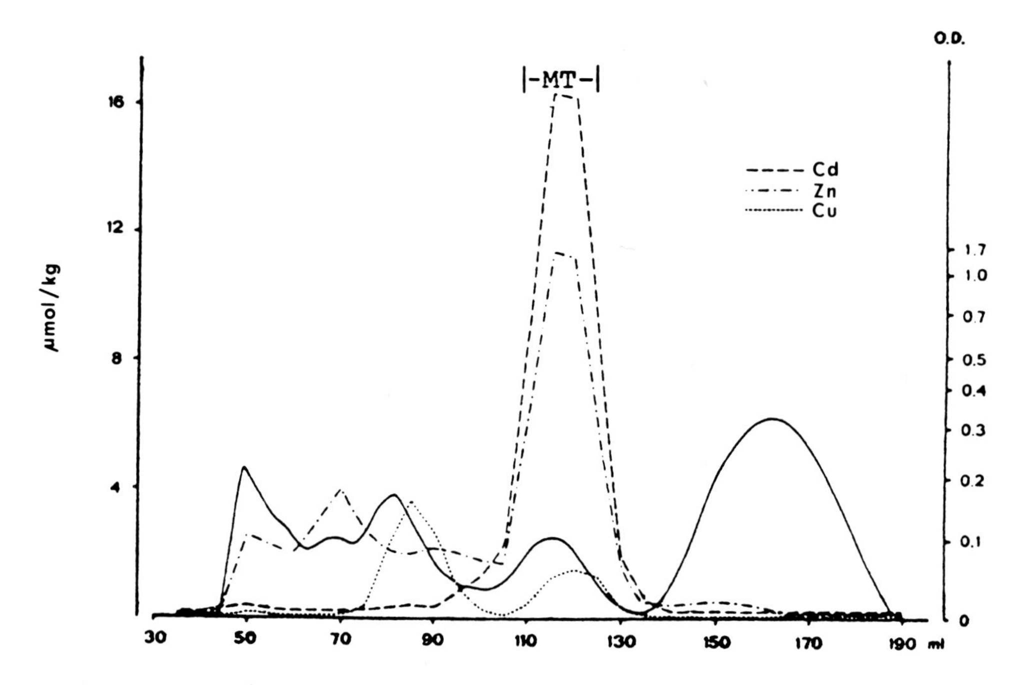

FIGURE 1. Gel filtration chromatography on G-75 Sephadex of homogenate from horse kidney cortex. Dotted lines show metal concentrations in different protein fractions. Solid lines show the optical density at 254 nm. (From Nordberg, M., Elinder, C.-G., and Rahnster, B., *Environ. Res.*, 20, 341—350, 1979. With permission.)

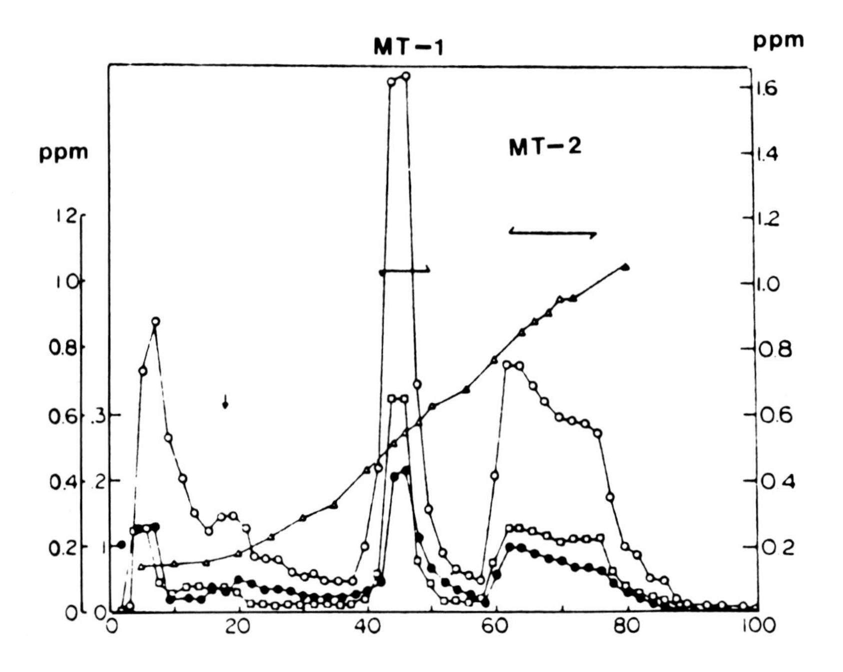

FIGURE 2. DEAE-cellulose column chromatography profile for low molecular weight cadmium peak (metallothionein) obtained from Sephadex G-75 chromatography on a homogenate prepared from cadmium-exposed rat livers (○) ppm Cd. (□) ppm Zn. (●) A_{250}. (△) Conductivity. (From Ridlington, J. W., Winge, D. R., and Fowler, B. A., *Biochim. Biophys. Acta*, 673, 177—183, 1981. With permission.)

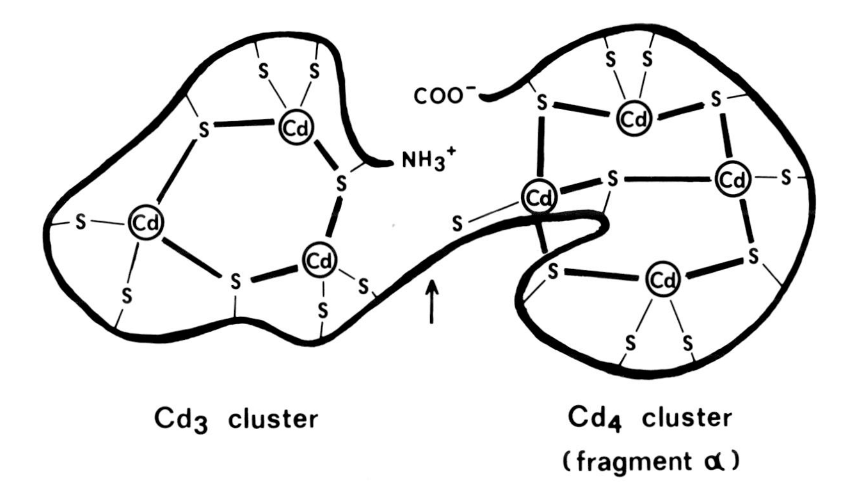

FIGURE 3. Schematic representation of the structure of (Cd_7-Th) from rat liver. (From Winge, D. R. and Miklossy, K.-A., *Arch. Biochem. Biophys.*, 214, 80—88, 1982b. With permission.)

much less prominent than the increased excretion of other low molecular weight proteins (see Chapter 9, Section I.).

There is a considerable variation in reported levels of metallothionein in plasma and urine of nonoccupationally exposed persons. Therefore, the true normal concentrations of metallothionein in urine and plasma are as yet not well established.

IV. METAL BINDING

Metallothionein was discovered because of its capacity to bind cadmium and for the possibility to exchange cadmium for other metals, such as mercury, copper, and zinc in vitro. From experiments on horse kidney metallothionein, Kägi and Vallee[48] were able to show that the bond between metallothionein and mercury was stronger than that between metallothionein and cadmium, which in turn was much stronger than the bond with zinc. However, metals (zinc, cadmium, mercury, and copper) in metallothionein can be replaced in vitro either by direct displacement or after acidification and reconstitution with the desired metal.[48,74] In rat kidney metallothionein, relative affinity for the metals lies in a descending order, Hg > Cu (depending on state of oxidation) > Cd > Zn.[32,46] The spectral locations of the specific absorption shoulders for zinc, cadmium, copper and mercury metallothionein differ, being 225, 250, 275, and 300 nm, respectively.[46,48,97] Probably, the same order of affinity applies for the different metals in metallothionein obtained from other tissues and species.

The molar ratio of the number of SH-groups in the protein to the number of zinc plus cadmium atoms is close to three,[72] but each metal ion is coordinated to four thiolate ligands.[110] Seven binding sites for cadmium and zinc atoms are available for each molecule of metallothionein. Only two SH-groups are considered to be necessary for mercury and copper, and thus one molecule of metallothionein can bind more of these atoms than those of cadmium or zinc. Boulanger and Armitage[6] found that human liver metallothionein saturated with cadmium ions consists of one four- and one three-metal cluster, which contain metal-tetrathiolate coordinations (Figure 3).[13,118]

Metallothionein isolated from kidneys of old horses contains mainly cadmium, lesser amounts of zinc, and only small amounts of copper,[49,75] whereas liver metallothionein from horses contains more zinc than cadmium,[27,49] In fetal and neonatal liver, copper and zinc are the major metal constituents; there is no cadmium present.[4,5,12,41,89,90]

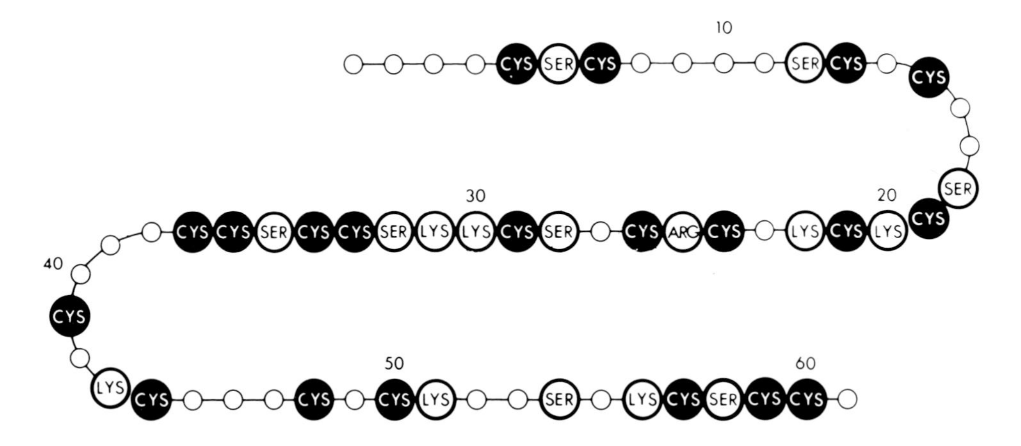

FIGURE 4. Distribution of cysteine, serine, and basic amino acids in horse kidney metallothionein-1B. Small open circles indicate other amino acid residues. (Adapted from Kojima, Y., Berger, C., Vallee, B. L., and Kägi, J. H. R., *Proc. Natl. Acad. Sci. U.S.A.*, 73, 3413—3417, 1976.)

In cadmium-exposed rats (s.c. 0.5 mg Cd per kg × 6 days/week × 25 weeks), it has been shown that when hepatic metallothionein containing cadmium and zinc is released into plasma, it contains mainly cadmium and copper. It is conceivable that zinc, in rat liver metallothionein released into serum, might be replaced by copper and that metallothionein under these particular experimental conditions might serve as a carrier protein for copper to the kidney.[99,101]

Human autopsy studies have shown that the zinc concentration increases in kidney cortex with increasing concentration of cadmium.[26,40,83] At relatively low concentrations of cadmium in kidney cortex, less than about 50 mg/kg, the increase in zinc in human kidney cortex is almost equimolar to the increase of cadmium. At high concentrations of cadmium in kidney cortex, the increase in zinc levels off and the relationship between cadmium and zinc concentrations in kidney cortex can more adequately be expressed by a second-order mathematical function.[24,44]

The relative content of zinc and cadmium in human kidney metallothionein has not been extensively studied, but studies on environmentally exposed horses[75] have shown that the increase of zinc in kidney cortex with increasing cadmium concentration can be explained by and related to the production of metallothionein which contains cadmium as well as zinc. It is most likely that even the zinc increase seen in human kidney cortex with increasing cadmium concentration can be explained by the formation of metallothionein-binding zinc and cadmium.

V. AMINO ACID SEQUENCE

Mammalian metallothionein contains 61 amino acids, of which 20 are cysteine; it has no aromatic amino acids or histidine.[46,50]

Metallothionein always contains one methionine, serine, and a number of basic amino acids. The protein contains no free SH-groups and disulfide bridges in vivo and always binds metals.

The amino acid sequence has been determined for metallothionein obtained from various mammals such as mice,[42,43] rats,[118] rabbits,[52] horses,[56,57] and also from humans.[53] Acetylmethionine and alanine have been identified as the N- and C-terminal amino acids of the protein. The predominating amino acid sequence in metallothionein is Cys-X-Cys.[50] Basic amino acids and serine are usually located in juxtaposition to cysteinyl residues. Figure 4 presents the amino acid sequence for horse kidney metallothionein.

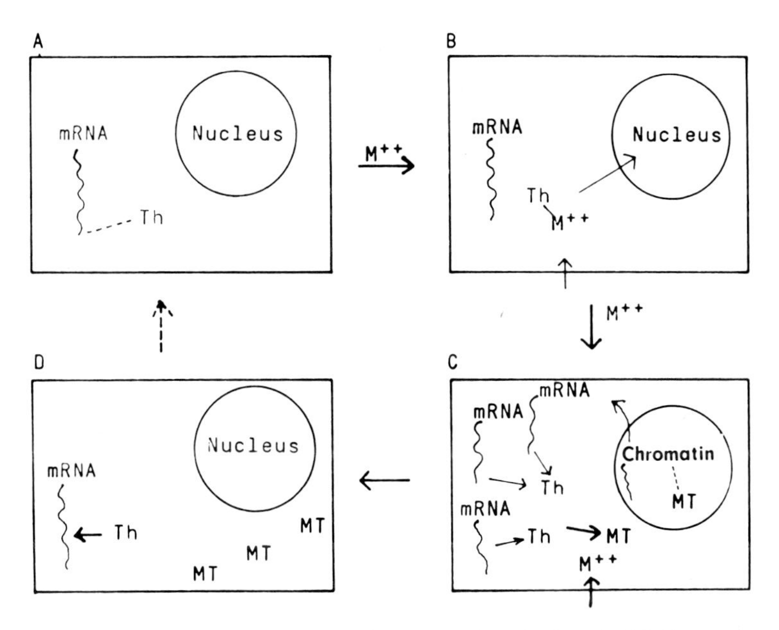

FIGURE 5. A scheme for the induced synthesis of metallothionein (MT) by metals. (A) Control cell. (B, C, D) Cells at various time periods after exposure to inducing metals (M^{2+}). Th, thionein; MT, thionein plus metal. The metal in MT will be either M^{2+} or previously present metal. (From Cherian, M. G. and Nordberg, M., *Toxicology*, 28, 1—15, 1983. With permission.)

The two polymorphic forms of metallothionein from the same animal and tissue may differ in amino acid composition, tertiary structure, as well as in affinity to different metal ions.[118] The isomers of horse kidney metallothionein (Figure 4) have variations in 7 amino acid positions and mouse metallothionein in 15 positions. The two major forms of metallothionein — MT-1 and MT-2 — are believed to be coded by two different cistrons.[72]

From an evolutionary aspect, metallothionein is considered to be an "old" protein. Metallothionein-like proteins are found in most living organisms, microorganisms, invertebrates, and vertebrates.[46,50]

VI. INDUCTION OF METALLOTHIONEIN SYNTHESIS, OCCURRENCE IN ORGANS, AND DEGRADATION

Parenteral administration of cadmium or zinc is the most effective means of inducing metallothionein synthesis. This was shown initially by Piscator[82] for cadmium and by Webb[112] for zinc. The most marked response is seen in the liver, but also in kidneys, pancreas, and to a lesser extent in other tissues, in which metallothionein synthesis may be induced.[17,61,67,72,78]

Shortly (2 to 3 hr) after parenteral administration of cadmium to rodents and mice, most of the cadmium in the liver is bound to high molecular weight proteins in the cytosol,[1,38,59] but already after 8 hr more than 80% of the cadmium present in liver cell cytoplasma is bound to metallothionein.[1,67] Although metallothionein is mainly present in the cytoplasm of renal and hepatic cells, induction of metallothionein results in its presence in cell nucleus also, as can be observed using immunohistochemical techniques.[3,23] It has been suggested[3,20] that there is a small amount of thionein (metallothionein without metal ions) present in the cytoplasma of the cells (Fiqure 5A). The thionein is bound to cytoplasmatic thionein mRNA and as a result of this binding the translation of mRNA to thionein is inhibited. When cadmium ions enter the cells, they will be picked up by the thionein, forming metallothionein and transported to the cell nucleus. At this stage there will be no inhibition of mRNA and translation to thionein will be started (Figure 5B). In the cell nucleus, metallothionein will possibly bind to DNA also and induce an increased production of thionein mRNA (Figure 5C). When all cadmium ions have been chelated by metallothionein, a small amount of thionein will again be capable of inhibiting cytoplasmatic mRNA (Figure 5D).

The synthesis of metallothionein in kidney cells is considerably slower than that in liver cells.[115] The tissue metallothionein level is mainly related to the tissue deposition of the inducing metal. Thus, the maximum induction of metallothionein was found in the pancreas after injection of zinc salts.[79] Taguchi and Nakamura[102] and Oullette et al.[80] have been able to identify metallothionein in intestinal cells from rats 24 hr after a large oral dose of cadmium, and von Post et al.[84] have identified low molecular proteins similar to metallothionein in rabbit lungs after cadmium inhalation. The latter group of authors also described another cadmium-binding protein in the rabbit lungs which was considered to be the major intracellular cadmium-binding protein in this organ.

Apart from exposure to metals (cadmium, zinc, mercury, and copper), a number of other factors may induce metallothionein synthesis in animals. It has, e.g., been shown that environmental stresses — such as starvation,[10] heat, and exercise[76] — increased metallothionein levels in rat liver. It is possible that the induction of metallothionein synthesis by various stress factors is mediated via a release of glucocorticoid hormones, since glucocorticoid administration also has been shown to be very effective in inducing metallothionein synthesis in rat liver.[28] However, there is a marked difference in the magnitude of metallothionein synthesis induced by metals and that induced by other stress factors. Metal induction is much more effective.[54]

In adult cadmium-exposed animals, liver and kidney are the major cadmium storage organs (Chapter 6, Section IV. A). More than 80% of the cadmium in these organs is bound to metallothionein[113] (Chapter 6, Section IV. B). Metallothionien is, however, also found in other tissues, usually in amounts proportional to the cadmium or zinc content.[46,78,79]

The biological half-time of zinc- and cadmium-metallothionein in liver and kidneys has been examined by injecting radiolabeled amino acids, such as ^{35}S-cysteine together with the inducing metal.[21,91,116] The radiolabeled amino acids are incorporated into the newly formed metallothionein. There is a fairly rapid loss of labeled amino acids from metallothionein in spite of the fact that cadmium or zinc ions are still bound to metallothionein. This can be taken as evidence of a constant synthesis of metallothionein in order to sequester cadmium or zinc ions which have been released from degraded metallothionein.[91] The biological half-time of zinc-metallothionein in liver and kidney from rats has been found to range from 11 to 65 hr and from 22 to 103 hr, respectively.[117] The biological half-time of cadmium-metallothionein appears to be somewhat longer, in the order of days.[17,30,46,91] Hepatic cadmium-metallothionein in rats has a biological half-time of 2.8 to 3.5 days,[46] while in kidneys the biological half-time is about 5 days.[92] This is still considerably shorter than the biological half-time for cadmium which is in the order of months or years (see Chapter 6, Section V. D, Table 7). Thus, a constant synthesis of metallothionein takes place in order to sequester the cadmium ions which have been released from the degraded metallothionein.[71,88,91]

VII. BIOLOGICAL ROLE OF METALLOTHIONEIN OTHER THAN IN RELATION TO CADMIUM

The biological function of metallothionein under physiological conditions is still not completely understood. Reviews summarizing the present knowledge of the functions of metallothionein have been presented.[2,22a,32,46,114]

A high content of zinc- and copper-containing metallothionein in newborn[12] makes it possible to assume that metallothionein serves as a storage protein for these metals in early life. Experiments on adult animals also indicated that metallothionein may serve as a reserve for zinc, releasing the metal when the animal is in need of zinc for the various cellular processes in the body.[117] It has been shown in vitro that zinc-metallothionein will serve as a zinc donator to zinc-requiring apoenzymes, thereby reestablishing enzyme activity.[60,108] Similar to zinc salts, the zinc bound to metallothionein can also be reutilized in vivo.[18]

Richards and Cousins[86,87] suggested that metallothionein plays a controlling role in the absorption of zinc from the gastrointestinal tract. It was proposed that during zinc overload, metallothionein formed in the intestinal cells would bind zinc and thereby prevent further absorption. Excess zinc would then be rejected upon the sloughing of epithelial cells. This hypothesis has been questioned to a certain extent by Foulkes[33] and Webb and Cain.[114] Nevertheless, Cousins[22,22a] considered metallothionein to have somewhat more than merely a storage function in view of the marked induction of metallothionein synthesis seen after the administration of glucocorticoids.

VIII. DISTRIBUTION AND EFFECT OF PARENTERALLY ADMINISTERED CADMIUM-CONTAINING METALLOTHIONEIN

Nordberg[66] gave mice subcutaneous injections of a mixture of cadmium chloride and cadmium-metallothionein. Although the total dose of cadmium was the same in all groups of mice (1.1 mg/kg), the mice that were given cadmium bound to metallothionein were protected from the testicular damage seen among mice given cadmium chloride. In contrast to the mice given cadmium chloride only, animals given higher doses of cadmium-metallothionein revealed histological evidence of renal tubular damage. Later studies have confirmed that cadmium-metallothionein is highly nephrotoxic.[21,34,35,69,111,116] The distribution of a single dose of cadmium salt differs considerably from that of cadmium-metallothionein. A couple of hours after cadmium salt was administered, about 50% of the dose was found in the liver and only about 10%, or less, in the kidney. However, when cadmium-metallothionein has been administered, up to 90% will be found in the kidneys 2 hr later.[21,73,93]

It has been shown that cadmium-metallothionein is freely filtered through the glomerulus and subsequently that the complex, similar to other low molecular weight proteins, is being reabsorbed from the tubular fluid[31,63] (see Chapter 6, Sections III. and IV.B). At low levels of cadmium metallothionein in the glomerular filtrate, the reabsorbtion is almost complete, whereas at high levels the uptake is saturated and the relative reabsorption is decreased.[33,45,63] Accordingly, high urinary excretion of cadmium-containing metallothionein is seen shortly after administration of larger doses of cadmium-metallothionein containing more than 0.1 mg Cd per kilogram body weight.[21,71,73,93]

IX. MECHANISM OF THE CADMIUM-INDUCED RENAL DYSFUNCTION

When cadmium is given in the form of cadmium-metallothionein, the LD_{50} is only about one tenth of that for inorganic cadmium salts.[69,116] After parenteral administration of cadmium-metallothionein to mice, Nordberg et al.[69] observed severe tubular damage and uremia at a renal cortex concentration of about 10 mg/kg wet weight. This should be compared to a concentration of about 100 to 300 mg/kg wet weight necessary to induce renal dysfunction after exposure to inorganic cadmium salts (Volume II, Chapter 9, Sections III.A and V.A). The mechanism underlying this phenomenon is, probably, glomerular filtration of cadmium-metallothionein and a subsequent efficient uptake from the tubular fluid into the tubular cells followed by a rapid degradation and release of cadmium from its protein ligand. The occurrence of nonmetallothionein-bound cadmium ions in the cells produces the toxic effects.

Basically, this hypothesis was proposed by Piscator in 1964,[82] by Nordberg et al. in 1971[67] and by Nordberg et al. in 1975.[69] During the last 10 years, confirming experimental data have become available (see also Chapter 6, Sections IV.B.1 and IV.B.2 and Figure 19).

Fowler and Nordberg[35] and Squibb et al.[94-96] have studied the fate of a single dose of cadmium-metallothionein given parenterally. After filtration through the glomerulus, cadmium-metallothionein is taken up by the tubular cells by pinocytosis. In lysosomes, cadmium-metallothionein is rapidly degraded and cadmium is released into the cytoplasm of the tubular cells. Squibb et al.[96] reported that 1 hr after intraperitoneal injection of cadmium-metallothionein (0.6 mg Cd per kilogram body weight) to rats, more than 95% of cadmium ions present in the cytoplasm of tubular kidney cells were nonmetallothionein bound and that these cadmium ions probably were associated with the profound signs of tubular toxicity seen shortly after the injection. Zinc-metallothionein (0.9 mg Zn per kilogram body weight) produced no signs of tubular renal toxicity.[96]

Tubular cells have a certain capacity for producing their own metallothionein which can bind cadmium and thereby prevent the toxic effects of cadmium ions.[15,94] Following large doses of cadmium-metallothionein, the cells cannot cope with all the cadmium being released and cell damage occurs. Chronic toxic effects of cadmium in the kidney are likely to occur when the tubular cell capacity for producing metallothionein is insufficient to sequester all the cadmium ions in the cell cytoplasm.[25,36,64,71,75]

It is probably the concentration of nonmetallothionein-bound cadmium ions that is the direct cause of toxicity, rather than the total concentration of cadmium.[96] During long-term exposure, the concentration of nonmetallothionein-bound cadmium is likely to be related to the total concentration of cadmium. After parenteral administration of cadmium-metallothionein, the situation is different. When large doses of cadmium-metallothionein are given, the input rate of cadmium into the tubular cells is much larger than the capacity of these cells to produce their own protecting metallothionein. Therefore, under this type of experimental condition, the toxicity will be related to the input and degradation rate of cadmium-metallothionein rather than to the total concentration of cadmium.

Figure 6 is a schematic illustration of two different alternatives during long-term cadmium exposure. In Figure 6A, a constant proportion of the cadmium ions is bound, or not bound, to metallothionein. There is no threshold in the metallothionein-producing capacity of the kidney. At a certain critical total concentration of cadmium in the kidney the amount or concentration of nonmetallothionein-bound cadmium will nevertheless become high enough to produce toxic effects.

In Figure 6B, a maximum metallothionein-producing capacity is assumed to exist. At low concentrations of cadmium almost all cadmium ions are bound to metallothionein, but at a high total cadmium concentration the kidney cells are not capable of producing enough metallothionein to bind the cadmium ions and, therefore, nonmetallothionein-bound cadmium ions will occur, causing toxicity. This latter alternative (Figure 6B) is supported by results from horses environmentally exposed to cadmium.[25] In renal metallothionein, obtained from horses with a high total concentration of cadmium, there was a high molar ratio between cadmium and zinc[75] and this was interpreted as an indication of a decreased capacity of the renal tubular cells to produce protective metallothionein.[25] On the other hand, experimental data from rabbits and monkeys reported by Nomiyama and Nomiyama[64] and Nomiyama et al.[65] suggest that there is a constant relationship between the total cadmium and metallothionein-bound cadmium in kidneys and no threshold (Figure 6A). The authors nevertheless concluded from their experiments that there was a critical concentration of nonmetallothionein-bound cadmium above which renal toxic effects developed.

It is evident that much more experimental work is needed before the mechanism underlying intracellular toxicity of cadmium in tubular cells is fully elucidated.

There are likely to be species differences with regard to the capacity of different animals to produce metallothionein in the renal cortex. Therefore, signs of renal toxicity may occur at different total concentrations of cadmium. In the case of human exposure, constitutional factors as well as age and simultaneous exposure to other nephrotoxic agents may influence renal metallothionein production capacity and thus the susceptibility of the kidneys to cadmium.

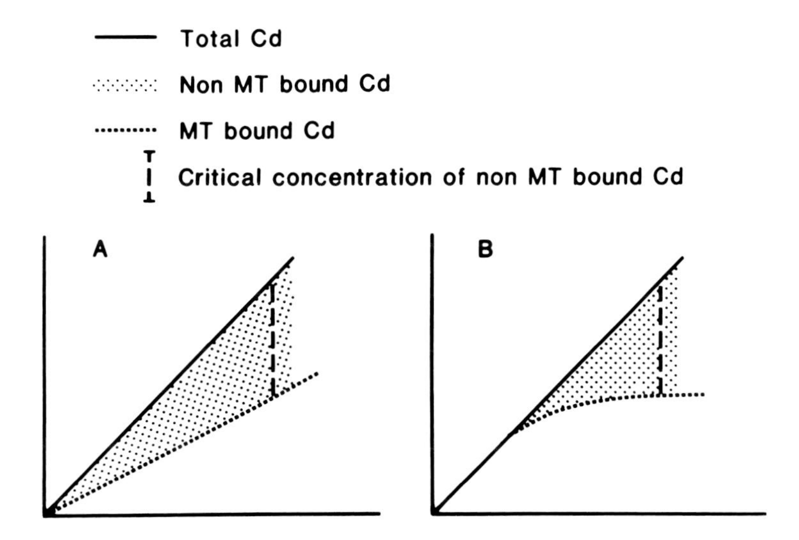

FIGURE 6. Schematic presentation of two possible relationships between the total concentration of cadmium in kidney cells and the proportion of cadmium bound, or not bound, to metallothionein, respectively. (A) No maxium metallothionein-producing capacity. (B) Assuming a maximum metallothionein-producing capacity.

In this chapter we do not discuss or draw conclusions about the numerical concentrations at which cadmium kidney damage develops. Here we discuss only the tentative involvement of metallothionein. The mechanism underlying cadmium nephrotoxicity and the total concentration of cadmium related to different types of renal effects are discussed in Chapter 9, Sections III.A.B.C and V.A.B.

X. SUMMARY AND CONCLUSIONS

Metallothionein is a low molecular weight protein which plays a major role in the metabolism and toxicity of cadmium. It sequesters cadmium in the cells and thereby acts as a detoxifying agent. At the same time, it may serve as a carrier of cadmium between tissues, mainly from the liver to the kidney where cadmium, which has been released from degraded metallothionein, exerts toxic effects. The biological role of metallothionein under physiological conditions is still not completely established. It is conceivable that metallothionein is involved in the metabolism and storage of zinc and possibly copper.

REFERENCES

1. **Andersen, O., Hägerstrand, I., and Nordberg, G. F.,** Effect of the chelating agent sodium tripolyphosphate on cadmium toxicity in mice, *Environ. Res.*, 29, 54—61, 1982.
2. **Bakka, A.,** Studies on Possible Functions of Metallothionein, Doctoral thesis, University of Oslo, Norway, 1983.
3. **Banerjee, D., Onosaka, S., and Cherian, M. G.,** Immunohistochemical localization of metallothionein in cell nucleus and cytoplasm of rat liver and kidney, *Toxicology*, 24, 95—105, 1982.
4. **Bell, J. U.,** A renal-hepatic comparison of metallothionein in the sheep fetus, *Toxicol. Lett.*, 4, 407—411, 1979a.

5. **Bell, J. U.,** Native metallothionein levels in rat hepatic cytosol during perinatal development, *Toxicol. Appl. Pharmacol.*, 50, 101—107, 1979b.
6. **Boulanger, Y. and Armitage, I. M.,** ^{113}Cd study of the metal cluster structure of human liver metallothionein, *J. Inorg. Biochem.*, 17, 147—153, 1982.
7. **Brady, F. O.,** The physiological function of metallothionein, *Trends Biochem. Sci.*, 7(4), 143—145, 1982.
8. **Brady, F. O. and Kafka, R. L.,** Radioimmunoassay of rat liver metallothionein, *Anal. Biochem.*, 98, 89—94, 1979.
9. **Bremner, I.,** Cadmium toxicity. Nutritional influences and the role of metallothionein, *World Rev. Nutr. Diet.*, 32, 165—197, 1978.
10. **Bremner, I. and Davies, N. T.,** The induction of metallothionein in rat liver by zinc injection and restriction of food intake, *Biochem. J.*, 149, 733—738, 1975.
11. **Bremner, I. and Young, B. W.,** Copper thionein in the kidneys of copper-poisoned sheep, *Chem. Biol. Interact.*, 19, 13—23, 1977.
12. **Bremner, I., Williams, R. B., and Young, B. W.,** Distribution of copper and zinc in the liver of the developing sheep foetus, *Br. J. Nutr.*, 38, 87—92, 1977.
13. **Briggs, R. W. and Armitage, J. M.,** Evidence for site-selective metal binding in calf liver metallothionein, *J. Biol. Chem.*, 257, 1259—1262, 1981.
14. **Bühler, R. H. O. and Kägi, J. H. R.,** Human hepatic metallothioneins, *FEBS Lett.*, 39, 229—234, 1974.
15. **Cain, K. and Holt, D. E.,** Studies of cadmium-thionein induced nephropathy: time course of cadmium-thionein uptake and degradation, *Chem Biol. Interact.*, 43, 223—237, 1983.
16. **Chang, C. C., Lauwerys, R., Bernard, A., Roels, H., Buchet, J. P., and Garvey, J. S.,** Metallothionein in cadmium-exposed workers, *Environ. Res.*, 23, 422—428, 1980.
17. **Chen, R. W. and Ganther, H. E.,** Relative cadmium binding capacity of metallothionein and other cytosolic fractions in various tissues of the rat, *Environ. Physiol. Biochem.*, 5, 378—388, 1975.
18. **Cherian, M. G.,** Studies on the synthesis and metabolism of zinc-thionein in rats, *J. Nutr.*, 107, 965—972, 1977.
19. **Cherian, M. G. and Goyer, R. A.,** Metallothioneins and their role in the metabolism and toxicity of metals, *Life Sci.*, 23, 1—10, 1978.
20. **Cherian, M. G. and Nordberg, M.,** Cellular adaptation in metal toxicology and metallothionein, *Toxicology*, 28, 1—15, 1983.
21. **Cherian, M. G., and Shaikh, Z. A.,** Metabolism of intravenously injected cadmium-binding protein, *Biochem. Biophys. Res. Commun.*, 65, 863—869, 1975.
22. **Cousins, R. J.,** Relationship of metallothionein synthesis and degradation to intracellular zinc metabolism, in *Biological Roles of Metallothionein*, Foulkes, E. C., Ed., Elsevier, Amsterdam, 1982, 251—260.
22a. **Cousins, R. J.,** Metallothionein — aspects related to copper and zinc metabolism, *J. Inher. Metab. Dis.*, 6(Suppl. 1), 15—21, 1983.
23. **Danielson, K. G., Ohi, S., and Huang, P. C.,** Immunochemical detection of metallothionein in specific epithelial cells of rat organs, *Proc. Natl. Acad. Sci. U.S.A.*, 79, 2301—2304, 1982.
24. **Elinder, C.-G.,** Early Effects of Cadmium Accumulation in Kidney Cortex, with Special Reference to Cadmium and Zinc Interactions. A Study of Humans, Horses and Laboratory Animals, Doctoral thesis, Karolinska Institute, Stockholm, 1979.
25. **Elinder, C.-G. and Nordberg, M.,** Critical concentration of cadmium estimated by studies on horse kidney metallothionein, in *Biological Roles of Metallothionein*, Foulkes, E. C., Ed., Elsevier, Amsterdam, 1982, 37—46.
26. **Elinder, C.-G., Piscator, M., and Linnman, L.,** Cadmium and zinc relationships in kidney cortex, liver and pancreas, *Environ. Res.*, 13, 432—440, 1977.
27. **Elinder, C.-G., Nordberg, M., Palm, B., and Piscator, M.,** Cadmium, zinc and copper in horse liver and in horse liver metallothionein: comparison with kidney cortex, *Environ. Res.*, 26, 22—32, 1981.
28. **Etzel, K. R., Shapiro, S. G., and Cousins, R. J.,** Regulation of liver metallothionein and plasma zinc by the glucocorticoid dexamethasone, *Biochem. Biophys. Res. Commun.*, 89(4), 1120—1126, 1976.
29. **Falck, F. Y., Jr., Fine, L. J., Smith, R. G., Garvey, J., Schork, A., England, B., McClatchey, K. D., and Linton, J.,** Metallothionein and occupational exposure to cadmium, *Br. J. Ind. Med.*, 40, 305—313, 1983.
30. **Feldman, S. L., Failla, M. L., and Cousins, R. J.,** Degradation of rat liver metallothioneins *in vitro*, *Biochim. Biophys. Acta*, 544, 638—646, 1978.
31. **Foulkes, E. C.,** Renal tubular transport of cadmium metallothionein, *Toxicol. Appl. Pharmacol.*, 45, 505—512, 1978.
32. **Foulkes, E. C., Ed.,** *Biological Roles of Metallothionein*, Vol. 9, Elsevier, Amsterdam, 1982a.
33. **Foulkes, E. C.,** Role of metallothionein in transport of heavy metals, in *Biological Roles of Metallothionein*, Foulkes, E. C., Ed., Elsevier, Amsterdam, 1982b, 131—140.
34. **Fowler, B. A. and Nordberg, G. F.,** The renal toxicity of cadmium metallothionein, in Int. Conf. Heavy Metals in the Environment, Abstr., Toronto, October 27 to 31, 1975.

35. **Fowler, B. A. and Nordberg, G. F.,** The renal toxicity of cadmium metallothionein: morphometric and X-ray microanalytical studies, *Toxicol. Appl. Pharmacol.*, 46, 609—623, 1978.
36. **Friberg, L., Piscator, M., Nordberg, G. F., and Kjellström, T.,** *Cadmium in the Environment*, 2nd ed., CRC Press, Boca Raton, Fla., 1974.
37. **Garvey, J. S. and Chang, C. C.,** Detection of circulating metallothionein in rats injected with zinc or cadmium, *Science*, 214, 805—807, 1981.
38. **Goering, P. L. and Klaassen, C. D.,** Altered subcellular distribution of cadmium following cadmium pretreatment: possible mechanism of tolerance to cadmium-induced lethality, *Toxicol. Appl. Pharmacol.*, 70, 195—203, 1983.
39. **Goyer, R. A., Fowler, B. A., Nordberg, G. F., Shepherd, G., and Moustafa, L., Eds.,** Proceedings of metallothionein and cadmium nephrotoxicity conference, North Carolina, 1983, *Environ. Health Perspect.*, 54, 1—295, 1984.
40. **Hammer, D. I., Colucci, A. V., Hasselblad, V., Williams, M. E., and Pinkerton, C.,** Cadmium and lead in autopsy tissues, *J. Occup. Med.*, 12, 956—963, 1973.
41. **Hartmann, H. J. and Weser, U.,** Copper-thionein from fetal bovine liver, *Biochim. Biophys. Acta*, 491, 211—222, 1977.
42. **Huang, I.-Y., Yoshida, A., Tsunoo, H., Nakajima, H.,** Mouse liver metallothioneins. Complete amino acid sequence of metallothionein-I., *J. Biol. Chem.*, 252, 8217—8221, 1977.
43. **Huang, I.-Y, Tsunoo, H., Kimura, M., Nakashima, H., and Yoshida, A.,** Primary structure of mouse liver metallothionein-I and -II, in *Metallothionein*, Kägi, J. H. R. and Nordberg, M., Eds., Birkhäuser Verlag, Boston, 1979, 169—172.
44. **Iwao, S., Tsuchiya, K., and Sugita, M.,** Variation of cadmium accumulation among Japanese, *Arch. Environ. Health*, 38, 156—162, 1983.
45. **Johnson, D. R. and Foulkes, E. C.,** On the proposed role of metallothionein in the transport of cadmium, *Environ. Res.*, 21, 360—365, 1980.
46. **Kägi, J. H. R. and Nordberg, M., Eds.,** *Metallothionein*, Birkhäuser Verlag, Boston, 1979.
47. **Kägi, J. H. R. and Vallee, B. L.,** Metallothionein: a cadmium- and zinc-containing protein from equine renal cortex. I, *J. Biol. Chem.*, 235, 3460—3465, 1960.
48. **Kägi, J. H. R. and Vallee, B. L.,** Metallothionein: a cadmium- and zinc-containing protein from equine renal cortex. II. Physicochemical properties, *J. Biol. Chem.*, 236, 2435—2442, 1961.
49. **Kägi, J. H. R., Himmelhoch, S. R., Whanger, P. D., Bethune, J. L., and Vallee, B. L.,** Equine hepatic and renal metallothioneins. Purification, molecular weight, amino acid composition and metal content, *J. Biol. Chem.*, 249, 3537—3542, 1974.
50. **Kägi, J. H. R., Vašák, M., Lerch, K., Gilg, D. E. O., Hunziker, P., Bernhard, W. R., and Good, M.,** Structure of mammalian metallothionein, *Environ. Health Perspect.*, 54, 93—103, 1984.
51. **Kimura, M., Otaki, N., Yoshiki, S., Suzuki, M., Horiuchi, N., and Suda, T.,** The isolation of metallothionein and its protective role in cadmium poisoning, *Arch. Biochem. Biophys.*, 165, 340—348, 1974.
52. **Kimura, M., Otaki, N. and Imano, M.,** Rabbit liver metallothionein tentative amino acid sequence of metallothionein-B, in *Metallothionein*, Kägi, J. H. R. and Nordberg, M., Eds., Birkhäuser Verlag, Boston, 1979, 163—168.
53. **Kissling, M. M. and Kägi, J. H. R.,** Amino acid sequence of human hepatic metallothioneins, in *Metallothionein*, Kägi, J. H. R. and Nordberg, M., Eds., Birkhäuser Verlag, Boston, 1979, 145—151.
54. **Klaassen, C. D.,** Induction of metallothionein by adrenocortical steroids, *Toxicology*, 20, 275—279, 1981.
55. **Kojima, Y. and Kägi, J. H. R.,** Metallothionein, *Trends Biochem. Sci.*, 3, 90—93, 1978.
56. **Kojima, Y., Berger, C., Vallee, B. L., and Kägi, J. H. R.,** Amino-acid sequence of equine renal metallothionein-1B, *Proc. Natl. Acad. Sci. U.S.A.*, 73, 3413—3417, 1976.
57. **Kojima, Y., Berger, C., and Kägi, J. H. R.,** The amino acid sequence of equine metallothioneins, in *Metallothionein*, Kägi, J. H. R. and Nordberg, M., Eds., Birkhäuser Verlag, Boston, 1979, 153—161.
58. **Kotsonis, F. N. and Klaassen, C. D.,** Metallothionein and its interactions with cadmium, in *Cadmium in the Environment, Part 2*, Nriagu, J. O., Ed., John Wiley & Sons, New York, 1981, 595—616.
59. **Leber, A. P. and Miya, T. S.,** A mechanism for cadmium- and zinc-induced tolerance to cadmium toxicity: involvement of metallothionein, *Toxicol. Appl. Pharmacol.*, 37, 403—414, 1976.
60. **Li, T.-Y., Kraker, J., Shaw, C. F., III, and Petering, D. H.,** Ligand substitution reactions of metallothionein with EDTA and apo-carbonic anhydrase, *Proc. Natl. Acad. Sci. U.S.A.*, 77, 6334—6338, 1980.
61. **Lucis, O. J., Shaikh, Z. A., and Embil, J. A.,** Cadmium as a trace element and cadmium binding components in human cells, *Experientia (Basel)*, 26, 109—110, 1970.
62. **Margoshes, M. and Vallee, B. L.,** A cadmium protein from equine kidney cortex, *J. Am. Chem. Soc.*, 79, 4813—4814, 1957.
63. **Nomiyama, K. and Foulkes, E. C.,** Reabsorption of filtered cadmium-metallothionein in the rabbit kidney, *Proc. Soc. Exp. Biol. Med.*, 156, 97—99, 1977.

64. **Nomiyama, K. and Nomiyama, H.,** Tissue metallothioneins in rabbits chronically exposed to cadmium, with special reference to the critical concentration of cadmium in the renal cortex, in *Biological Roles of Metallothionein,* Foulkes, E. C., Ed., Elsevier, Amsterdam, 1982, 47—67.
65. **Nomiyama, K., Nomiyama, H., Akahori, F., and Masaoka, T.,** Cadmium health effects in monkeys with special reference to the critical concentration of cadmium in the renal cortex, in *Cadmium 81, Edited Proc. 3rd Int. Cadmium Conf. Miami, Feburary 3 to 5, 1981,* Cadmium Association, London, 1982, 151—156.
66. **Nordberg, G. F.,** Effects of acute and chronic cadmium exposure on the testicles of mice. With special reference to protective effects of metallothionein, *Environ. Physiol. Biochem.,* 1, 171—187, 1971.
67. **Nordberg, G. F., Piscator, M., and Lind, B.,** Distribution of cadmium among protein fractions of mouse liver, *Acta Pharmacol. Toxicol.,* 29, 456—470, 1971.
68. **Nordberg, G. F., Nordberg, M., Piscator, M., and Vesterberg, O.,** Separation of two forms of rabbit metallothionein by isoelectric focusing, *Biochem. J.,* 126, 491—498, 1972.
69. **Nordberg, G. F., Goyer, R. A., and Nordberg, M.,** Comparative toxicity of cadmium-metallothionein and cadmium chloride on mouse kidney, *Arch Pathol.,* 99, 192—197, 1975.
70. **Nordberg, G. F., Garvey, J. S., and Chang, C. C.,** Metallothionein in plasma and urine of cadmium workers, *Environ. Res.,* 28, 179—182, 1982.
71. **Nordberg, M.,** Studies on metallothionein and cadmium, *Environ. Res.,* 15, 381—404, 1978.
72. **Nordberg, M. and Kojima, Y.,** Metallothionein and other low molecular weight metal-binding proteins, in *Metallothionein,* Kägi, J. H. R. and Nordberg, M., Eds., Birkhäuser Verlag, Boston, 1979, 41—124.
73. **Nordberg, M. and Nordberg, G. F.,** Distribution of metallothionein-bound cadmium and cadmium chloride in mice: preliminary studies, *Environ. Health Perspect.,* 12, 103—108, 1975.
74. **Nordberg, M., Trojanowska, B., and Nordberg, G. F.,** Studies on metal-binding proteins of low molecular weight from renal tissue of rabbits exposed to cadmium or mercury, *Environ. Physiol. Biochem.,* 4, 149—158, 1974.
75. **Nordberg, M., Elinder, C.-G., and Rahnster, B.,** Cadmium, zinc and copper in horse kidney metallothionein, *Environ. Res.,* 20, 341—350, 1979.
76. **Oh, S. H., Deagen, J. T., Whanger, P. D., and Weswig, P. H.,** Biological function of metallothionein. V. Its induction in rats by various stresses, *Am. J. Physiol.,* 234, E282—E285, 1978.
77. **Olafson, R. W. and Sim, R. G.,** An electrochemical approach to quantitation and characterization of metallothioneins, *Anal. Biochem.,* 100, 343—351, 1979.
78. **Onosaka, S. and Cherian, M. G.,** The induced synthesis of metallothionein in various tissues of rats in response to metals. I. Effect of repeated injection of cadmium salts, *Toxicology,* 22, 91—101, 1981.
79. **Onosaka, S. and Cherian, M. G.,** The induced synthesis of metallothionein in various tissues of rats in response to metals. II. Influence of zinc status and specific effect on pancreatic metallothionein, *Toxicology,* 23, 11—20, 1982.
80. **Ouellette, A. J., Aviles, L., Burnweit, C. A., Frederick, D., and Malt, R. A.,** Metallothionein mRNA induction in mouse small bowel by oral cadmium and zinc, *Am. Physiol. Soc.,* 243, C396—C403, 1982.
81. **Piotrowski, J. K., Bolanowska, W., and Sapota, A.,** Evaluation of metallothionein content in animal tissues, *Acta Biochim. Pol.,* 20, 207—215, 1973.
82. **Piscator, M.,** Cadmium in the kidneys of normal human beings and the isolation of metallothionein from liver of rabbits exposed to cadmium (in Swedish), *Nord. Hyg. Tidskr.,* 45, 76—82, 1964.
83. **Piscator, M. and Lind, B.,** Cadmium, zinc, copper and lead in human renal cortex, *Arch. Environ. Health,* 24, 426—431, 1972.
84. **Post, von, C. T., Squibb, K. S., Fowler, B. A., Gardner, D. E., Illing, J., and Hook, G. E. R.,** Production of low molecular weight cadmium-binding proteins in rabbit lung following exposure to cadmium chloride, *Biochem. Pharmacol.,* 31(18), 2969—2975, 1982.
85. **Pulido, P., Kägi, J. H. R., and Vallee, B. L.,** Isolation and some properties of human metallothionein, *Biochemistry,* 5, 1768—1777, 1966.
86. **Richards, M. P. and Cousins, R. J.,** Mammalian zinc homeostasis: requirement for RNA and metallothionein synthesis, *Biochem. Biophys. Res. Commun.,* 64, 1215—1223, 1975.
87. **Richards, M. P. and Cousins, R. J.,** Metallothionein and its relationship to the metabolism of dietary zinc in rats, *J. Nutr.,* 106, 1591—1599, 1976.
88. **Ridlington, J. W., Winge, D. R., and Fowler, B. A.,** Long-term turnover of cadmium metallothionein in liver and kidney following a single low dose of cadmium in rats, *Biochim. Biophys. Acta,* 673, 177—183, 1981.
89. **Riordan, J. R. and Richards, V.,** Human fetal liver contains both zinc- and copper-rich forms of metallothionein, *J. Biol. Chem.,* 255, 5380—5383, 1980.
90. **Rydén, L. and Deutsch, H. F.,** Preparation and properties of the major copper-binding component in human fetal liver. Its identification as metallothionein, *J. Biol. Chem.,* 253, 519—524, 1978.
91. **Shaikh, Z. A.,** Metallothionein as a storage protein for cadmium: its toxicological implications, in *Biological Roles of Metallothionein,* Foulkes, E. C., Ed., Elsevier, Amsterdam, 1982, 69—76.

92. **Shaikh, Z. A. and Smith, J. C.,** The biosynthesis of metallothionein in rat liver and kidney after administration of cadmium, *Chem. Biol. Interact.*, 15, 327—336, 1976.
93. **Shaikh, Z. A. and Hirayama, K.,** Metallothionein and metabolism of cadmium, in *Kadmium-Symposium*, Anke, M. and Schneider, H.-J., Eds., Wissenschaftliche Beiträge der Friedrich-Schiller-Universität, Jena, 1979, 95—101.
94. **Squibb, K. S., Ridlington, J. W., Carmichael, N. G., and Fowler, B. A.,** Early cellular effects of circulating cadmium-thionein on kidney proximal tubules, *Environ. Health Perspect.*, 28, 287—296, 1979.
95. **Squibb, K. S., Pritchard, J. B., and Fowler, B. A.,** Renal metabolism and toxicity of metallothionein, in *Biological Roles of Metallothionein*, Foulkes, E. C., Ed., Elsevier, Amsterdam, 1982, 181—192.
96. **Squibb, K. S., Pritchard, J. B., and Fowler, B. A.,** Cadmium-metallothionein nephropathy: relationships between ultrastructural/biochemical alterations and intracellular cadmium binding, *J. Pharmacol. Exp. Ther.*, in press.
97. **Suzuki, K. T.,** Copper content in cadmium-exposed animal kidney metallothioneins, *Arch. Environ. Contam. Toxicol.*, 8, 255—268, 1979.
98. **Suzuki, K. T.,** Direct connection of high-speed liquid chromatograph (equipped with gel permeation column) to atomic absorption spectrophotometer for metalloprotein analysis. Metallothionein, *Anal. Biochem.*, 102, 31—34, 1980.
99. **Suzuki, K. T.,** Induction and degradation of metallothionein and their relation to the toxicity of cadmium, in *Biological Roles of Metallothionein*, Foulkes, E. C., Ed., Elsevier, Amsterdam, 1982, 215.
100. **Suzuki, K. T. and Maitani, T.,** Metal-dependent properties of metallothionein, *Biochem. J.*, 199, 289—295, 1981.
101. **Suzuki, Y.,** Metal-binding properties of metallothionein in extracellular fluids and its role in cadmium-exposed rats in *Biological Roles of Metallothionein*, Foulkes, E. C., Ed., Elsevier, Amsterdam, 1982, 25—67.
102. **Taguchi, T. and Nakamura, K.,** Isolation and properties of cadmium-binding protein induced in rat small intestine by oral administration of cadmium, *J. Toxicol. Environ. Health*, 9, 401—409, 1982.
103. **Tohyama, C. and Shaikh, Z. A.,** Cross-reactivity of metallothioneins from different origins with rabbit anti-rat hepatic metallothionein antibody, *Biochem. Biophys. Res. Commun.*, 84, 907—913, 1978.
104. **Tohyama, C. and Shaikh, Z. A.,** Metallothionein in plasma and urine of cadmium exposed rats determined by a single-antibody radioimmunoassay, *Fundam. Appl. Toxicol.*, 1, 1—7, 1981.
105. **Tohyama, C., Shaikh, Z. A., Ellis, K. J., and Cohn, S. H.,** Metallothionein excretion in urine upon cadmium exposure: its relationship with liver and kidney cadmium, *Toxicology*, 22, 181—191, 1981a.
106. **Tohyama, C., Shaikh, Z. A., Nogawa, K., Kobayashi, E., and Honda, R.,** Elevated urinary excretion of metallothionein due to environmental cadmium exposure, *Toxicology*, 20, 289—297, 1981b.
107. **Tohyama, C., Shaikh, Z. A., Nogawa, K., Kobayashi, E., and Honda, R.,** Urinary metallothionein as a new index of renal dysfunction in "Itai-itai" disease patients and other Japanese women environmentally exposed to cadmium, *Arch. Toxicol.*, 50, 159—166, 1982.
108. **Udom, A. O. and Brady, F. O.,** Reactivation *in vitro* of zinc requiring apoenzymes by rat liver zinc thionein, *Biochem. J.*, 187, 329—335, 1980.
109. **Vander Mallie, R. J. and Garvey, J. S.,** Production and study of antibody produced against rat cadmium thionein, *Immunochemistry*, 15, 857—868, 1978.
110. **Vašák, M. and Kägi, J. H. R.,** Spectroscopic properties of metallothionein, in *Metal Ions in Biological Systems*, Vol. 15, Sigel, H., Ed., Marcel Dekker, New York, 1983, 213—273.
111. **Vostal, J. J. and Cherian, M. G.,** Effects of cadmium metallothionein on the renal tubular transport of sodium, *Fed. Proc.*, 33, 519, 1974.
112. **Webb, M.,** Protection by zinc against cadmium toxicity, *Biochem. Pharmacol.*, 21, 2767—2771, 1972.
113. **Webb, M., Ed.,** *The Chemistry, Biochemistry and Biology of Cadmium*, Elsevier, Amsterdam, 1979.
114. **Webb, M. and Cain, K.,** Functions of metallothionein, *Biochem. Pharmacol.*, 31(2), 137—142, 1982.
115. **Webb, M. and Daniel, M.,** Induced synthesis of metallothionein by pig kidney cells *in vitro* in response to cadmium, *Chem. Biol. Interact.*, 10, 269—276, 1975.
116. **Webb, M. and Etienne, A. T.,** Studies on the toxicity and metabolism of cadmium-thionein, *Biochem. Pharmacol.*, 26, 25—30, 1977.
117. **Whanger, P. D. and Ridlington, J. W.,** Role of metallothionein in zinc metabolism, in *Biological Roles of Metallothionein*, Foulkes, E. C., Ed., Elsevier, Amsterdam, 1982, 263—277.
118. **Winge, D. R. and Miklossy, K.-A.,** Domain nature of metallothionein, *J. Biol. Chem.*, 257(7), 3471—3476, 1982a.
119. **Winge, D. R. and Miklossy, K.-A.,** Differences in the polymorphic forms of metallothionein, *Arch. Biochem. Biophys.*, 214, 80—88, 1982b.

Chapter 5

NORMAL VALUES FOR CADMIUM IN HUMAN TISSUES, BLOOD, AND URINE IN DIFFERENT COUNTRIES

Carl-Gustaf Elinder

TABLE OF CONTENTS

I. INTRODUCTION

Ten years ago, Friberg et al.[28] reviewed the literature on normal values of cadmium in different human tissues. Since then much additional data have been made available. In this chapter it is not our intention to cover all of the extensive literature on normal values of cadmium in different tissues. Emphasis has been given to readily available materials such as blood, urine, and hair which may be used for screening and monitoring programs and for the evaluation of exposure, as well as to kidney and liver, the two organs containing the highest concentrations of cadmium. Another section in this chapter is devoted to cadmium in muscle, bone, and fatty tissue, and the final section to other tissues. With the exception of cadmium concentration in urine, we will restrict the presentation to deal with cadmium concentration in tissues and fluids of persons not occupationally exposed to cadmium. A more extensive review presenting data on cadmium levels in human tissues has been published.[10] The review by Cherry[10] presents results of normal cadmium levels in 52 human tissues, including blood, urine, and hair.

II. BLOOD

Cadmium is usually measured in whole blood. Most of the cadmium in blood is bound to the red blood cells (Chapter 6) and the concentration of cadmium in plasma and serum is often too low to be accurately measured. A great number of studies have presented normal values for cadmium in blood based on samples taken from persons in the general population. However, several of the early measurements have resulted in falsely high values. Friberg et al.[28] in 1974 summarized that there were at that time no accurate studies available and that the normal concentration range was not yet known. The normal concentration of cadmium in blood was, nevertheless, considered to be less than 10 μg/ℓ. Szadkowski[95] using atomic absorption after extraction, and Ediger and Coleman[14] using atomic absorption with the Delves cup technique and deuterium backgound correction, were the first investigators to report normal levels of cadmium which still appear essentially valid. Szadkowski[95] reported a mean of 3.5 μg/ℓ in 18 "normals" and Ediger and Coleman[14] found a median concentration of 0.6 μg/ℓ, range 0.2 to 2, in venous blood taken from 50 industrial workers not occupationally exposed to cadmium. Ulander and Axelson[107] using the same Delves cup technique as Ediger and Coleman,[14] were able to distinguish between smokers and nonsmokers. Male smokers had a mean cadmium level in blood of 2.3 μg/ℓ, range 0.6 to 6.1, while nonsmokers had a mean of 0.6, range 0.3 to 1.2.

The pronounced influence of smoking habits on blood cadmium concentration has later been confirmed in a series of reports which appeared within the last 10 years. Blood cadmium is also influenced by sex and age,[18] but this effect is usually totally overshadowed by the impact of smoking habits. Among nonsmokers, a slight increase in the blood cadmium concentration with age can usually be seen and nonsmoking females tend to have somewhat higher blood cadmium levels compared to males.[18]

A number of studies providing data on the cadmium concentration in blood of nonoccupationally exposed persons are summarized in Table 1. It may be concluded that nonsmokers in countries where dietary cadmium intake is 10 to 20 μg/day have a median cadmium concentration in whole blood in the order of 0.4 to 1.0 μg/ℓ, whereas smokers have a median concentration of 1.4 to 4.5 μg/ℓ.

It should be pointed out that the results presented in Table 1 are not readily comparable, as analytical quality assurance programs were used in a few of the studies only. For smokers, the variation in results obtained in various studies can, to a great extent, be explained by differences in the degree of smoking. For nonsmokers, however, one would not expect to find such a wide variation in the average cadmium concentration as the daily intake figure for cadmium (see Chapter 3, Table 7) and the cadmium levels in kidney cortex (Table 2), with the exception of Japan, are similar. It is, therefore, likely that the variation in blood cadmium levels of nonsmokers in different countries to some extent reflects analytical problems rather than true differences. Some recently published results are also in great error since quality control aspects were not considered by the researchers. There are even reports published in the 1980s which present data on cadmium in blood 5 to 20 times higher than generally reported concentrations.[47,81]

In 1980, WHO/UNEP launched a global biological monitoring program for the assessment of human exposure to heavy metals. The program was initiated to enable valid comparisons of blood cadmium levels for different countries.[26,108,113] In the program blood samples were obtained from similar groups of volunteers (teachers) in each participating country. Different analytical techniques were used. Figure 1 presents the median blood cadmium for smokers and nonsmokers in ten participating countries. The data in this figure should be comparable and reliable since the monitoring was coupled to an ambitious quality assurance program to ensure accurate results. Therefore, the variations seen between nonsmoking groups in different countries, seen in Figure 1, are probably indicative of true differences in exposure.

The cadmium level in whole blood from smokers is related to the degree of smoking (Figure 2) and a dose-related increase in blood cadmium levels with increasing number of cigarettes smoked per day has been seen in several studies.[18,66,72]

III. URINE

Normal urine contains various salts, especially sodium chloride which may interfere with analysis. Analysis of cadmium in urine with atomic absorption spectrophotometry without previous extraction or deuterium background correction will thus easily result in falsely high values due to nonspecific absorption from interfering salts. Early measurements using atomic absorption and excluding the above procedures have, therefore, given "normal" values of cadmium in urine which are several magnitudes higher than the actual concentration.[67-69,89,102] From more recent and reliable reports including analytical quality assurance, it is now evident that the cadmium concentration in urine from humans not excessively exposed to cadmium is usually around or below 1 $\mu g/\ell$.[43]

Data on the urinary excretion of a substance can be presented in various ways, the most straightforward being as micrograms per liter. The drawback is that different urine samples frequently have a variable degree of dilution and a diluted sample cannot readily be compared with a concentrated one. A common way of making urine samples more comparable is to adjust the results to refer to a certain specific gravity,[20] corresponding to a certain degree of dilution, or to use creatinine as a reference metabolite, quoting micrograms of Cd per gram of creatinine. Adult males excrete about 1.7 g creatinine per 24 hr and females about 1.0 g.[42] This difference should be borne in mind when data for the two sexes are compared.

It has been shown that urinary concentration of cadmium increases with age[46,104] and, furthermore, that smokers have higher averages of cadmium in urine compared to nonsmokers of the same age.[17,43,86]

In Figure 3, average cadmium concentrations in urine from male volunteers in Tokyo, Dallas, and Stockholm are given in relation to age. Geometric averages have been used in the figure, since the distribution of urinary cadmium concentrations in any age group is closer to a log-normal frequency distribution than a normal distribution.[17,43,56] The Japanese and Swedish data in the figure have been adjusted to a specifc gravity of 1.020. Data on specific gravity were not available for the American study. In another U.S. city (Chicago) the geometric averages of urinary cadmium concentrations in different age groups ranged from 0.3 to 0.9 $\mu g/\ell$.[56] In 1983, Kowal and Zirkes[55] reported on the concentration of cadmium in urine samples collected from almost 1000 persons (males and females), aged from 20 to

Table 1
CADMIUM IN BLOOD (μg/ℓ) IN HUMANS NOT OCCUPATIONALLY EXPOSED TO CADMIUM

Country	Year of publication	Authors	Sex	Age	Smoking habits	Mean[a]	Range or ±SD	Analytical procedure	Quality assurance[b]
Belgium	1980	Buchet et al.	M + F	7—10	—	0.5—0.8	0.1—3.1	AAS, flameless	–
	1980	Lauwerys et al.	F	About 80	—	1.2	0.2—4.5	AAS, flameless	–
F.R.G.	1981	Manthey et al.	M	Adults	Smokers	2.5	0.5—6.4	AAS, flameless	+
				Adults	Nonsmokers	0.4	0.1—0.7	AAS, flameless	
	1983	Brockhaus et al.	F	60—65	Smokers	1.3—1.8		AAS, flameless	+ +
				60—65	Nonsmokers	0.4		AAS, flameless	
			M	60—65	Smokers	1.0—1.9		AAS, flameless	
				60—65	Nonsmokers	0.4—0.5		AAS, flameless	
France	1983	Moreau et al.	M	25—55	Smokers	1.3		AAS, flameless	–
				25—55	Nonsmokers	0.4		AAS, flameless	
Japan	1979	Saito et al.	M	—	—	3.6	0.4—9.4	AAS, Zeeman background correction	–
	1983	Watanabe et al.	M	40—59	Smokers	4.4	1.0—10.0	AAS, flameless	–
				40—59	Nonsmokers	3.2	0.5—9.5	AAS, flameless	
			F	40—50	Nonsmokers	3.7	1.0—9.5	AAS, flameless	
Netherlands	1977	Zielhuis et al.	F	Adults	Smokers	0.6—0.7	0.2—4.4	AAS, flameless	–
			F	Adults	Nonsmokers	0.4	0.2—2.5	AAS, flameless	
Sweden	1974	Ulander and Axelson	M	20—55	Smokers	2.3	0.6—6.1	AAS, Delves cup	+
				20—55	Nonsmokers	0.6	0.3—1.2	AAS, Delves cup	
			F	20—55	Smokers	2.0	0.5—7.6	AAS, Delves cup	
				20—55	Nonsmokers	0.5	0.2—1.0	AAS, Delves cup	
	1983a	Elinder et al.	M	18—72	Smokers	1.5	0.2—7.3	AAS, flameless	+ +
				19—70	Nonsmokers	0.2	0.2—1.2	AAS, flameless	
			F	20—70	Smokers	1.3	0.3—3.9	AAS, flameless	
				23—71	Nonsmokers	0.3	0.2—1.2	AAS, flameless	
U.K.	1976	Beevers et al.	M + F	45—64	Smokers	3.3	±2.0	AAS, extraction	–
				45—64	Nonsmokers	1.8	±0.9	AAS, extraction	
	1978	Ward et al.	M + F	16—51	Smokers	4.5	±2	AAS, extraction	–
				16—51	Nonsmokers	2.2	±0.7	AAS, extraction	

U.S.	1973	Ediger and Coleman	M	Adult	—	0.6	0.2—20	AAS, Delves cup	−
	1977	Baker et al.	M + F	1—5	—	2.0	±1.6	AAS, flameless	−
	1978	Wysowski et al.	M + F	1—4	—	0.4	0.4—3.7	AAS, Delves cup	−
				5—9	—	0.4	0.4—1.6	AAS, Delves cup	
				10—19	Smokers	3.1	2.3—3.2	AAS, Delves cup	
				10—19	Nonsmokers	0.7	0.4—4.2	AAS, Delves cup	
				20—39	Smokers	1.8	0.4—3.6	AAS, Delves cup	
				20—39	Nonsmokers	0.4	0.4—1.0	AAS, Delves cup	
				40—	Smokers	3.4	0.5—6.9	AAS, Delves cup	
				40—	Nonsmokers	0.5	0.4—2.7	AAS, Delves cup	
	1979	Kowal et al.	M + F	4—69	Current smokers	1.4	0.4—3.3	AAS, extraction	
					Former smokers	0.9	0.3—2.3	AAS, extraction	
					Nonsmokers	0.8	0.2—3.3	AAS, extraction	
	1982	Tulley and Lehmann	M	Adults	Smokers	1.5	±1.0	AAS, flameless	+ +
				Adults	Nonsmokers	0.6	±0.2	AAS, flameless	
			F	Adults	Smokers	1.0	±0.6	AAS, flameless	
				Adults	Nonsmokers	0.8	±0.4	AAS, flameless	

[a] Arithmetic means or medians are used in the table.

[b] + +, valid quality assurance data reported; +, probably valid quality assurance but detailed data not reported; −, no valid quality assurance information reported.

Table 2
AVERAGE CADMIUM CONCENTRATION IN RENAL CORTEX AND LIVER AT AGES 40 TO 59, SMOKERS AND NONSMOKERS COMBINED (ARITHMETIC AVERAGES ARE USED IF NOT OTHERWISE STATED)

			mg/kg wet weight			
Country	Area	Sex	Renal cortex	Liver	Ratio liver/ renal cortex concentration	Ref.
Australia	Brisbane	M + F	33			70
	Perth	M + F	35	2.2	0.06	91
Belgium	Liège	M + F	39[a]			108
Canada	Ontario	M	44[b]			60
		F	78[b]			
China[c]	Beijing	M + F	17[a]			108
Denmark	Copenhagen	M	36			76
		F	35			
Finland	Helsinki	M + F	22	0.8	0.03	109
France	Strasbourg	M	26	1.3		31
		F	14	1.1		
F.R.G.	Stuttgart	M + F	12	1.0	0.09	22
G.D.R.	Jena	M	22			1
		F	11			
India	Different areas	M + F	20[a]			108
Israel	Jerusalem	M + F	24[a]			108
Japan	Kanazawa	M + F	85	10	0.12	40
	Kobe	M + F	60	5		48
	Tokyo	M + F	100	5.7	0.06	105
		M	65	4.5	0.07	50, 106
		F	73	5.4	0.07	
	Different areas	M	60	4	0.07	41
		F	70	7	0.10	
	Akita	M	119			54
		F	140			
Norway	Bergen	M + F	25	2.4	0.10	94
Romania	Different areas	M	15—27	1.3—3		4
		F	9—36	0.8—1.3		
Sweden	Stockholm	M + F	30			80
		M	21	0.7	0.03	15, 50
		F	28	0.8	0.03	
U.S.	Large cities	M + F	42	2.6		87
	North Carolina	M	32	3.2	0.10	34
		F	27			
	Cincinnati	M	36	1.8		32
		F	31	2.5		
	Dallas	M	25	1.3	0.05	45, 50
	Baltimore	M + F	25[a]			108
Yugoslavia	Zagreb	M + F	32[a]			6, 108

[a] All data from Vahter[108] report are given as geometric averages, which on average is 85% of the corresponding arithmetic ones. Thus, the arithmetic average can be approximated as 1.18 times the geometric average.

[b] Results by LeBaron et al.[60] may not be readily comparable with the other results, since they found an unusually high ratio between the cadmium concentration in cortex and medulla, about 3, whereas most other investigators have reported a ratio of about 2.

[c] People's Republic of China.

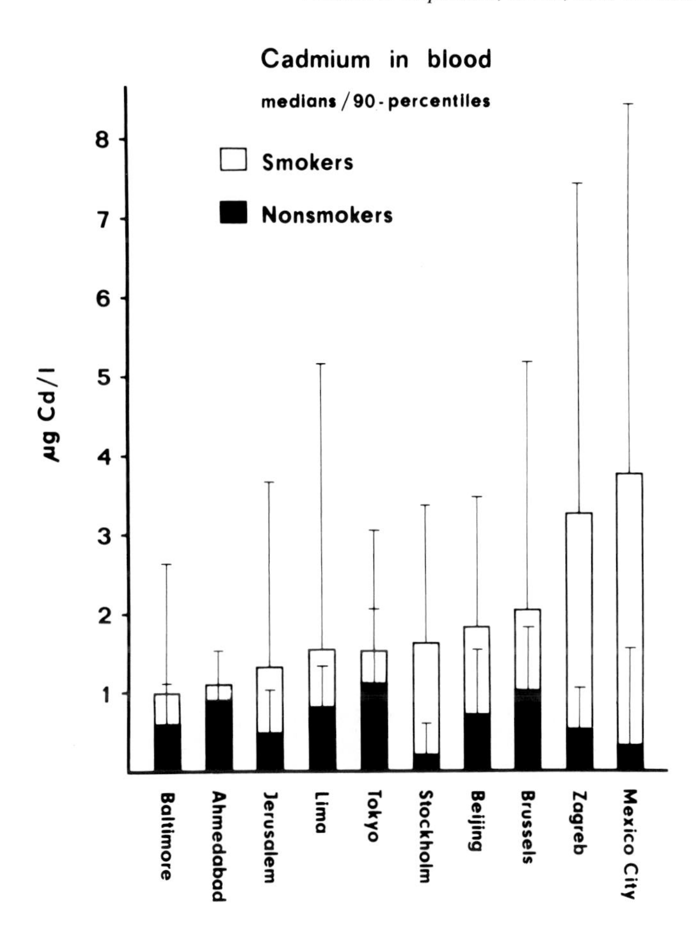

FIGURE 1. Concentrations of cadmium in blood (median values with 90 percentiles indicated) of teachers in nine different cities and of a randomly selected group of people living in Stockholm. (Modified from Vahter, M., Ed., Report prepared for United Nations Environment Programme and World Health Organization by National Swedish Institute of Environmental Medicine and Department of Environmental Hygiene, Karolinska Institute, Stockholm, 1982.)

74, living in 9 states of the U.S. The specific gravity-adjusted urinary cadmium concentration increased almost linearly from the age group 20 to 29 to the one comprised of persons aged 60 to 69. At age 20 to 29, the geometric average urinary cadmium concentration was 0.4 μg/ℓ and at age 60 to 69, it was 1.2 μg/ℓ.

In order to estimate the 24-hr urinary excretion of cadmium instead of concentration only, Elinder et al.[17] measured the urinary cadmium concentration in relation to creatinine. The 24-hr urinary excretion of cadmium was estimated by multiplying the average urinary concentration per gram of creatinine according to age and sex by values given in earlier published data on 24-hr urinary excretion of creatinine from persons of the same sex and age.[11,29,42] The geometric average 24-hr excretion increases with age and is higher among smokers than nonsmokers (Figure 4).

Several observations indicate that the urinary excretion of cadmium is mainly related to the body burden or kidney burden of cadmium, and only to a limited extent related to recent exposure. This general statement only holds true when the kidneys are functioning normally and are not damaged by, e.g., cadmium. If exposure is very excessive, the urinary excretion of cadmium may sometimes increase even without kidney dysfunction.[58]

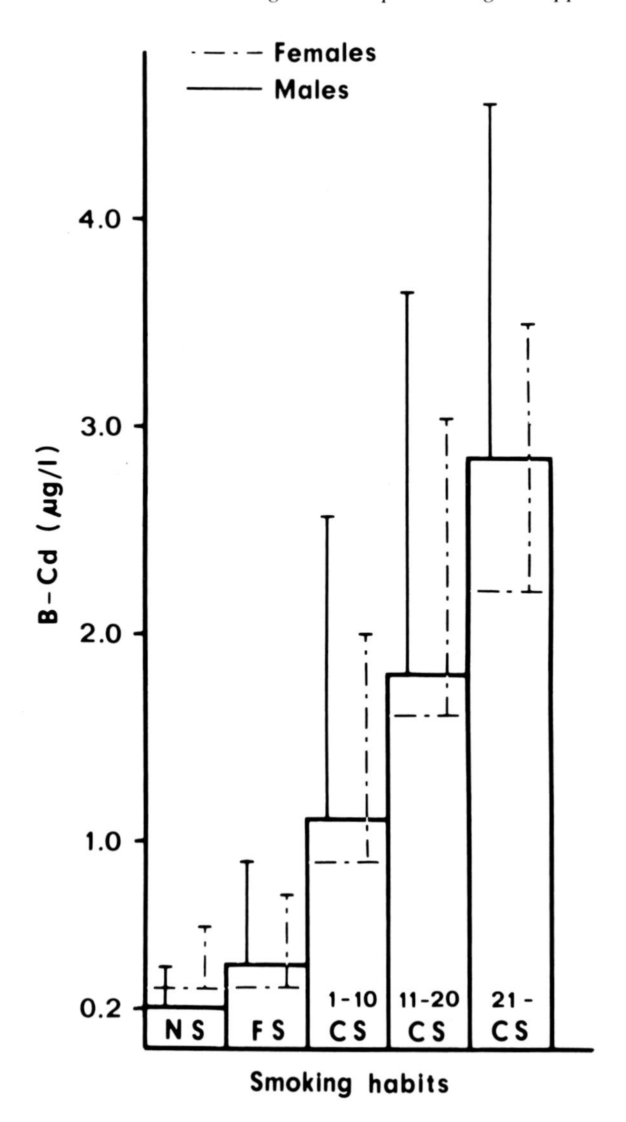

FIGURE 2. Median blood cadmium levels (μg/ℓ) among males and females in different smoking categories. Vertical bars denote the 90 percentiles. NS, nonsmokers; FS, former smokers; CS, 1-10, current smokers, smoking 1 to 10 cigarettes daily; CS, 11-20, current smokers, smoking 11 to 20 cigarettes daily; CS, >20, current smokers, smoking more than 20 cigarettes daily. (From Elinder, C.-G., Friberg, L., Lind, B., and Jawaid, M., *Environ. Res.*, 30, 233—253, 1983a. With permission.)

Normally, the urinary excretion of cadmium increases with age in a way similar to that of body or kidney burden, and does not change with age in a similar way to the daily intake of cadmium.[17,49] Smokers have higher urinary cadmium levels compared to nonsmokers, the difference being of the same order as that for cadmium levels in tissues. Long-term studies of not excessively occupationally exposed workers have shown that urinary levels of cadmium usually do not increase much despite the fact that blood cadmium levels increase dramatically only a short time (months) after occupational exposure commences.[51,57] After longer periods of exposure to cadmium which have led to an increase in body burden of cadmium, a significant relationship between urinary cadmium and duration of occupational exposure is evident.[35,57]

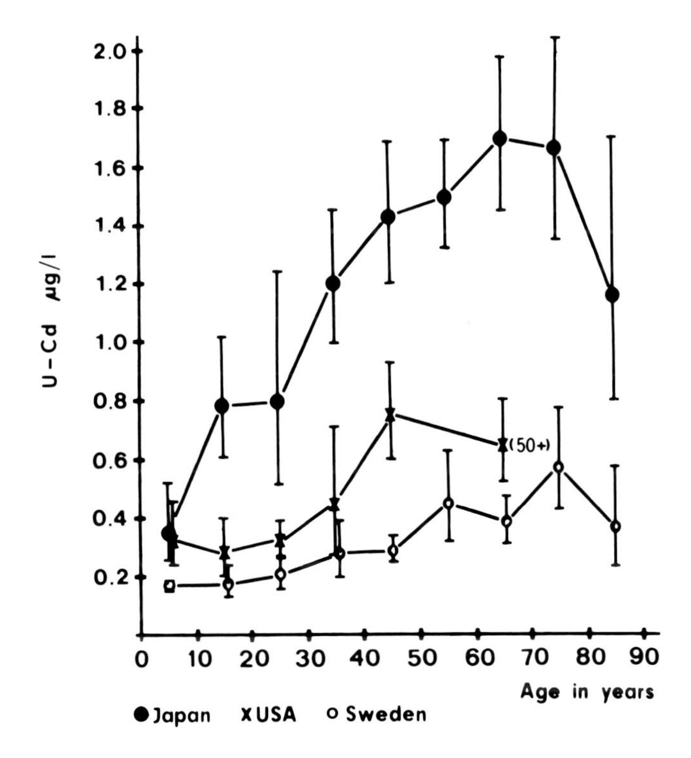

FIGURE 3. Geometric average of cadmium concentration in urine, obtained from male volunteers in Japan, U.S., and Sweden. The Japanese and American data are on smokers and nonsmokers combined, whereas Swedish values are based on nonsmokers only. (From Kjellström, T., *Environ. Health Perspect.*, 28, 169—197, 1979. With permission.)

In Japan,[39,52,96,111] and in Belgium,[8,83] people living in cadmium-contaminated or -polluted areas have higher urinary levels of cadmium than people living in nonpolluted areas.

When a cadmium-induced renal dysfunction is present, the urinary excretion of cadmium usually increases markedly and is no longer directly proportional to the body or renal burden of cadmium (see Chapter 6, Sections V.A.3 and VI.C).

IV. HAIR

Hair is a readily available material which has been analyzed by several investigators, e.g., Schroeder and Nason,[88] Hammer et al.,[33] Nishiyama,[73] Petering et al.,[77] Jervis et al.,[44] Gross et al.,[32] Oleru,[75] Rosmanith et al.,[84] Baker et al.,[3] Pihl and Parkes,[78] and Kowal et al.[56] Average concentrations have been in the order of 0.5 to 2 mg/kg. However, the interpretation of results of analysis of cadmium in hair is difficult since external contamination both from dust and from metals in hair lotions, hair sprays, etc. may occur. Nishiyama and Nordberg[74] showed that cadmium adsorbed on the hair, due to external contamination, was virtually impossible to remove even after using various washing procedures and could, therefore, not be distinguished from endogenous cadmium. Elevated concentrations of cadmium have been found in hair samples obtained from children living in the vicinity of cadmium-polluting industries (Figure 5)[3,33] and hair samples collected from cadmium workers.[21] This is related to external contamination of the hair.

Cadmium in hair neither reflects the body burden of cadmium, nor recent exposure.[21] In contrast to body burden of cadmium, which increases with age, levels of cadmium in hair tend to decrease with age.[32,56] Furthermore, smokers have not been shown to have higher cadmium concentrations in hair than nonsmokers.[56]

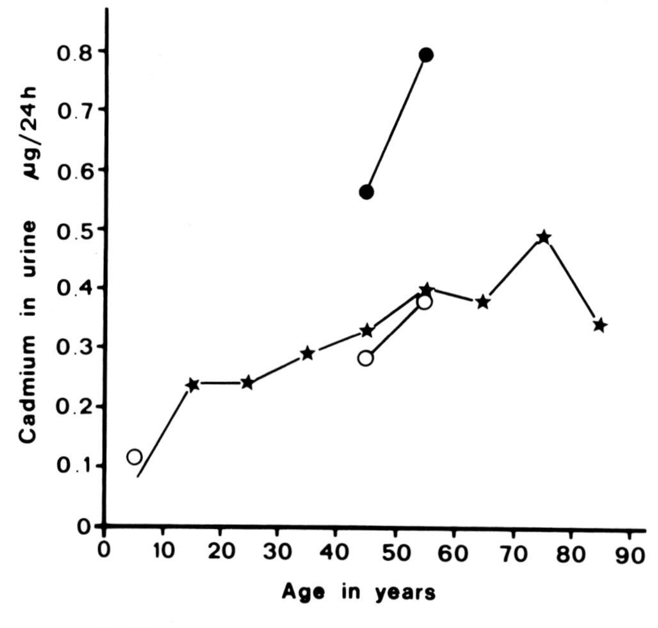

FIGURE 4. Geometric average of estimated cadmium excretion per 24 hr at different ages in Swedish volunteers. (From Elinder, C. G., Kjellström, T., Linnman, L., and Pershagen, G., *Environ. Res.*, 15, 473—484, 1978. With permission.)

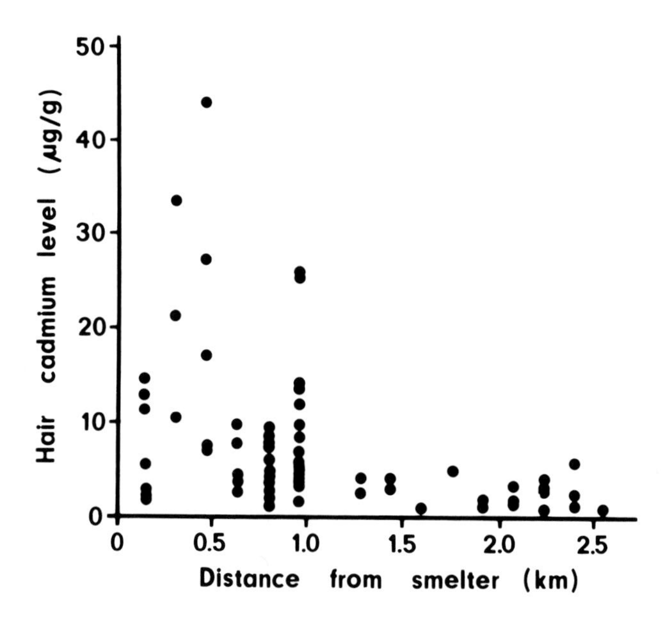

FIGURE 5. Hair cadmium levels among children aged 1 to 5, living at different distances from a zinc smelter. (From Baker, E. L., Hayes, C. G., Landrigan, P. J., Handke, J. L., Leger, R. T., Housworth, W. J., and Harrington, J. M., *Am. J. Epidemiol.*, 106(4), 261—273, 1977. With permission.)

V. LIVER AND KIDNEY

In contrast to blood and urine, analysis of cadmium in liver and kidney from adult humans will give accurate results using most methods as the concentrations usually are relatively high. The basic work in this field was done in the early 1960s by Tipton, Schroeder, Perry, and co-workers. They showed that cadmium occurred in comparably high concentrations in liver and kidney and that the concentrations in these tissues increased with age. Later on their results were confirmed by a large number of investigators[28] with data being reported from most parts of the developed world (Table 2). The tissue concentrations have been reported in various ways, e.g., milligrams per kilogram wet weight, dry weight, and ash weight. Some authors have determined cadmium in whole kidney and other concentrations in the renal cortex. To obtain approximate wet weight values for renal cortex, ash values can be multiplied by about 0.013[16,100] and dry weight values by about 0.20.[16,79,109] For liver, the conversion factor from dry to wet weight is somewhat higher, about 0.28.[16,79]

A. Distribution of Cadmium within the Kidney

Cadmium concentration is higher in the kidney cortex than in the medulla. Livingston[65] analyzed cadmium in eight serial sections of one kidney, from the outer layer of cortex inward to the inner medulla. The concentration of cadmium was found to decrease from the outer layers to the inner, and the cadmium concentration in outer cortex was about twice that in medulla. A ratio of about 2:1 between the cadmium content in cortex and medulla has been found by most investigators.[2,30,40,94] LeBaron et al.[60] and Svartengren et al.,[93] however, arrived at a slightly higher ratio, 2.9:1 and 2.7:1, respectively. It is clear that different ratios between cortex and medulla are easily obtained as a result of different sampling methods. Standardized sampling methods are, therefore, desirable.

In human beings, about 65 to 75% of the whole kidney weight is cortex and the remaining 25 to 35% is medulla and collecting system.[38,93] In other mammals the proportion may be different.[90] These relationships are crucial when whole kidney cadmium concentrations are recalculated to cortex concentrations in animals or humans. If it is assumed[25] that the ratio between cadmium concentration in cortex and medulla is usually 2:1, and that 75% of the kidney weight is cortex, an approximate estimate of the concentration in kidney cortex can be arrived at by multiplying whole kidney values by 1.15 (calculated from $a \times 0.75 + b \times 0.25 = c$, where (a) is the kidney cortex concentration of cadmium, which is twice the medulla concentration (b), and (c) is the whole kidney concentration. Thus, $a \times 0.75 + a \times 0.25/2 = c$, which will be the same as $a = 1.15 \times c$). More precise information has been provided by Svartengren et al.[93] They examined whole kidneys obtained from 20 males aged 30 to 59 and found that cortex, medulla, and the remainder (mainly the renal pelvis) of the kidney constituted 65, 27 and 8% of the whole kidney weight, respectively. The average cadmium concentration was 18.4, 6.9, and 3.2 mg/kg wet weight. The average cadmium concentration in the whole kidney was 14.4 mg/kg.

Friberg et al.[27,28] used the constant 1.5 to estimate renal cortex concentration from whole kidney values. This was based on a very limited amount of data available at that time. In view of the more detailed information that is now at hand,[93] a conversion factor of about 1.25 seems to be more appropriate (see Chapter 6, Section IV.B.2).

B. Variation with Age, Sex, Smoking Habits, and Residency

The cadmium concentration in liver and kidney from fetuses and neonates is very low but increases markedly with age.[9,32,36] Figures 6 and 7 present age-specific data on cadmium in liver and kidney cortex samples obtained from autopsies in Japan, U.S., and Sweden. The results should be comparable since the analytical methods were cross-checked.[50]

Table 2 presents the mean cadmium concentration in renal cortex and liver at ages of 40

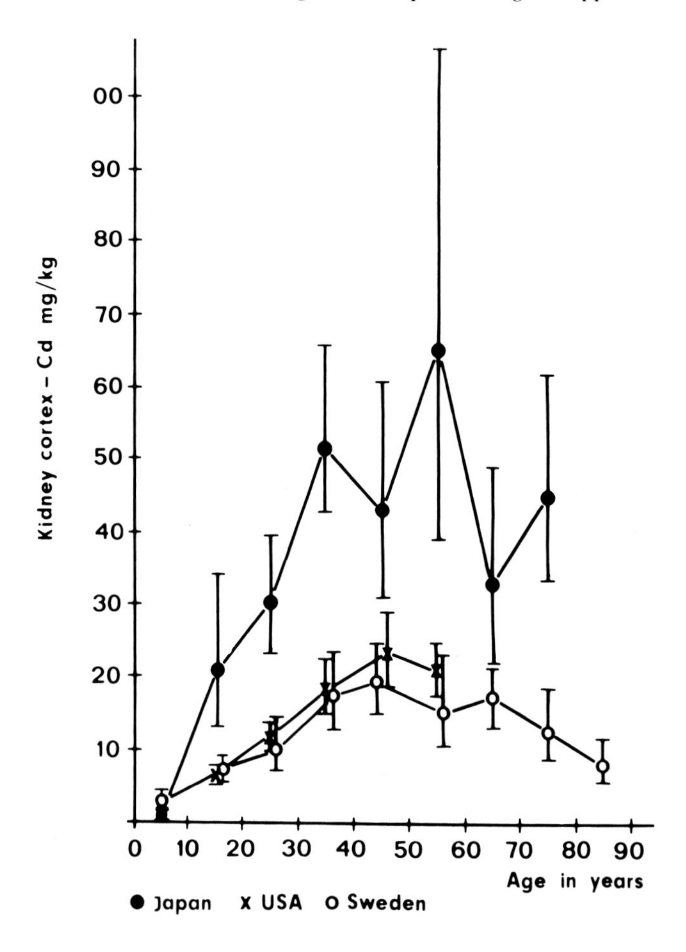

FIGURE 6. Geometric average cadmium concentrations in male kidney cortex in three countries: (●) Japan; (X) U.S.; (○) Sweden. (From Kjellström, T., *Environ. Health Perspect.*, 28, 169—197, 1979. With permission.)

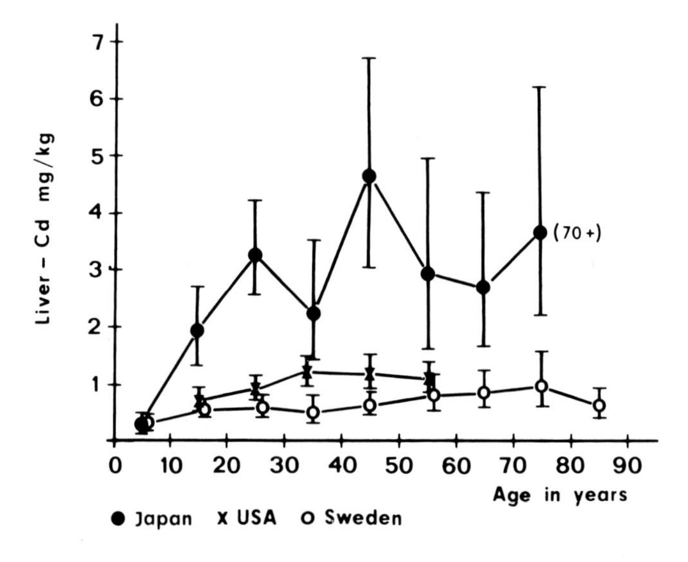

FIGURE 7. Geometric average cadmium concentrations in liver of men in three countries: (●) Japan; (X) U.S.; (○) Sweden (From Kjellström, T., *Environ. Health Perspect.*, 28, 169—197, 1979. With permission.)

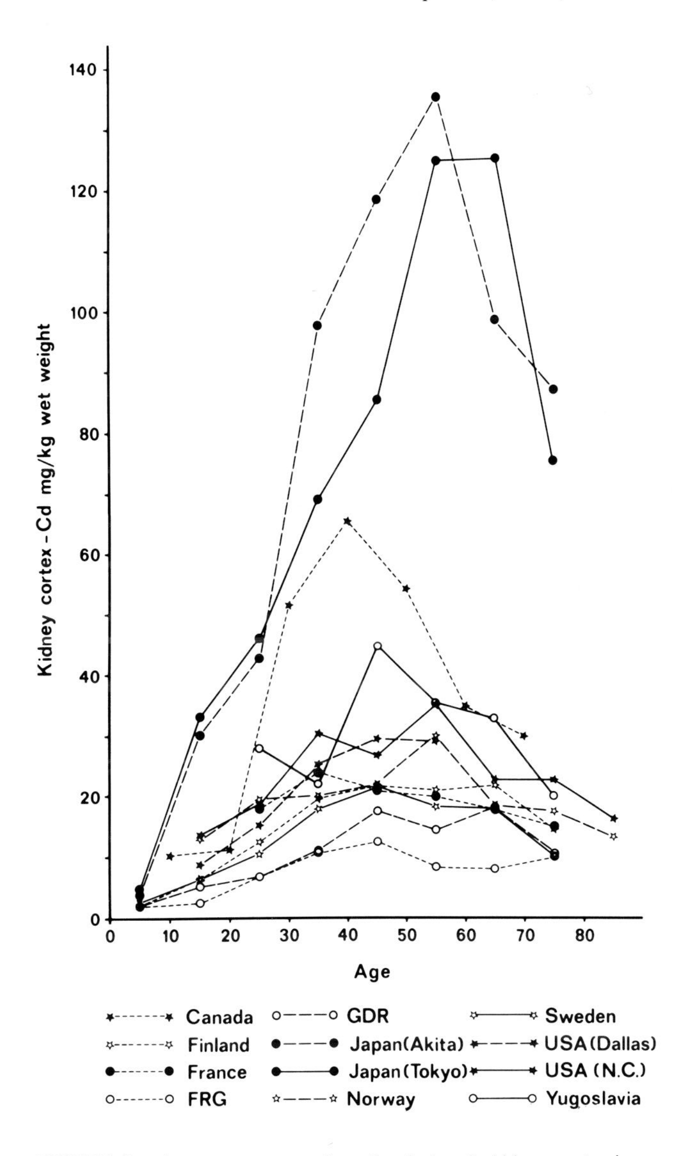

FIGURE 8. Average concentration of cadmium in kidney cortex in relation to age. Results from 12 different studies in 10 countries are summarized. The averages are based on smokers, nonsmokers, females, and males combined. For some studies, the kidney cortex concentrations in wet weight have been recalculated from ash or dry weight values or from whole kidney values. (1) Canada;[60] (2) Finland;[109] (3) France;[31] (4) F.R.G.;[22] (5) G.D.R.;[1] (6) Japan, Akita;[54] (7) Japan, Tokyo;[105] (8) Norway;[94] (9) Sweden;[15] (10) U.S., Dallas;[56] (11) U.S., North Carolina;[34] (12) Yugoslavia.[6]

to 59 years. The data are from different countries, as reported by different investigators. When reports included cadmium concentration in whole kidney only, the concentration in kidney cortex was estimated by multiplying by 1.25. In the same way, wet weight concentration was calculated from data on dry or ash weight concentration using the conversion factors mentioned earlier. The mean values given in the table are based on both smokers and nonsmokers taken jointly, since data on smoking habits were not available for most of the studies. Figure 8 summarizes data on age-related changes in the renal cortex concentration of cadmium as it has been reported in 12 different autopsy studies from 10 countries.

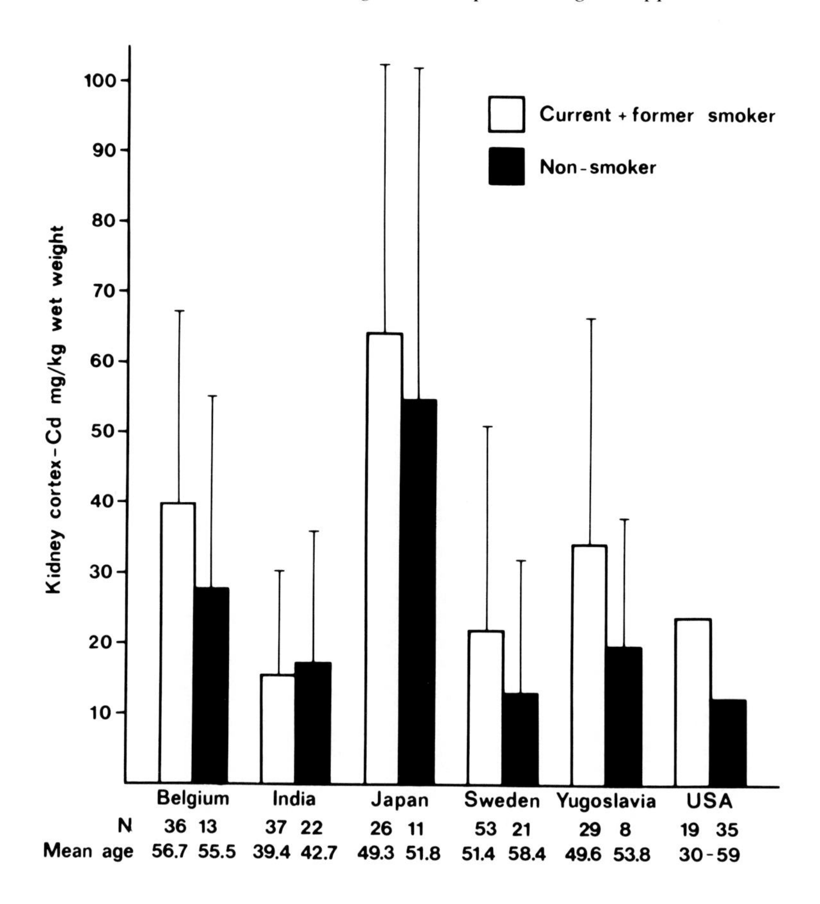

FIGURE 9. Concentration of cadmium in kidney cortex (geometric mean values) in relation to smoking habits among subjects (30 to 69 years of age) studied in Belgium, India, Japan, and Yugoslavia.[108] Also included in the figure are data from Sweden and U.S. (subjects aged 30 to 59).[50]

In most countries, the average cadmium concentration in renal cortex of middle-aged people is around 20 to 40 mg/kg wet weight. In Japan, the average cadmium concentration in kidney cortex of middle-aged people is around 100 mg/kg. An important exception from this general statement is the comparably high levels reported by LeBaron et al.[60] from Canada.

Due to the fact that cigarettes contain a certain amount of cadmium which is inhaled during smoking, smokers have higher tissue levels of cadmium than nonsmokers.[34,61,62] In a Swedish autopsy study,[15] middle-aged smokers were found to have cadmium levels in kidney cortex about 10 mg/kg higher than those for nonsmokers of the same age. In liver, the corresponding increase in cadmium level was about 0.5 mg/kg. Similar data were obtained from the U.S. by Kowal et al.[56]

Figure 9 presents data on the cadmium concentration in kidney cortex of smokers and nonsmokers in six countries. Middle-aged smokers and former smokers in five of the countries had cadmium concentrations in kidney cortex about 10 to 15 mg/kg higher than those for nonsmokers of similar age, but in India there was no difference. This could be due to the low cadmium concentration in Indian cigarettes.[19]

The variations in average kidney cortex cadmium concentration between different countries may thus be partially explained by differences in smoking habits. However, it is unlikely that differences in smoking habits will explain the differences found between means of cadmium in kidney cortex, which are larger than 10 mg/kg. As there are great differences also between nonsmokers in different countries (Figure 9), factors other than smoking must be of decisive importance for differences between countries.

Another possible reason for variable results may be that the different investigators analyzed divergent parts of the kidney cortex. As mentioned earlier, the outer part of the cortex has the highest cadmium concentration[65] and investigators analyzing superficial kidney cortex samples will obtain higher concentrations than those analyzing the whole cortex. Different sample preparation techniques can, however, hardly explain variations greater than 10 mg/kg. In the WHO/UNEP[113] study (Figure 9), similar methods for sample preparation and collection were used.

Therefore, differences in average kidney cortex concentration exceeding 20 mg/kg (Table 2), including possible divergency caused by variable smoking habits in the analyzed group and by sampling methods, have to be related to differences in exposure.

In most countries the average level of cadmium in kidney cortex ranges from 20 to 40 mg/kg, whereas in Japan the levels are considerably higher. A relationship between high daily intake of cadmium in Japan (Chapter 3, Section IV.) and elevated tissue levels is apparent. Likewise, the average liver-cadmium level of middle-aged persons ranges from 0.7 to 2.6 mg/kg in countries other than Japan. In Japanese autopsy studies, average cadmium levels in liver have been in the order of 5 to 10 mg/kg.

It should be emphasized that there is a large interindividual variation in kidney cortex concentrations even among persons of the same age. Using a larger number of samples, the range usually exceeds one order of magnitude, i.e., subjects with cadmium concentrations in kidney cortex lower than 5 mg/kg, as well as persons with more than 50 mg/kg, may be found in the same age group. Elinder et al.[15] noticed that the frequency distribution for different cadmium levels in any specific age group was closer to a log-normal, than to a linear-normal distribution. In log-normal distributions, geometric averages and standard deviation should be used for estimating means and normal ranges. The finding of a log-normal frequency distribution of cadmium levels in human tissues has been confirmed in other studies, e.g., Tsuchiya and Iwao,[103] Kowal et al.,[56] Vuori et al.,[109] and Vahter.[108] Therefore, geometric means have been used in Figures 5 and 6. In Table 2 and Figure 8, arithmetic averages have been used since geometric means were not always available. Due to the skewed distribution of cadmium concentration in liver and kidney, as well as in other tissues, geometric averages are always lower than the corresponding arithmetic ones. Kjellström[50] has provided data on the geometric and arithmetic average cadmium concentration in liver and kidney for different age groups in Japan, U.S., and Sweden. It can be seen from these data that as a rule, the geometric average cadmium concentration in kidney cortex is about 85% of the corresponding arithmetic average. For liver, the ratio between geometric and arithmetic average is about 80%. It is noteworthy that the ratios between geometric and arithmetic averages do not appear to vary with the cadmium concentration in organs.

A limited number of measurements have also been made on liver and kidney samples obtained from occupationally exposed workers and from environmentally exposed Japanese farmers. Such data are presented in Volume II, Chapter 9.

VI. MUSCLE, BONE, AND FAT

The cadmium concentration in muscle is of special interest since muscle constitutes a large proportion of the body mass, about 40% of the body weight of an adult man according to the International Commission on Radiation Protection.[38] Furthermore, autopsy data indicate that the biological half-time of cadmium in muscle is very long, exceeding that in kidney and liver.[49,50] In a mathematical model of cadmium metabolism,[51] it was necessary to assume a longer half-time in "other tissues" than in liver and kidney in order to accurately

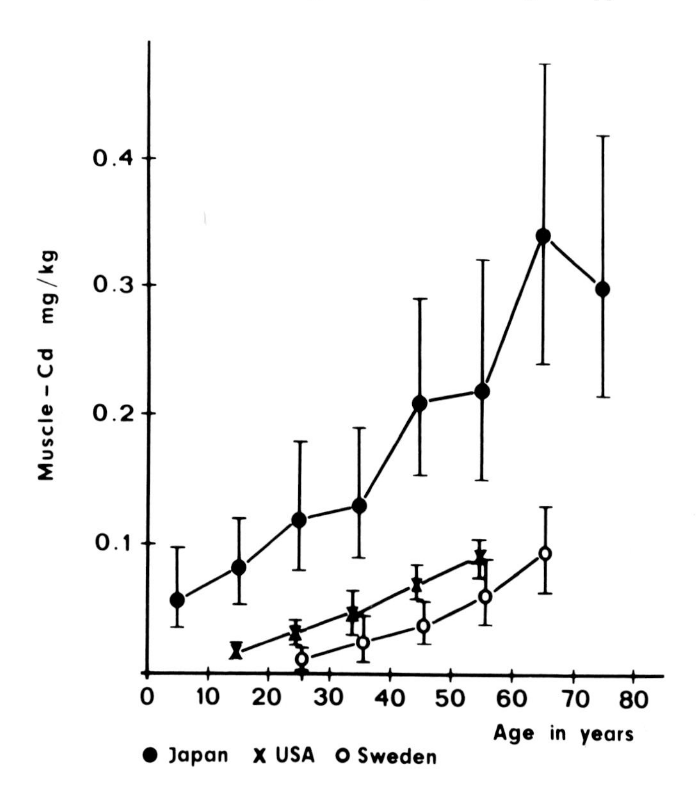

FIGURE 10. Geometric average cadmium concentrations in male muscle in three countries: (●) Japan; (X) U.S.; (○) Sweden. (From Kjellström, T., *Environ. Health Perspect.*, 28, 169—197, 1979. With permission.)

simulate the observed accumulation curves (see Chapter 7, Section IV.B). For the future it is possible that biopsies from muscle tissue can be used as a measure of the body burden, or accumulated exposure of cadmium. Figure 10 presents data on cadmium in abdominal wall muscle specimens, collected from autopsies performed in Japan, U.S., and Sweden.[50] In all three countries there was a continuous increase in cadmium concentration with age. Similar data have also been reported from Finland.[109] Differences between the countries follow the same pattern as that for liver and kidney cortex. Generally, the average cadmium concentration in muscle is below 0.1 mg/kg wet weight in the U.S. and Sweden, whereas in Japan a geometric mean of 0.3 mg/kg was reached at the age of 60 to 69.

In bone, the cadmium concentration is probably quite variable depending on which part of the osseous system is being analyzed.[64] In compact bone, Lindh et al.[63] found a median value of 0.041 mg/kg wet weight. Knuuttila et al.[53] reported an average of 0.22 mg/kg dry weight in cancellous bones obtained from Finnish autopsies.

There is very little data available on cadmium in fat, another major constituent of the body (10%). Available data indicate that the cadmium concentration in fatty tissue is in the same order as that in muscle tissue[50] or slightly lower.[92]

VII. OTHER TISSUES

There are several studies of cadmium in pancreas. The concentrations are usually below 2 mg/kg wet weight in samples obtained from humans living in the U.S. and Europe.[15,100,109] From Japan, averages of 2 to 3 mg/kg have been reported.[92] The cadmium concentration in pancreas increases with age in a similar way to the increase of cadmium in liver and muscle.[15,56,109] The concentrations are also higher among smokers than among nonsmokers. From Sweden, Elinder et al.[15] reported an average of 0.3 mg/kg in middle-aged nonsmokers (40 to 50 years) and of 0.7 mg/kg in smokers of the same age group. Kowal et al.[56] reported 0.3 and 0.6 mg/kg for nonsmokers and smokers, respectively, in the U.S.

In lung, the cadmium concentration is usually below 1 mg/kg wet weight. Reported averages have ranged from 0.1 to 0.7 mg/kg.[30,62,71,92,101,109] In most studies based on a sufficient number of samples, the cadmium concentrations in the lung increased with age. Lewis et al.[61] noticed that smokers had about twice the cadmium concentration in lungs (0.56 mg/kg) compared to nonsmokers (0.33 mg/kg).

The normal concentration of cadmium in the thyroid has not been investigated very thoroughly. In thyroid obtained from normal donors, the cadmium concentration has been found to be in the order of 30 to 50 mg/kg ash weight, which corresponds to about 0.3 to 0.5 mg/kg wet weight.[23,100]

Cadmium concentration in the prostate increases with age in a mode similar to that found in other organs. Hienzsch et al.[37] reported an average of 0.2 to 0.6 mg/kg wet weight among boys aged less than 20. Among men aged 21 to 90, the average ranged from 1.3 to 3.1 mg/kg, with the highest average reached in the age group 51 to 60.

In human placenta, the average cadmium concentration has been reported to be in the order of 20 to 150 μg/kg dry weight, which corresponds to about 3 to 24 μg/kg on a wet weight basis.[12,13,82,97] Thürauf et al.[98] reported almost ten times higher values, but since deuterium background correction was not reported as being used during the analysis, the actual concentration may well be lower. Piscator,[79] using a spectrographic method, arrived at an average of 10 μg/kg wet weight. Roels et al.[82] and Copius-Peereboom et al.[12] noticed that women smoking 15 to 25 cigarettes per day had almost twice the cadmium concentration (83 μg/kg dry weight) compared to nonsmoking women (46 μg/kg dry weight).

Cadmium has also been determined in several other human tissues such as adrenal gland, aorta, brain, heart, intestine, skin, spleen, teeth, thymus, urinary bladder, and uterus.[10] Apart from the spleen and adrenal gland, which may sometimes contain concentrations up to 1 mg/kg wet weight, concentrations in most other tissues are well below 1 mg/kg dry weight.[4,10,24,92,99,109]

VIII. SUMMARY AND CONCLUSIONS

The cadmium concentration in blood and urine from persons not occupationally exposed is low and difficult to measure accurately. The blood cadmium concentration is heavily influenced by a person's smoking habits. In most countries the median blood cadmium concentration for nonsmokers is in the order of 0.4 to 1 μg/ℓ, whereas smokers have a median value of about 1.4 to 4.2 μg/ℓ.

The urinary excretion of cadmium is also influenced by smoking habits, but not to the same extent as blood. It increases with age and reaches a median concentration of about 0.5 to 1 μg/ℓ at the age of 70. A higher urinary excretion is seen among Japanese people and this is related to the higher dietary intake and body burden of cadmium in this country. The cadmium concentration in hair is usually around 1 mg/kg and heavily affected by different sources of external contamination.

Liver and kidney contain the highest concentrations of cadmium in the human body. The concentration is influenced by age, smoking habits, and residency. At the age of 50, the average liver concentration is around 1.5 mg/kg and the kidney cortex concentration usually ranges between 20 to 40 mg/kg. In Japan, the average level of cadmium in liver and kidney cortex is about two to five times higher.

In other tissues such as muscle, bone and fat, the cadmium concentration is usually below 1 mg/kg wet weight.

REFERENCES

1. **Anke, M. and Schneider, H.-J.,** Trace element content in human kidneys in relation to age and sex (in German), *Z. Urol.*, 67, 357—362, 1974.
2. **Anke, M., Partschefeld, M., Grün, M., and Groppel, B.,** The capacity of different parts of the body to reflect Cd-exposure, in *Kadmium-Symposium, August 1977, Jena,* Veröffentlichung der Friedrich-Schiller-Universität, Jena, 1979, 242—249.
3. **Baker, E. L., Hayes, C. G., Landrigan, P. J., Handke, J. L., Leger, R. T., Housworth, W. J., and Harrington, J. M.,** A nationwide survey of heavy metal absorption in children living near primary copper, lead and zinc smelters, *Am. J. Epidemiol.*, 106(4), 261—273, 1977.
4. **Balazova, G., Rippel, A., and Rosival, L.,** Cadmium concentration in human autopsy material (in German), in *Kadmium-Symposium, August 1977, Jena,* Veröffentlichung der Friedrich-Schiller-Universität, Jena, 1979, 302—306.
5. **Beevers, D. G., Campbell, B. C., Goldberg, A., Moore, M. R., and Hawthorne, V. M.,** Blood-cadmium in hypertensives and normotensives, *Lancet*, 4, 1222—1224, 1976.
6. **Blanuša, M., Kralj, Z., and Bunarevič, A.,** Cadmium and other trace metals in normal human renal cortex, in *Heavy Metals in the Environment, Heidelberg, Sept. 1983,* CEP Consultants Ltd., Edinburgh, 1983.
7. **Brockhaus, A., Freier, I., Ewers, U., Jermann, E., and Dolgner, R.,** Levels of cadmium and lead in blood in relation to smoking, sex, occupation, and other factors in an adult population of the FRG, *Int. Arch. Occup. Environ. Health*, 52, 167—175, 1983.
8. **Buchet, J. P., Roels, H., Lauwerys, R., Bruaux, P., Clays-Thoreau, F., Lafontaine, A., and Verduyn, G.,** Repeated surveillance of exposure to cadmium, manganese and arsenic in school-age children living in rural, urban and nonferrous smelter areas in Belgium, *Environ. Res.*, 22, 95—108, 1980.
9. **Casey, C. E. and Robinson, M. F.,** Copper, manganese, zinc, nickel, cadmium and lead in human foetal tissues, *Br. J. Nutr.*, 39, 639—646, 1978.
10. **Cherry, W. H.,** Distribution of cadmium in human tissues, in *Cadmium in the Environment*, Part 2, Nriagu, J. O., Ed., John Wiley & Sons, New York, 1981, 69—536.
11. **Clark, L. C., Thompson, H. L., Beck, E. I., and Jacobson, W.,** Excretion of creatine and creatinine by children, *Am. J. Dis. Child.*, 81, 744—783, 1951.
12. **Copius Peereboom, J. W., de Voogt, P., van Hattum, B., van der Velde, W., and Copius Peereboom-Stegeman, J. H. J.,** The use of the human placenta as a biological indicator for cadmium exposure, in *Int. Conf. Management and Control of Heavy Metals in the Environment, London, September 1979,* CEP Consultants Ltd., Edinburgh, 1979, 8—10.
13. **Copius Peereboom-Stegeman, J. H. J., Velde, van der, W. J., and Dessing, J. W. M.,** Influence of cadmium on placental structure, *Ecotoxicol. Environ. Saf.*, 7, 79—86, 1983.
14. **Ediger, R. D. and Coleman, R. L.,** Determination of cadmium in blood by a Delves cup technique, *At. Absorpt. Newsl.*, 12, 3, 1973.
15. **Elinder, C.-G., Kjellström, T., Friberg, L., Lind, B., and Linnman, L.,** Cadmium in kidney cortex, liver, and pancreas from Swedish autopsies, *Arch. Environ. Health,* 31, 292—302, 1976.
16. **Elinder, C.-G., Piscator, M., and Linnman, L.,** Cadmium and zinc relationships in kidney cortex, liver and pancreas, *Environ. Res.*, 13, 432—440, 1977.
17. **Elinder, C.-G., Kjellström, T., Linnman, L., and Pershagen, G.,** Urinary excretion of cadmium and zinc among persons from Sweden, *Environ. Res.*, 15, 473—484, 1978.
18. **Elinder, C.-G., Friberg, L., Lind, B., and Jawaid, M.,** Lead and cadmium levels in blood samples from the general population of Sweden, *Environ. Res.*, 30, 233—253, 1983a.
19. **Elinder, C.-G., Kjellström, T., Lind, B., Linnman, L., Piscator, M., and Sundstedt, K.,** Cadmium exposure from smoking cigarettes. Variations with time and country where purchased, *Environ. Res.*, 32, 220—227, 1983b.
20. **Elkins, H. B. and Pagnotto, L. D.,** The specific gravity adjustment in urinanalysis, *Arch. Environ. Health,* 18, 996—1001, 1969.
21. **Ellis, K. J., Yasumura, S., and Cohn, S. H.,** Hair cadmium content: is it a biological indicator of the body burden of cadmium for the occupationally exposed worker?, *Am. J. Ind. Med.*, 2, 323—330, 1981.
22. **Fischer, H. and Weigert, P.,** Investigations on the lead and cadmium content in human organs (in German), *Oeff. Gesundheitsw.*, 37, 732—737, 1975.
23. **Forssén, A.,** Inorganic elements in the human body. I. Occurrence of Ba, Br, Ca, Cd, Cs, Cu, K, Mn, Ni, Sn, Sr, Y and Zn in the human body, *Ann. Med. Exp. Biol. Fenn.*, 50, 99—162, 1972.
24. **Fosse, G. and Berg-Justesen, N. P.,** Cadmium in deciduous teeth of Norwegian children, *Int. J. Environ. Stud.*, 11, 17—27, 1977.
25. **Friberg, L.,** Cadmium and the kidney, *Environ. Health Perspect.*, 54, 1—11, 1984.

26. **Friberg, L. and Vahter, M.,** Assessment of exposure to lead and cadmium through biological monitoring: results of a UNEP/WHO global study, *Environ. Res.*, 30, 95—128, 1983.
27. **Friberg, L., Piscator, M., and Nordberg, G., Eds.,** *Cadmium in the Environment*, CRC Press, Boca Raton, Fla., 1971.
28. **Friberg, L., Piscator, M., Nordberg, G. F., and Kjellström, T.,** *Cadmium in the Environment*, 2nd ed., CRC Press, Boca Raton, Fla., 1974.
29. **Gault, M. H. and Cockcroft, D. W.,** Creatinine: clearance and age, *Lancet*, 1, 612—613, 1975.
30. **Geldmacher-v. Mallinckrodt, M. and Opitz, O.,** Diagnosis of cadmium intoxication of the normal cadmium content in human organs and body fluids (in German), *Arbeitsmed. Sozialmed. Arbeitshyg.*, 3, 276—279, 1968.
31. **Gretz, M. and Laugel, P.,** A study on the level of human contamination by cadmium (results of a regional survey) (in French), *Toxicol. Eur. Res.*, 4(2), 63—70, 1982.
32. **Gross, S. B., Yeager, D. W., and Middendorf, M. S.,** Cadmium in liver, kidney, and hair of humans, fetal through old age, *J. Toxicol. Environ. Health*, 2, 153—167, 1976.
33. **Hammer, D. I., Finklea, J. F., Hendricks, R. H., Shy, C. M., and Horton, R. J. M.,** Hair trace metal levels and environmental exposure, *Am. J. Epidemiol.*, 93, 84—91, 1971.
34. **Hammer, D. I., Calocci, A. V., Hasselblad, V., Williams, M. E., and Pinkerton, C.,** Cadmium and lead in autopsy tissues, *J. Occup. Med.*, 15(12), 956—963, 1973.
35. **Hassler, E., Lind, B., and Piscator, M.,** Cadmium in blood and urine related to present and past exposure. A study of workers in an alkaline battery factory, *Br. J. Ind. Med.*, 40, 420—425, 1983.
36. **Henke, G., Sachs, H. W., and Bohn, G.,** Cadmium determination in the liver and kidneys of children and juveniles by means of neutron activation analysis (in German), *Arch. Toxicol.*, 26, 8—16, 1970.
37. **Hienzsch, E., Schneider, H.-J., Anke, M., Hennig, A., and Groppel, B.,** The cadmium-, zinc-, copper- and maganese-level of different organs of human beings without considerable Cd-exposure in dependence on age and sex, in *Kadmium-Symposium, August 1977, Jena*, Veröffentlichung der Friedrich-Schiller-Universität, Jena, 1979, 276—282.
38. International Commission on Radiological Protection, Report of the Task Group on Reference Man, Rep. No. 23, ICRP, Pergamon Press, Oxford, 1975.
39. **Ishizaki, A.,** On the so-called Itai-itai disease (in Japanese), *J. Jpn. Med. Soc. (Nihon Ishikai Zasshi)*, 62, 242—248, 1969.
40. **Ishizaki, A., Fukushima, M., and Sakamoto, M.,** On the accumulation of cadmium in the bodies of Itai-itai patients, *Jpn. J. Hyg.*, 25, 86, 1970.
41. **Iwao, S., Tsuchiya, K., and Sugita, M.,** Variation of cadmium accumulation among Japanese, *Arch. Environ. Health*, 38, 156—162, 1983.
42. **Jackson, S.,** Creatinine in urine as an index of urinary excretion rate, *Health Phys.*, 12, 843—850, 1966.
43. **Jawaid, M., Lind, B., and Elinder, C.-G.,** Determination of cadmium in urine by extraction and flameless atomic-absorption spectrophotometry. Comparison of urine from smokers and non-smokers of different sex and age, *Talanta*, 30, 509—513, 1983.
44. **Jervis, R. E., Tiefenbach, B., and Chattopadhyay, A.,** Determination of trace cadmium in biological materials by neutron and pluton activation analyses, *Can. J. Chem.*, 52, 3008—3020, 1974.
45. **Johnson, D. E., Prevost, R. J., Tillery, J. B., and Thomas, R. E.,** The distribution of cadmium and other metals in human tissue, Environ. Prot. Tech. Serv. 232, U.S. Environmental Protection Agency, Washington, D.C., 1977.
46. **Katagiri, Y., Tati, M., Iwata, H., and Kawai, M.,** Concentration of cadmium in urine by age (in Japanese), *Med. Biol.*, 82, 239—243, 1971.
47. **Khera, A. K., Wibberley, D. G., Edwards, K. W., and Waldron, H. A.,** Cadmium and lead levels in blood and urine in a series of cardiovascular and normotensive patients, *Int. J. Environ. Stud.*, 14, 309—312, 1980.
48. **Kitamura, S., Sumino, K., and Kamatani, N.,** Cadmium concentrations in livers, kidneys and bones of human bodies (in Japanese), *Jpn. J. Public Health*, 17 (Abstr. 507) 177, 1970.
49. **Kjellström, T.,** Accumulation and Renal Effects of Cadmium in Man. A Dose-Response Study, Doctoral thesis, Karolinska Institute, Stockholm, 1977.
50. **Kjellström, T.,** Exposure and accumulation of cadmium in populations from Japan, the United States and Sweden, *Environ. Health Perspect.*, 28, 169—197, 1979.
51. **Kjellström, T. and Nordberg, G. F.,** A kinetic model of cadmium metabolism in the human being, *Environ. Res.*, 16, 248—269, 1978.
52. **Kjellström, T., Shiroishi, K., and Evrin, P.-E.,** Urinary β_2-microglobulin excretion among people exposed to cadmium in the general environment. An epidemiological study in cooperation between Japan and Sweden, *Environ. Res.*, 13, 318—344, 1977.
53. **Knuuttila, M., Olkkonen, H., Lammi, S., and Alhava, E. M.,** Cadmium content of human cancellous bone, *Arch. Environ. Health*, 37(5), 290—294, 1982.

54. **Kobayashi, S.,** Effect of ageing on the concentration of cadmium, zinc and copper in human kidney, *Jpn. J. Public Health,* 30, 27—34, 1983.
55. **Kowal, N. E. and Zirkes, M.,** Urinary cadmium and β_2-microglobulin: normal values and concentration adjustment, *J. Toxicol. Environ. Health,* 11, 607—624, 1983.
56. **Kowal, N. E., Johnson, D. E., Kraemer, D. F., and Pahren, H. R.,** Normal levels of cadmium in diet, urine, blood and tissues of inhabitants of the United States, *J. Toxicol. Environ. Health,* 5, 995—1014, 1979.
57. **Lauwerys, R., Buchet, J.-P., and Roels, H.,** The relationship between cadmium exposure or body burden and the concentration of cadmium in blood and urine in man, *Int. Arch. Occup. Environ. Health,* 36, 275—285, 1976.
58. **Lauwerys, R., Roels, H., Regniers, M., Buchet, J.-P., Bernard, A., and Goret, A.,** Significance of cadmium concentration in blood and in urine in workers exposed to cadmium, *Environ. Res.,* 20, 375—391, 1979.
59. **Lauwerys, R., Roels, H., Bernard, A., and Buchet, J. P.,** Renal response to cadmium in a population living in a nonferrous smelter area in Belgium, *Int. Arch. Occup. Environ. Health,* 45, 271—274, 1980.
60. **LeBaron, G. J., Cherry, W. H., and Forbes, W. F.,** Studies of trace-metal levels in human tissues. IV. The investigation of cadmium levels in kidney samples from 61 Canadian residents, in *Trace Substances in Environmental Health — XI,* Hemphill, D. D., Ed., University of Missouri, Columbia, 1977, 44—54.
59. **Lauwerys, R., Roels, H., Bernard, A., and Buchet, J. P.,** Renal response to cadmium in a population living in a nonferrous smelter area in Belgium, *Int. Arch. Occup. Environ. Health,* 45, 271—274, 1980.
60. **LeBaron, G. J., Cherry, W. H., and Forbes, W. F.,** Studies of trace-metal levels in human tissues. IV. The investigation of cadmium levels in kidney samples from 61 Canadian residents, in *Trace Substances in Environmental Health — XI,* Hemphill, D. D., Ed., University of Missouri, Columbia, 1977, 44—54.
61. **Lewis, G. P., Coughlin, L., Jusko, W., and Hartz, S.,** Contribution of cigarette smoking to cadmium accumulation in man, *Lancet,* 1, 291—292, 1972a.
62. **Lewis, G. P., Jusko, W. J., Coughlin, L. L., and Hartz, S.,** Cadmium accumulation in man: influence of smoking, occupation, alcoholic habit and disease, *J. Chronic Dis.,* 25, 717—726, 1972b.
63. **Lindh, U., Brune, D., Nordberg, G., and Wester, P. O.,** Level of antimony, arsenic, cadmium, copper, lead, mercury, selenium, silver, tin and zinc in bone tissue of industrially exposed workers, *Sci. Total Environ.,* 16, 109—116, 1980.
64. **Lindh, U., Brune, D., Nordberg, G., and Wester, P. O.,** Levels of cadmium in bone tissue (femur) of industrially exposed workers — a reply (to the letter to the editor), *Sci. Total Environ.,* 20, 3—11, 1981.
65. **Livingston, H. D.,** Measurement and distribution of zinc, cadmium and mercury in human kidney tissue, *Clin. Chem.,* 18, 67—72, 1972.
66. **Manthey, J., Stoeppler, M., Morgenstern, W., Nüssel, E., Opherk, D., Weintraut, A., Wesch, H., and Kübler, W.,** Magnesium and trace metals: risk factors for coronary heart disease?, *Circulation,* 64(4), 722—729, 1981.
67. **McKenzie, J. M.,** Urinary excretion of zinc, cadmium, sodium, potassium, and creatinine in ninety-six students, *Proc. Univ. Otago Med. Sch.,* 50, 16—18, 1972a.
68. **McKenzie, J. M.,** Variation in urinary excretion of zinc and cadmium, *Proc. Univ. Otago Med. Sch.,* 50, 15—16, 1972b.
69. **Mertz, D. P., Koschnick, R., and Wilk, G.,** The renal excretion of cadmium in normotensive and hypertensive humans (in German), *Z. Klin. Chem. Biochem.,* 10, 21—24, 1972.
70. **Miller, G. J., Wylie, M. J., and McKeown, D.,** Cadmium exposure and renal accumulation in an Australian urban population, *Med. J. Aust.,* 10, 20—23, 1976.
71. **Molokhia, M. M. and Smith, H.,** Trace elements in the lung, *Arch. Environ. Health,* 15, 745—750, 1967.
72. **Moreau, T., Lellouch, J., Orssaud, G., Claude, J. R., Juguet, B., and Festy, B.,** Blood cadmium levels in a general male population with special reference to smoking, *Arch. Environ. Health,* 38(3), 163—167, 1983.
73. **Nishiyama, K.,** in *Cadmium in the Environment,* Friberg, L., Piscator, M., and Nordberg, G., Eds., CRC Press, Boca Raton, Fla., 1971, 59.
74. **Nishiyama, K. and Nordberg, G.,** Adsorption and elution of cadmium on hair, *Arch. Environ. Health,* 25, 92—96, 1972.
75. **Oleru, U. G.,** Kidney, liver, hair and lungs as indicators of cadmium absorption, *Am. Ind. Hyg. Assoc. J.,* 37, 617—621, 1976.
76. **Østergaard, K.,** Concentrations of cadmium in renal tissue in smokers and non-smokers, *Norr. Kr. Laeg.,* 139(17), 989—991, 1977.
77. **Petering, H. G., Yeager, D. W., and Witherup, S. O.,** Trace metal content of hair. II. Cadmium and lead of human hair in relation to age and sex, *Arch. Environ. Health,* 27, 327—330, 1973.
78. **Pihl, R. O. and Parkes, M.,** Hair element content in learning disabled children, *Science,* 198, 204—206, 1977.

79. **Piscator, M.,** in *Cadmium in the Environment*, Friberg, L., Piscator, M., and Nordberg, G., Eds., CRC Press, Boca Raton, Fla., 1971.
80. **Piscator, M. and Lind, B.,** Cadmium, zinc, copper and lead in human renal cortex, *Arch. Environ. Health*, 24, 426—431, 1972.
81. **Revis, N. W. and Zinsmeister, A. R.,** The relationship of blood cadmium level to hypertension and plasma norepinephrine level: a Romanian study, *Proc. Soc. Exp. Biol. Med.*, 167, 254—260, 1981.
82. **Roels, J., Hubermont, G., Buchet, J. P., and Lauwerys, R.,** Placental transfer of lead, mercury, cadmium and carbon monoxide in women. III. Factors influencing the accumulation of heavy metals in the placenta and the relationship between metal concentrations in the placenta and in maternal and cord blood, *Environ. Res.*, 16, 236—247, 1978.
83. **Roels, H. A., Lauwerys, R. R., Buchet, J.-P., and Bernard, A.,** Environmental exposure to cadmium and renal function of aged women in three areas of Belgium, *Environ. Res.*, 24, 117—130, 1981.
84. **Rosmanith, J., Einbrodt, H. J., and Ehm, W.,** The interaction between lead, cadmium and zinc in children in an industrial area, *Staub-Reinhalt. Luft.*, 36(2), 55—62, 1976.
85. **Saito, K., Sasaki, T., Sato, Y., and Yasuda, H.,** Distribution of trace metals in snowfall and human blood in the northern region of Japan, in *Trace Substances in Environmental Health — XIII*, Hemphill, D. D., Ed., University of Missouri, Columbia, 1979, 68—79.
86. **Schaller, K. H. and Zober, A.,** Renal excretion of metals toxicologically relevant to occupationally exposed persons (in German), *Aerztl. Lab.*, 28, 209—214, 1982.
87. **Schroeder, H. A. and Balassa, J. J.,** Abnormal trace metals in man: cadmium, *J. Chronic Dis.*, 14, 236—258, 1961.
88. **Schroeder, H. A. and Nason, A. P.,** Trace metals in human hair, *J. Invest. Dermatol.*, 53, 71—78, 1969.
89. **Schroeder, H. A., Nason, A. P., Tipton, I. H., and Balassa, J. J.,** Essential trace metals in man: zinc. Relation to environmental cadmium, *J. Chronic Dis.*, 20, 179—210, 1967.
90. **Sperber, I.,** *Studies on the Mammalian Kidney*, Zool. Bidrag 22, Almquist & Wiksell, Uppsala, Sweden, 249—450, 1942—44.
91. **Spickett, J. T. and Lazner, J.,** Cadmium concentrations in human kidney and liver tissues from Western Australia, *Bull. Environ. Contam. Toxicol.*, 23, 627—630, 1979.
92. **Sumino, K., Hayakawa, K., Shibata, T., and Kitamura, S.,** Heavy metals in normal Japanese tissues, *Arch. Environ. Health*, 30, 487—494, 1975.
93. **Svartengren, M., Elinder, C.-G., Lind, B., and Friberg, L.,** Distribution and concentration of cadmium in human kidney, *Environ. Res.*, in press.
94. **Syversen, T. L. M., Stray, T. L., Syversen, G. B., and Ofstad, J.,** Cadmium and zinc in human liver and kidney, *Scand. J. Clin. Lab. Invest.*, 36, 251—256, 1976.
95. **Szadkowski, D.,** Cadmium — an ecological disturbance at the working place (in German), *Med. Monatsschr.*, 26, 553—556, 1972.
96. **Takebatake, E., Keshino, M., and Matsuo, I.,** Urinary cadmium concentration of population in Sasu area, Tsushima, Nagasaki, Japan (in Japanese with English summary), *J. Hyg. Chem.*, 18, 41, 1972.
97. **Thieme, R., Schramel, P., and Kurz, E.,** Trace element concentration in the human placenta in a strong contaminated environment (in German), *Geburtshilfe Frauenheilkd.*, 37, 756—761, 1977.
98. **Thürauf, J., Schaller, K.-H., Engelhardt, E., and Gossler, K.,** Cadmium content in the human placenta (in German), *Int. Arch. Occup. Environ. Health*, 36, 19—27, 1975.
99. **Thürauf, J., Schaller, K. H., and Weltle, D.,** Cadmium content in different human organs (in German), in *Kadmium-Symposium, Kongress-Bericht*, Wissenschaftliche Beiträge der Friedrich-Schiller-Universität, Jena, 1979, 296—302.
100. **Tipton, I. H. and Cook, M. J.,** Trace elements in human tissue. II. Adult subjects from the United States, *Health Phys.*, 9, 103—145, 1963.
101. **Tipton, I. H. and Shafer, J. J.,** Statistical analysis of lung trace element levels, *Arch. Environ. Health*, 8, 58—67, 1964.
102. **Tipton, I. H. and Stewart, P. L.,** Pattern of elemental excretion in long-term balance studies. II, in *Internal Dosimetry*, Synder, W. S., Ed., Ann. Prog. Rep., ORNL-4446, Health Physics Division, Oak Ridge, Tenn., 1970, 303—304.
103. **Tsuchiya, K. and Iwao, S.,** Interrelationships among zinc, copper, lead and cadmium in food, feces and organs of humans, *Environ. Health Perspect.*, 25, 119—124, 1978.
104. **Tsuchiya, K., Seki, Y., and Sugita, M.,** Organ and tissue cadmium concentration of cadavers from accidental deaths, in Proc. 17th Int. Congr. Occup. Health, available through the Secretariat, Av. Rogue Saenz. Pena, 110-2 piso, Oficio 8, Buenos Aires, 1972.
105. **Tsuchiya, K., Seki, Y., and Sugita, M.,** Cadmium concentrations in the organs and tissues of cadavers from accidental deaths, *Keio J. Med.*, 25, 83—90, 1976.
106. **Tulley, R. T. and Lehman, H. P.,** Method for the simultaneous determination of cadmium and zinc in whole blood by atomic absorption spectrophotometry and measurement in normotensive and hypertensive humans, *Clin. Chim. Acta*, 122, 189—202, 1982.

107. **Ulander, A. and Axelson, O.,** Measurement of blood-cadmium levels, *Lancet,* 1, 682—683, 1974.
108. **Vahter, M., Ed.,** Assessment of human exposure to lead and cadmium through biological monitoring, Report prepared for United Nations Environment Programme and World Health Organization by National Swedish Institute of Environmental Medicine and Department of Environmental Hygiene, Karolinska Institute, Stockholm, 1982.
109. **Vuori, E., Huunan-Seppälä, A., Kilpiö, J. O., and Salmela, S. S.,** Biologically active metals in human tissues. II. The effect of age on the concentration of cadmium in aorta, heart, kidney, liver, lung, pancreas and skeletal muscle, *Scand. J. Work Environ. Health,* 5, 16—22, 1979.
110. **Ward, R. J., Fisher, M., and Tellez-Yudilevich, M.,** Significance of blood cadmium concentrations in patients with renal disorders or essential hypertension and the normal population, *Ann. Clin. Biochem.,* 15, 197—200, 1978.
111. **Watanabe, H.,** A study of health effect indices in populations in cadmium-polluted areas (in Japanese), presented at Meeting on Research on Cadmium Poisoning, Tokyo, March 25, 1973.
112. **Watanabe, T., Koizumi, A., Fujita, H., Kumai, M., and Ikeda, M.,** Cadmium levels in the blood of inhabitants in nonpolluted areas in Japan with special references to aging and smoking, *Environ. Res.,* 31, 472—483, 1983.
113. WHO/UNEP, Pilot project on assessment of human exposure to pollutants through biological monitoring, report on a planning meeting on quality control, World Health Organization, Geneva, 1979.
114. **Wysowski, D. K., Landrigan, P. J., Ferguson, S. W., Fontaine, R. E., Tsongas, T. A., and Porter, B.,** Cadmium exposure in a community near a smelter, *Am. J. Epidemiol.,* 107, 27—35, 1978.
115. **Zielhuis, R. L., Struik, E. J., Herber, R. F. M., Sallé, H. J. A., Verberk, M. M., Posma, F. D., and Jager, J. H.,** Smoking habits and levels of lead and cadmium in blood in urban women, *Int. Arch. Occup. Environ. Health,* 39, 53—58, 1977.

Chapter 6

KINETICS AND METABOLISM

Gunnar F. Nordberg, Tord Kjellström, and Monica Nordberg

TABLE OF CONTENTS

I. INTRODUCTION

The kinetics and metabolism of cadmium as dealt with in this chapter include uptake, absorption, transport, distribution within the body, excretion, and accumulation, as well as binding to proteins in tissues and biological fluids. The data have been derived from experimental studies on animals and human beings, and from studies of populations exposed to different levels of cadmium. Studies on autopsy materials are included when the data can be interpreted in terms of kinetics and metabolism of cadmium. No attempt is made to refer to all publications containing data of relevance to this chapter because a large number of reports provide only confirmatory data on earlier findings.

Data on tissue concentrations of populations exposed to "normal" levels of cadmium in the general environment or to cigarette smoking have been reviewed in Chapter 5, which will be referred to when discussing the descriptive data for human beings. Data from Chapter 4 will be used when discussing the occurrence and significance of metallothionein-bound cadmium in various tissues.

As was pointed out in Chapter 2, in many studies the accuracy of the analytical data on cadmium has not been verified. Experimental research may involve the measurement of cadmium at higher concentrations than does epidemiological research, but the need for quality control in analysis is still inherent. Much of the experimental data referred to in this chapter were obtained by radioactive isotope techniques. In general, these techniques pose fewer problems in achieving accurate and sensitive analysis than do other chemical techniques. Two particular isotopes have mainly been used in the studies: ^{109}Cd and ^{115m}Cd. In some early studies, the isotope ^{115}Cd was referred to and we have retained this designation, even though it is likely that the isotope used was ^{115m}Cd.

The terminology and definitions of terms used in this chapter have been adopted by international scientific meetings on metal toxicology.[259,261] The results of a great deal of research on cadmium metabolism have appeared since the publication of our earlier general review.[97] More recent reviews have dealt with different specific aspects of cadmium metabolism.[85,98,102,172,185,200]

II. UPTAKE AND ITS DETERMINANTS

There is abundant evidence from human data that cadmium is found in most internal organs in concentrations that increase with level of cadmium intake and age (see Chapter 5). The main absorption routes are the respiratory and gastrointestinal tracts, however, limited skin penetration can take place when soluble cadmium compounds are applied in solution onto the skin.[127,235] In guinea pigs, 1.8% of cadmium applied to the skin in this way was absorbed within 5 hr.[235] In rabbits painted repeatedly with a $CdCl_2$ solution, 0.4 to 0.6% was absorbed within 3 weeks and in mice the absorption in 1 week was 0.2 to 0.8%.[127] This type of skin exposure situation is uncommon in humans, therefore in this section we will discuss the other absorption routes only.

A. Via the Respiratory Tract

Human cadmium exposure via inhalation (Chapter 3) will be in the form of an aerosol containing very small particles (cadmium fume) or larger particles (cadmium dust). Many animal experiments on cadmium use these types of exposure. The inhaled particles may be deposited in the respiratory tract or exhaled.[258]

The size, shape, and density of the particles, as they occur in the respiratory tract, determine in which part of the respiratory tract the particles will be deposited. Together these characteristics of an airborne particle are often termed "equivalent aerodynamic diameter" or "mass median aerodynamic diameter".[27]

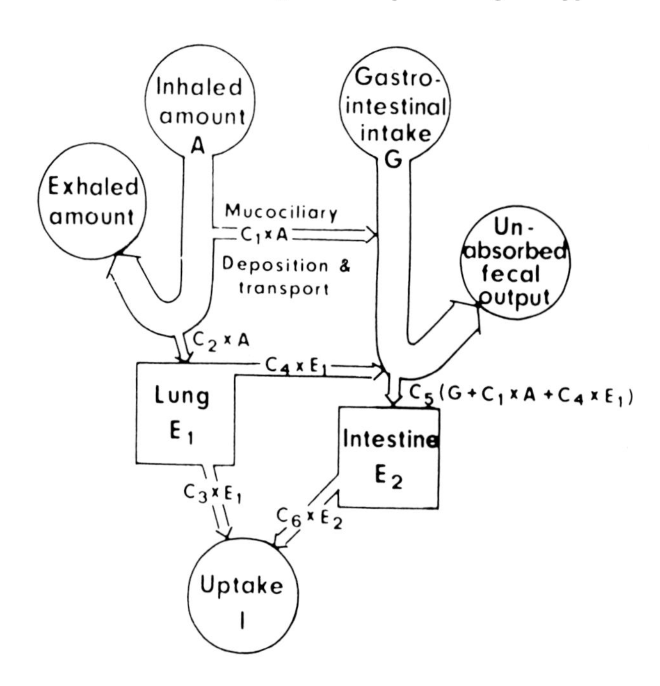

FIGURE 1. Model for the total (systemic) absorption (uptake I) resulting from inhalation or gastrointestinal intake of a metal compound. The proportion (C_1) of the inhaled amount (A) that will be deposited on the mucociliary escalator and transferred to the GI tract varies with particle size (see Table 1). This is also the case regarding the proportion (C_2) deposited in the peripheral parts of the lung. The amount retained in these parts (E_1) will be dependent largely on in vivo solubility and the extent of phagocytosis, processes which govern the coefficients C_3 and C_4. The proportion (C_5) of the amount of metal present in the intestinal lumen that will be taken up in the intestinal mucosa (E_2) depends on a number of factors such as solubility and particle size. Retention in the mucosa is dependent on binding characteristics for the metal compound in intestinal cells, and C_6 therefore may vary depending on, e.g., the amount of metallothionein present in the mucosal cell. (From Nordberg, G. F. and Kjellström, T., *Environ. Health Perspect.*, 28, 211—217, 1979. With permission.)

Absorption of highly soluble cadmium compounds may occur through the epithelium of the airways, but for most compounds and most particle sizes, the major absorption to the blood and the lymph takes place from the alveoli. Thus, alveolar deposition is of considerable importance for respiratory absorption.

Particles deposited in the nasopharyngeal tract are eliminated from the respiratory tract by "blowing the nose" or swallowing. Those deposited in the tracheobronchial tract are transported by mucociliary clearance up to the pharynx and are then spat out or swallowed.

Clearance of particles deposited in the alveoli also occurs. Macrophages may phagocytize the particles[27] and to some degree transport them to the ciliated parts of the small bronchi, from which the macrophages and particles can be transported by mucociliary clearance. Some of the macrophages are eliminated to the lymphatic system. Dissolution of the particles and transport to the blood stream is a third method of alveolar clearance. The alveolar clearance to lymph and blood constitutes the absorbed part of the cadmium deposited in the alveoli. These pathways for the absorption of cadmium via the respiratory tract and the interrelationship to the absorption via the gastrointestinal tract are depicted in Figure 1.

It should be pointed out that the cadmium deposited in the upper and lower respiratory tract is possibly of importance for the development of some local effects such as obstructive lung disease, anosmia, and lung cancer (Volume II, Chapters 8, 11, and 12).

1. Respiratory Deposition in Animals

Only limited data on respiratory deposition of specific cadmium compounds are available

Table 1
ABSORPTION AFTER INHALATION OF AN AEROSOL OF CADMIUM COMPOUND: CALCULATION OF RESPIRATORY (r) AND TOTAL ABSORPTION (t) INTO THE BODY AS A FUNCTION OF TWO DIFFERENT RATES OF ALVEOLAR ABSORPTION AND DIFFERENT PARTICLE SIZES FOR A SPECIFIC DEPOSITION AND CLEARANCE MODEL[259]

Particle size (MMAD) (μm)	Alveolar deposition (%)	Tracheobronchial-nasopharyngeal deposition (%)	Absorption (%) into body when alveolar absorption is 100% r	t	50% r	t
0.1	50	9	50	50.4	25	26.7
0.5	30	16	30	30.8	15	16.6
2.0	20	43	20	22.2	10	12.6
5.0	10	68	10	13.4	5	8.6
10.0	5	83	5	9.2	2.5	6.8

Note: Gastrointestinal absorption is assumed to be 5%. MMAD, mass median aerodynamic diameter.

for animals or humans. In some animal experiments, cadmium chloride aerosols are used for the exposure situation. It is possible that these mists do not accurately represent the conditions of humans exposed to dust or fume.

The experiments using solid particle aerosols show that cadmium is deposited in the lung, is cleared slowly, and is absorbed to a certain degree. This implies that the mechanisms described in the preceding section do exist.

In an early study by Barrett et al.,[13] several animal species were exposed to cadmium oxide fume for 10 to 30 min. The total doses varied up to above 15,000 min × mg CdO/m^3 as indicated by the LD_{50} after 7 to 28 days (Volume II, Chapter 8, Section I.A). The percentage retention of the inhaled dose varied between 5 and 20%, as measured at autopsy.

Boisset et al.[20] gave rats five consecutive 30-min exposures to a cadmium oxide aerosol at 1-day intervals. The dose (280 min × mg/m^3), was sufficient to cause lung damage (Volume II, Chapter 8). It was reported that 12% of the inhaled dose was deposited in the lungs, but it was not shown how this value was arrived at.

In other studies[19,93,105,204,205] animals were exposed to cadmium oxide aerosols and cadmium was found in the lungs. However, quantitative calculations of deposition cannot be made. Some studies have revealed deposition of cadmium after exposure to cadmium chloride aerosol (liquid),[25,110,204,205] but only one of these gives quantitative estimates of deposition. Syrian hamsters were exposed to ^{115m}Cd-labeled $CdCl_2$ aerosol with a mass median aerodynamic diameter (MMAD) of 1.7 μm.[110] The concentrations were 1.8 and 16.5 mg/m^3 and on average 19% +/− 1.65% (SEM) of the dose was deposited in the lungs after 2 hr. How the solubility of the liquid compound affects deposition values is not known.

In view of the lack of human data, estimates based on the lung deposition and retention model[258] may provide the best approximation. For humans breathing at a moderate work rate (20 ℓ/min), the deposition in the alveolar compartment is estimated to vary from about 5% for particles with a MMAD of 10 μm to about 50% for particles with a MMAD of 0.1 μm (Table 1).

2. Respiratory Clearance in Animals

As mentioned above, the lung clearance of cadmium involves both the mucociliary clearance of the bronchioli, bronchi, and trachea, as well as alveolar clearance (which includes the absorbed cadmium). Most studies on clearance have only measured total lung clearance, which appears to be biphasic. There is a slow component with a half-time of about 2 months and a fast component with a half-time of 1 day. About half of the deposited cadmium is cleared in the fast phase. This may represent the tracheobronchial clearance as the particles deposited in the most peripheral of the ciliated airways are usually eliminated from the lung within 24 hr.[4,26] The first study of lung clearance[109] involved exposing dogs for 30 min to a cadmium aerosol obtained by nebulizing a 25% $CdCl_2$ solution. The cadmium air concentrations varied between 280 and 360 mg/m^3. The doses were so high that many of the dogs died, which may affect the validity of the clearance data.

Lung clearance was rapid during the first 2 weeks[109] with a half-time of about 5 days. The long-term clearance appears to be very slow as no further decrease was evident 10 weeks after exposure. At this time the cadmium concentration in the lung had decreased to about one third of the initial concentration. The slower long-term elimination in this study compared to more recent studies is probably related to the existence of severe pathological changes in the lung caused by the very high exposure (Volume II, Chapter 8).

Henderson et al.[110] exposed Syrian hamsters to a radioactive cadmium chloride aerosol (Section II.A.1). About one third of the cadmium deposited in the lung had been cleared within 24 hr, and 3 weeks later about 40% remained in the lung. This indicates that the half-time of the slow component is longer than 3 weeks.

Oberdörster et al.[205] studied lung deposition and clearance of cadmium in rats (Figure 2). Rats were exposed via inhalation to a $^{115m}CdCl_2$ aerosol with 660 μg Cd per cubic meter for 1 hr or via intratracheal instillation to 8 μg $^{115m}CdCl_2$ in 0.3 $m\ell$ NaCl. There was a biexponential clearance pattern after both types of exposure (Figure 2). The slow component half-times were 61 days (inhalation) and 66 days (instillation). The fast component half-times were 1.1 days (inhalation) and 0.7 days (instillation). About 50% of the deposited cadmium was cleared in the fast component (Figure 2). As mentioned earlier, the fast component has half-times similar to those expected for the tracheobronchial clearance[3] and it is likely that the slow component represents alveolar clearance. In a preceding similar study, rats were exposed to aerosols of CdO and $CdCl_2$ and the slow component (alveolar) clearance half-time was 67 days for both compounds.[204]

Boisset et al.[20] exposed young rats to five consecutive exposures of a cadmium oxide aerosol. The total dose was 280 min $\times$ mg/m^3. Some degree of permanent lung damage was observed and the half-time in the lung was estimated to be 56 days. Another study of rats exposed to a 0.5% $CdCl_2$ aerosol[25] revealed pathological changes in the lungs. About half of the deposited cadmium in the lungs was cleared within 27 days.

One study[18] indicates that the slow component clearance may be particularly slow for cadmium pigments with poor water solubility. Rats exposed by inhalation to aerosols of cadmium pigments (''cadmium red'' and ''cadmium yellow'') retained 100% of the amount deposited in the lung even after 30 days. On the other hand, by that time about 40% of the deposited amount of cadmium carbonate or cadmium fume had been absorbed and distributed to other tissues. The slow component clearance half-times for cadmium carbonate and fume were similar to those reported for cadmium chloride and oxide mentioned above.

It should be pointed out that both types of clearance do contribute to the uptake of cadmium. The tracheobronchial clearance will transport cadmium to the gastrointestinal tract and a day or two after inhalation, a proportion of the cadmium will be absorbed from the intestines. The alveolar clearance incorporates pulmonary absorption, but this is influenced by the solubility of the cadmium compound. In addition, binding of cadmium to proteins (e.g., metallothionein) in the lung cells may influence the clearance, absorption, and toxicity of cadmium to the lungs[51,116,205,217,278] (Volume I, Chapter 4 and Volume II, Chapter 8).

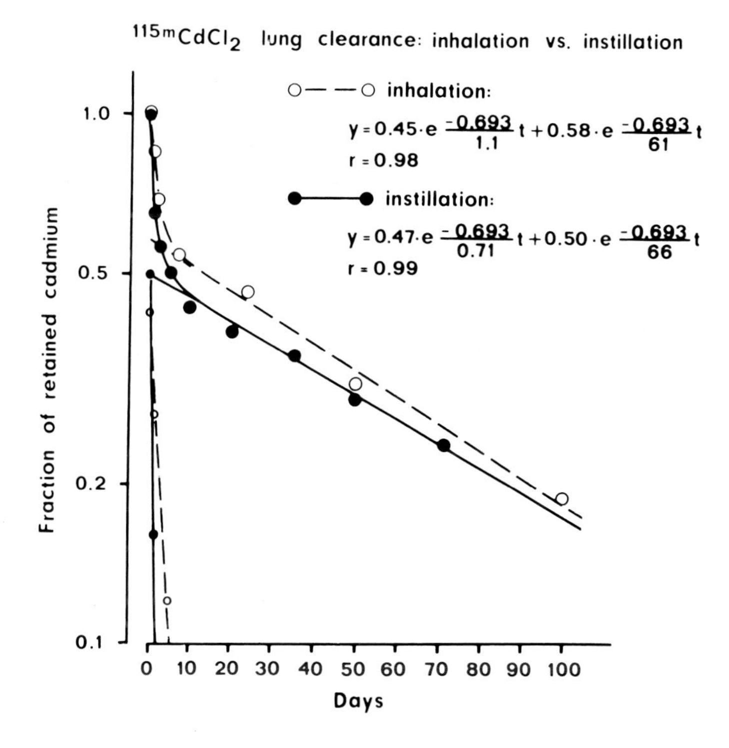

FIGURE 2. Lung clearance of $CdCl_2$ (as ^{115m}Cd) in rats after 1-hr inhalation (660 μg Cd/m^3, – – – – – –) and after intratracheal instillation (8 μg Cd in 0.3 mℓ, ——). Each data point represents the mean of five rats. Lung cadmium content on day 0 was set at 1. After extrapolation of the slow clearance curve to time 0, the fast phase could be calculated by subtracting the extrapolated curve from original values (days 0 to 5). The two straight lines in the lower left hand part of the figure are the resulting initial fast clearance curves. The intercepts of the slow phase with the y–axis indicate that more cadmium was deposited in the peripheral lung by inhalation than by instillation. (From Oberdörster, G., Oldiges, H., and Zimmerman, B., *Zentralbl. Bakteriol. (B)*, 170, 35—43, 1980. With permission.)

3. Respiratory Absorption in Animals

Several of the studies referred to in the previous two sections report data on cadmium in tissues other than the lung, which indicates that cadmium is absorbed after inhalation. Friberg et al.[97] calculated the total inhaled amount of cadmium in the study of dogs by Harrison et al.[109] They also calculated the total retained amount in kidneys, liver, and other internal organs and estimated that the overall retention was 40% of the inhaled amount a few days after exposure. The respiratory absorption must have been higher. A similar calculation[97] based on the study of rabbits inhaling cadmium-iron dust,[93] suggests an overall retention in the body of 30% of the inhaled dose after some months of chronic exposure.

As only a proportion of the inhaled dose is deposited in the lung and only 50% of the deposited amount is available for alveolar absorption (Section II.A.2), the absorption of the cadmium in the alveoli may be very high, even though it takes considerable time (Section II.A.2).

The studies of Syrian hamsters exposed to a $CdCl_2$ aerosol[110] showed about 25 to 35% of the initial lung burden in liver, kidneys, and skull 3 weeks after exposure. At this time the lung still contained about 50% of the initial lung burden.

One day after rats were exposed to a single inhalation of CdO,[204] 5% of the initial lung burden was found in the liver and kidneys, whereas after 100 days these tissues contained

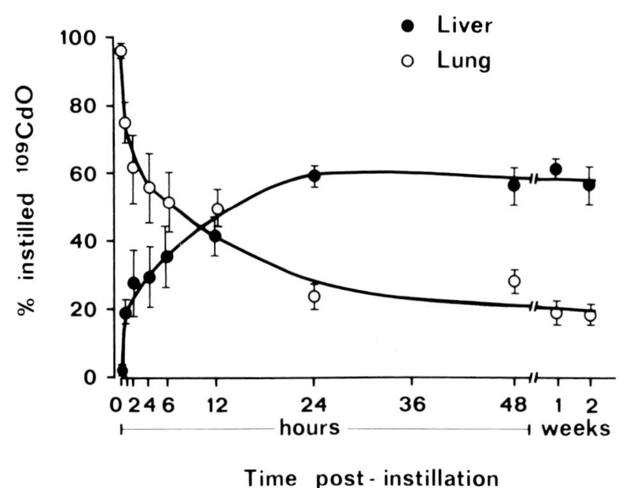

FIGURE 3. Lung and liver burdens of ^{109}CdO as a function of time after intratracheal instillation. Values are means ± SD of three to six samples. (From Hadley, J. G., Conklin, A. S., and Sanders, C. L., *Toxicol. Appl. Pharmacol.*, 54, 156—160, 1980. With permission.)

9%. After inhalation exposure to $CdCl_2$, the corresponding figures were 37 and 49%. The absorption of the soluble cadmium compound is faster and greater than for CdO. In rats exposed to $CdCl_2$ via intratracheal instillation,[205] 60% of cadmium instilled was found in kidneys and liver 2 weeks after instillation (Section II.A.2). Thus, the absorption appears to be even greater after instillation.

Already 1 day after exposure a considerable proportion (5 to 40%) of the initial lung burden is found in the liver, kidneys, and skull,[110,204,205] which indicates that both after exposure to $CdCl_2$ and CdO a significant absorption of cadmium must occur during the first fast component clearance of cadmium from the lung. Some of this absorption may be the result of rapid absorption through the epithelium of the airways, including the alveoli, and a certain amount may be due to absorption in the gastrointestinal tract of cleared cadmium.

After single inhalation exposure[110] or single instillation exposure[204,205] liver levels are maintained at high levels for weeks. After single injection exposure[15,104,188] the liver level decreased rapidly during the first 16 to 100 days. This difference in liver levels between the respiratory and injection exposure routes indicates that the absorption from the lung continues during the slow component phase of lung clearance.

Hadley et al.[105] exposed rats to an intratracheal instillation of 15 μg ^{109}CdO slurry (particle size 1.0 μm) in physiological saline. The lung levels decreased rapidly while the liver levels increased (Figure 3). After two weeks about 60% of the instilled dose was in the liver and 8% was in the kidneys, so the absorption was at least about 70%. In this study only small amounts of cadmium (less than 5% of instilled dose) were found in the feces during the first days. The gastrointestinal route of absorption could, therefore, not have contributed much of the cadmium absorbed to the liver during the first 24 hr (Figure 3).

In the study of rats exposed to a CdO aerosol[20] (Section II.A.1), an accumulation of cadmium in liver and kidneys was seen during 6 to 8 weeks after exposure. It was estimated[20] that about 60% of the cadmium deposited in the lungs (which was 12% of the inhaled dose) was absorbed. The overall absorption in this study was thus 7.2% of inhaled dose.

Clearly, the alveolar absorption and the overall absorption vary, depending on the type of exposure, being higher after intratracheal instillation than after inhalation of an aerosol. The absorption of a $CdCl_2$ aerosol is higher than a CdO aerosol. Considering that about 50% of the deposited cadmium aerosol may be deposited in the tracheobronchial section of the respiratory tract (Section II.A.2), it appears that for different cadmium compounds the alveolar absorption may vary between 50 and 100% of the amount deposited in the alveoli. The absorption is a slow process that continues for many weeks after a single inhalation exposure.

4. In Humans

No empirical data on respiratory cadmium absorption in human beings are available. Based on data on the increased body burden of cadmium among smokers[149] and estimates of the total inhaled amount of cadmium from the cigarette smoke, Friberg et al.[97] calculated the long-term retention in the body to be 27 to 54%. The actual absorption would be higher than the retention. The MMAD of cadmium aerosol in cigarette smoke would be 0.1 μm or less.

Elinder et al.,[61] also used data from autopsies to estimate the respiratory absorption of cadmium inhaled from cigarettes. It was found that an average middle-aged Swedish smoker, who had smoked one pack of cigarettes per day over a period of 25 years, had a cadmium concentration in kidney cortex which on average was 15 mg/kg higher than a nonsmoker of the same age. Based on the proportion of the total cadmium in the body that is accumulated in kidneys, the biological half-time of cadmium in the body, and the total amount of cadmium assumed to be inhaled from 2 to 5 years of smoking, the respiratory absorption was estimated to be about 45%.

Other data of relevance are the concentrations of cadmium in blood of 473 adult Swedish smokers and nonsmokers.[64] There was a very strong association between smoking habits and average blood cadmium level. The nonsmokers had on average 0.2 μg Cd per liter blood and those who smoked 20 cigarettes per day had on average about 2 μg Cd per liter blood. Based on the same cadmium intake data as those used by Elinder et al.,[61] such a great difference between smokers and nonsmokers would indicate an almost complete absorption of all the cadmium inhaled from cigarettes. It is possible that, in addition, the cadmium absorbed via the respiratory route is retained to a higher degree in the blood than the cadmium absorbed via the gastrointestinal route.[64] The latter passes through the liver on its way to the main blood stream (Section II.B). Thompson and Klaassen[263] have shown that in rats with intravenous exposure to cadmium, the biliary excretion is higher if the exposure is via the portal vein (Section V.B.3). Although these data would support a different intertissue distribution for the two exposure routes, more data are needed in order to quantitatively take account of such a possible difference. Data from humans as well as from animals were used by Kjellström and Nordberg,[132] and Nordberg and Kjellström[187] in their model of cadmium kinetics (Chapter 7). A possible difference in Cd distribution depending on exposure route was not taken into account in this quantitative model which used 10% tracheobronchial deposition and 40% alveolar deposition for the type of fine particles that occur in cigarette smoke. The corresponding values for factory dust (containing larger particles) was 70% deposited tracheobronchially and 13% deposited alveolarly. The absorption fraction from the alveolar compartment was taken to be 90% (Chapter 7).

To a great extent, the overall absorption depends on particle size of the aerosol. Estimates of respiratory absorption of an aerosol of a metal compound were made by the Task Group on Metal Accumulation.[259] For cadmium, the validity of the estimates generally has been supported by the animal and human data reviewed above.

Table 1 shows estimates[259] of the respiratory and total absorption of cadmium (expressed as a percentage of inhaled amount) after inhalation of an aerosol of a compound with relatively low solubility, e.g., CdO. It is assumed that ventilation is moderate and that the cadmium aerosol is deposited and cleared in the respiratory tract as are particles in general.[258] It is also assumed that since the cadmium compound is of relatively low solubility, the particles deposited on the ciliated epithelium will be entirely transferred to the gastrointestinal tract.

It can be seen from Table 1 that under these assumptions, respiratory absorption will vary between 2.5 to 50%, depending on particle size and alveolar absorption. The corresponding total absorption of inhaled cadmium would be 6.8 to 50.4% (Table 1) the assumption of a 5% gastrointestinal absorption (Section II.B.6).

5. Conclusions

In animal experiments, the reported lung deposition (excluding the upper respiratory tract) of cadmium compound aerosols varies in the range 5 to 20%, and estimates based on a lung physiology model for humans give values for alveolar deposition in the range 5 to 50%. The deposition in the upper respiratory tract was estimated at 9 to 83% in the model (particle size 0.1 to 10 μm). According to animal experiments with CdO or $CdCl_2$ aerosols, about half of the deposited cadmium is cleared in a fast component with a half-time of about 1 day and the rest is cleared in a slow component with a half-time of about 30 to 60 days. The slow component may represent the alveolar clearance and a part of the fast component may represent the tracheobronchial clearance.

The theoretical model of lung physiology predicted a variation of respiratory absorption depending on particle size and solubility between 2.5 and 50% of the inhaled cadmium amount. This would correspond to a total absorption of about 7 to 50% of inhaled cadmium. In general, the limited empirical human and animal data which are available support these predictions. It also appears that the absorption of an aerosol of soluble cadmium compounds is greater than the absorption of a cadmium oxide aerosol.

B. Via the Gastrointestinal Tract

Mechanisms similar to those described for respiratory absorption could be applied to gastrointestinal absorption. A large proportion of the ingested cadmium passes the gastrointestinal tract without being taken up into the mucosal cells. Because the uptake proportion is so low, fecal cadmium has been used in many studies as a measure of human cadmium intake (Chapter 3).

Feces also contain cadmium excreted via bile (Section V.B.3), and some of the cadmium in bile is likely to be reabsorbed. The different pathways for gastrointestinal absorption and excretion of cadmium are depicted in Figure 4.

The cadmium incorportated into the mucosal cells can be retained in these cells or "cleared" to blood, lymph, or back to the lumen of the gastrointestinal tract. The kinetics of these relationships has not been studied in great detail. Most studies of gastrointestinal absorption have merely compared the dose given with the amount retained in the body shortly after dosing. This approach gives only an estimate of the overall absorption.

1. Retention Studies of Animals

A number of studies indicate that the retention, or overall absorption, of cadmium nitrate, chloride, or sulfate given orally is about 1 to 2%. The absorption of cadmium stearate is about half of the absorption of sulfate. The absorption in monkeys may be higher than in smaller animals.

The studies of rats and mice exposed to a single oral dose of radioactive cadmium chloride or nitrate have all shown a long-term systemic absorption of about 1 to 3% (Table 2). These studies will not be described in detail here.

It is of interest to note the difference in absorption after exposure via stomach tube, inhalation, and intravenous injection (Figure 5). The gastrointestinal absorption was about 2% and the respiratory absorption about 40% as assessed from the whole body retention.

Differences in retention after oral exposure to cadmium sulfate and cadmium stearate were found by Schmidt and Gohlke.[228] They gave rats 15 mg Cd per kilogram body weight of the compounds by stomach tube twice a week for 7 weeks. Cadmium levels in liver after

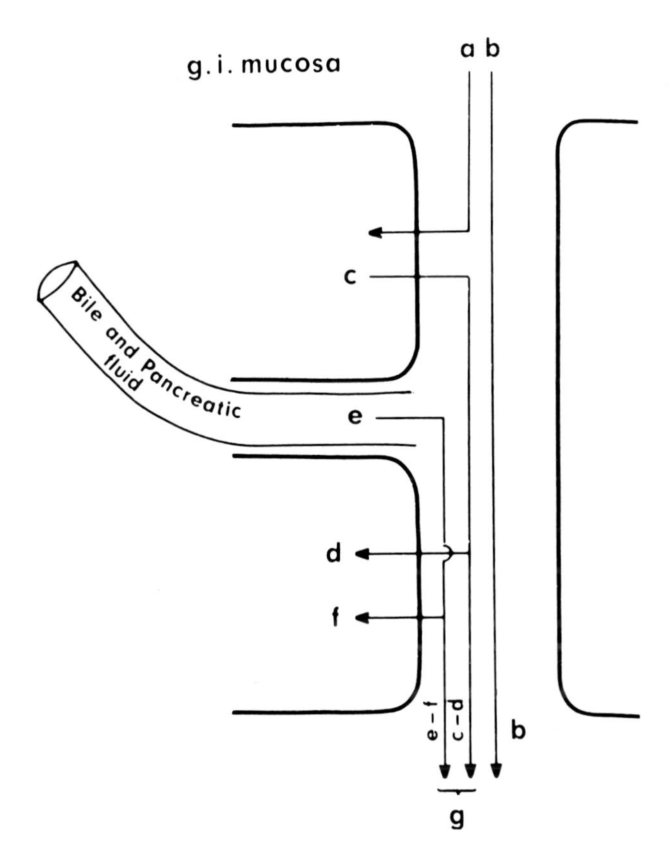

FIGURE 4. Routes for ingested cadmium. (a) Cadmium ingested and absorbed. (b) Cadmium ingested but passing unabsorbed through gastrointestinal tract. (c) Excretion of cadmium through intestinal mucosa. (d) Reabsorption of cadmium excreted through intestinal mucosa. (e) Excretion of cadmium through bile and pancreatic fluid. (f) Reabsorption of cadmium from bile and pancreatic fluid (enterohepatic circulation). (g) (e − f) + (c − d) = net gastrointestinal excretion. (g + b) Fecal content of cadmium. (Modified from Camner, P., Clarkson, T. W., and Nordberg, G. F., *Handbook on the Toxicology of Metals*, Friberg, L., Nordberg, G. F., and Vouk, V., Eds., Elsevier, Amsterdam, 1979, 65—97.

3, 7, and 16 weeks in rats that had received cadmium stearate were only 44, 32, and 51%, respectively, of the levels found in animals given cadmium as sulfate. It appears that the absorption of stearate may be half that of sulfate. The reason for this may be that cadmium stearate is not ionized in the gut to the same extent as is cadmium sulfate. On the other hand, compounds which can be expected to be similarly ionized would be absorbed to the same extent. Moore et al.[165] found similar organ retention values in rats after single oral exposures to ^{115m}Cd as acetate, chloride, or sulfate.

Nordberg et al.[191] studied the retention of cadmium in monkeys (*Saimiri sciureus*) by means of whole body measurements after ingestion of $^{115}CdCl_2$ through stomach tube. Two monkeys were given 1.7 mg Cd per kilogram body weight and two other monkeys, 0.17 mg Cd per kilogram body weight. The monkeys had retained 2.5 to 3.2% (average 2.9%) of the ingested amount 10 days after ingestion. In four additional monkeys exposed to a single dose of only 1 μg/kg body weight,[184] the average whole body retention after 10 days was 1% (range 0.5 to 1.5%) of the dose administered.

Suzuki et al.[250] gave ^{109}Cd by stomach tube as a single exposure to two Macaca irus

Table 2
GASTROINTESTINAL ABSORPTION OF SOLUBLE CADMIUM COMPOUNDS IN STUDIES OF SMALL ANIMALS

Type of animal	Exposure type	Compound and dose	Duration of experiment	Absorption (%)	Ref.
Rat	Stomach tube	$^{115}Cd(NO_3)_2$ 6.6 mg/kg body weight	24 hr	1—2	55
	Stomach tube	$^{115}CdCl_2$	4 days	3 (WBR)	166
			10 days	2.3 (WBR)	
Mouse	Stomach tube	$^{109}CdCl_2$	4 hr	0.5—8	49
	Stomach tube	^{109}Cd	2 days	0.5—3	220
	Stomach tube	$^{109}CdCl_2$	11 hr	4.5—12	248
		0.08 mg	164 hr	1—2.3	
	Oral, single	$^{109}CdCl_2$	15 days	1.6 (WBR)	180
	Oral, single	$^{115m}CdCl_2$	24 hr	7.3 (WBR)	206
			48 hr	2.7 (WBR)	
	Stomach tube	$^{109}CdCl_2$	4 hr	16 (including gut)	272
		24 μg Cd	3—10 days	2 (WBR)	

Note: WBR, Whole body retention.

monkeys, which weighed 3.60 and 3.13 kg. The night before the experiments, the monkeys were given no liquids. In the morning, 105 μCi carrier-free $^{109}CdCl_2$ was given to monkey A after which he was allowed to drink and eat freely. Monkey B received 57 μCi $^{109}CdCl_2$ in the same fashion, but in this case mixed with nonradioactive cadmium chloride, such that 1 mg of cadmium was given in 10 mℓ solution (concentration 100 mg Cd per liter). Monkey A was killed after 19 days, and monkey B after 25 days, and the cadmium content in the various organs was measured by scintillation counting. During the experiment, feces and urine were collected each day and cadmium content measured.

After 19 days, 12% of the dose was retained and about half of this retention still remained in the gastrointestinal tract mucosa. The cadmium concentration in the epithelium of the small intestine was about 33 times higher than the average concentration in the whole body and in the duodenal epithelium it was more than 50 times higher. The only other organ in which such high cadmium concentrations were found was the kidney. The cadmium concentration in the kidney was about 20 times higher than the average concentration in whole body.[250] These data have been published in English by Suzuki and Taguchi.[247] A rapid accumulation of cadmium in duodenal mucosa of quail exposed to 0.08 mg Cd per kilogram diet was observed by Fox.[88] Cadmium accumulated in this tissue at levels several times greater than the dietary cadmium concentration.

Suzuki et al.[249] had reported earlier a similar experiment on another Macaca irus monkey. In this study, the retention rate in whole body except for the gastrointestinal tract was 0.65%. The new data by Suzuki et al.[250] and Suzuki and Taguchi[247] indicate that the uptake of cadmium from the gastrointestinal tract is a slow process, taking several days or even weeks, and that the eventual retention is about 6%.

The gastrointestinal absorption of $^{109}CdCl_2$ (measured as whole body retention in percent of administered dose 5 days after exposure) in mice exposed by gavage to different cadmium exposure levels[71] was 0.5% for the lowest dose (1 μg Cd/kg body weight), 1.4% for intermediate doses (15 to 750 μg/kg), and 3.2% for the highest dose (37.5 mg/kg). Dose-dependent differences in organ distribution were also seen. It should be pointed out that in other studies at dose levels similar to the highest dose levels given in this study, morphological changes in the gastrointestinal mucosa have been seen (Volume II, Chapter 11, Section I.B).

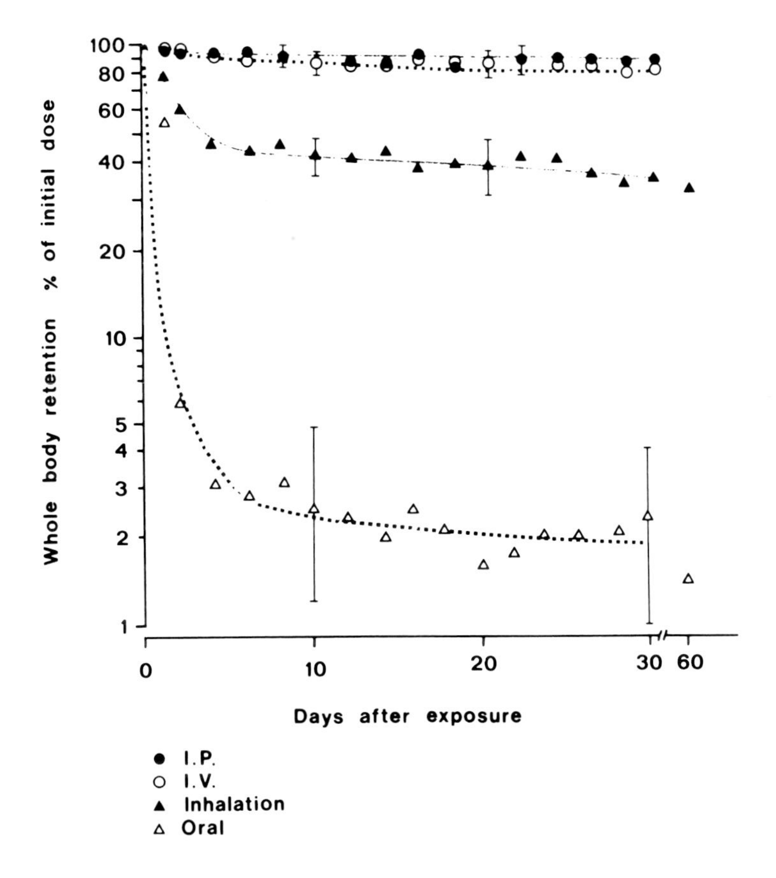

FIGURE 5. Whole body retention of ^{115}Cd following a single exposure to rats by different routes of administration (From Moore, W., Jr., Stara, J. F., and Crocker, W. C., *Environ. Res.*, 6, 159—164, 1973a. With permission.)

In conclusion, the gastrointestinal absorption is likely to be in the range 1 to 10% with higher values for monkeys and large animals than for rodents. The dietary composition and relative dose may, however, also influence these figures.

2. Influence of Age and Pretreatment on Gastrointestinal Absorption of Cadmium in Animals

Matsusaka et al.[159] studied whole body retention of $^{115m}CdCl_2$ after single and repeated peroral infusions, via gastric tube, to adult mice and young mice of the age 7, 14, and 21 days. Cadmium retention after single exposure (Figure 6) is higher in the younger mice. As shown in the same figure, the retention of $CdCl_2$ injected intravenously is almost 100%. Because the retention patttern of the 7-day-old mice appeared to differ from the others, Matsusaka et al. repeated the experiment on 20 such mice and killed them 1, 5, 10, and 20 days after single exposure. The cadmium content of the intestines was deducted from whole body and it could be shown that the actual gastrointestinal absorption in these mice was 8.1 to 9.2%. Repeated exposure of adult mice and 7-day-old mice over a period of 10 days showed that the excretion of infused cadmium was much slower from day 7 to day 15 after birth. A variation of the gastrointestinal handling of cadmium in mice at different ages may be the cause of the different retention patterns.

Kello and Kostial[123] demonstrated that neonatal animals absorbed cadmium to a much greater extent than adult animals. One important factor for this difference in absorption was considered to be the milk diet consumed by neonatal animals.[124,226] Engström and Nordberg[69] found a higher gastrointestinal absorption in adult mice given a milk diet compared to animals

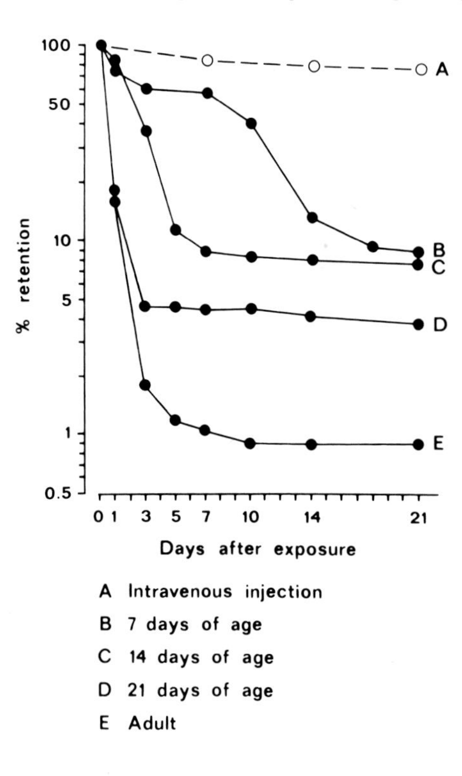

FIGURE 6. Whole body retention after a single peroral exposure to ^{115m}Cd in mice of different ages. Five animals in each group. (From Matsusaka, N., Tanaka, M., Nishimura, Y., Yugama, A., and Kobayashi, H., *Med. Biol.*, 85, 275—279, 1972. With permission.)

on a pellet diet. The chemical composition of the pellets or the milk may influence the absorption rate of cadmium (Section II.B.3).

In another study,[72] 1-month-old mice absorbed 5.2% , 3-month-old mice 2.9%, and 6-month-old mice 2.1% of a single dose (4 to 5 days after dosing) (Table 3). All animals were maintained on pelleted food in this study. In other experiments described in the same report,[72] the absorption was much higher for 3-month-old mice than for 8-month-old mice given the same dose (Table 3). In view of these observations concerning the influence of age and dosage, it is conceivable that some of the differences in the studies referred to in the previous section can be explained by these factors.

In another experiment,[72] a group of 10 male adult mice were given 75 mg Cd per liter in drinking water for 130 days. Another group was given the same treatment for 13 days and a third group was given no pretreatment. All groups were then given a single oral dose of $^{109}CdCl_2$ (15 μg Cd). The whole body retention 10 days after the single dose was about 1% in the pretreated groups and 0.4% in the nonpretreated group (statistically significant) (Table 3). Most of the added retention in the pretreated groups was due to an increased uptake of cadmium to the liver.[72]

3. Influence of Dietary Factors on Gastrointestinal Absorption in Animals

A number of studies of trace element interactions have shown that a low intake of calcium, iron, zinc, and copper increased the gastrointestinal absorption of several metals including cadmium.[12,80,89,197,260] On the other hand, a high cadmium ingestion may reduce the gastrointestinal absorption of calcium, copper, and iron[8,14,54,92,103,106] (see also Volume II, Chapters 10 and 11). In this section, several interactions of importance to the gastrointestinal absorption of cadmium will be discussed. Interactions with other aspects of cadmium metabolism are dealt with later in this chapter (Section VII).

a. Calcium

Several studies have shown that a low intake of calcium in the diet can increase whole body retention of cadmium, most likely due to an increased gastrointestinal absorption. Correspondingly, the uptake of calcium decreases when the ingestion of cadmium is high (Volume II, Chapter 10).

Larsson and Piscator[144] gave female rats on low and high calcium diets 25 mg/ℓ of cadmium as the chloride in drinking water for 1 and 2 months. They found that the rats on the low calcium diet had accumulated about 50% more cadmium in liver and kidney than the rats on the high calcium diet. This indicates a higher absorption in the former group, as there is no reason to believe that there would be a difference in excretion.

Kobayashi et al.[138] gave rice containing 0.1 to 0.6 mg/kg of cadmium to mice on low and normal calcium intake. After 70 weeks of exposure, animals on a low calcium diet had liver and kidney concentrations of cadmium 1.5 to 5 times higher than mice on a normal calcium diet. Piscator and Larsson[216] gave rats on low and normal calcium diets cadmium in drinking water in concentrations ranging from 0 to 10 mg/ℓ for 1 year. Calcium-deficient animals retained about twice as much cadmium in liver and kidney as those on a normal calcium diet.

In a study by Washko and Cousins,[277] the absorption of ^{109}Cd increased about 100% in rats given a low calcium diet (0.1% calcium) compared to rats given a diet with normal calcium content (0.6% calcium). Similar results in rats were reported by other authors.[123,124,152]

It has been suggested that water hardness is of importance for human uptake of metals, e.g., cadmium from drinking water. However, in a study by Schroeder et al.[229] there was no such effect on cadmium accumulation in rats given cadmium in "hard" and "soft" water for more than a year. The dietary intake of calcium was adequate, so the difference in calcium intake between the two groups was very small. Elinder et al.[63] found differences in renal cadmium accumulation in horses. These differences correlated with water hardness. One may speculate that in the latter case, water hardness was related to the total calcium intake of the horses, whereas this was not the case in the study by Schroeder et al.[229]

b. Vitamin D

There are conflicting reports on the effect of vitamin D on cadmium absorption. It appears that vitamin D itself does not increase cadmium absorption if the calcium intake is adequate.

Worker and Migicovsky[283] found that the uptake of cadmium in the tibia after an oral dose of ^{115}Cd was greater in rachitic chicks being treated with vitamin D than in untreated chicks. As there was no difference between similar groups in uptake in bone when ^{115}Cd was injected, they concluded that this difference was due to the effect of vitamin D on the intestinal absorption of cadmium. The chicks were on a low calcium diet[283] which could explain the increased cadmium uptake.

The influence of cholecalciferol (vitamin D3) on the absorption of cadmium has been studied by Cousins and Feldman.[50] In one experiment, male chicks were given a vitamin D-deficient diet containing 1% calcium for 28 days. On the 27th day of life, one half of the group received oral doses of 2000 I.U. of cholecalciferol. Eighteen hours later each chick received 5 μCi of ^{109}Cd orally. After a 24-hr fast, the chicks were killed. The cadmium content of liver and kidney in chicks given vitamin D was higher than in nontreated animals, but the difference was not significant.

In another experiment, chicks were given the vitamin D-deficient diet for 23 days, and then for 4 days half of the group received 600 I.U. of cholecalciferol per kilogram of diet.

Table 3
COMPARATIVE DATA ON THE GASTROINTESTINAL ABSORPTION AND BIOLOGICAL HALF-TIME OF ^{109}Cd IN MICE OF DIFFERENT AGES (I) AND THOSE RECEIVING CADMIUM IN DRINKING WATER BEFORE (II) AND AFTER (III) EXPOSURE TO A SINGLE ORAL DOSE OF RADIOLABELED CADMIUM

Experiment number and follow-up time	Age of animal at gavage (sex)	Oral dose of radiolabeled Cd (μg)	Pretreatment (P) or treatment with Cd after dosing (A)	Absorption (%) of radiolabeled dose (day after dosing)	Biological half-time of radiolabeled Cd (days)
Experiment I, 5 months	1 month (M)	9.59 ± 1.04	—	5.2 ± 1.0 (5)	72 ± 6
	3 months (M)	20.25 ± 2.09	—	2.9 ± 1.8 (4)	79 ± 4
	6 months (M)	21.79 ± 2.67	—	2.1 ± 1.1 (4)	72 ± 8
Experiment II, 5 months	8 months (M)	12.87 ± 1.38	—	0.40 ± 0.13 (7)	78 ± 12
	8 months (M)	13.06 ± 0.73	P—13 days	0.87 ± 0.30 (7)	145 ± 34
	8 months (M)	13.19 ± 1.27	P—130 days	0.96 ± 0.30 (7)	292 ± 113
Experiment III, 18 months	3 months (M)	12.47 ± 1.29	—	1.55 ± 0.22 (14)	231 ± 25[a] (14—510)
	3 months (F)	12.00 ± 1.46	—	1.62 ± 0.25 (14)	245 ± 24[b] (14—510)
	3 months (M)	12.69 ± 2.02	A—18 months	1.84 ± 0.73 (14)	580 ± 253 (14—510)
	3 months (F)	13.49 ± 1.72	A—18 months	1.50 ± 0.36 (14)	644 ± 144 (14—510)

[a] Three phases were distinguished: I (14—120) 96 ± 12; II (120—270) 200 ± 136; III (270—510. 460 ± 100.
[b] Three phases were distinguished: I (14—120) 155 ± 30; II (120—270) 228 ± 36; III (270—520) 330 ± 100.

From Engström, B. and Nordberg, G. F., *Acta Pharmacol. Toxicol.*, 45, 315—324, 1979b. With permission.

On the fifth day, all chicks received 5 μCi of ^{109}Cd orally. After a 24-hr fast, the chicks were killed. This experiment, likewise, revealed no significant difference between the groups with regard to liver and kidney levels of cadmium. Cousins and Feldman[50] concluded that vitamin D does not directly influence the uptake of oral cadmium into the liver and kidney.

The effects of cadmium exposure on vitamin D metabolism are reviewed in Volume II, Chapter 10.

c. Protein

Protein in the diet or the protein-binding of cadmium may influence cadmium absorption. Fitzhugh and Meiller[79] mentioned briefly that the toxicity of cadmium was increased by a low protein diet. Suzuki et al.[248] gave mice low and high protein diets for 24 hr before and after an oral dose of $^{115}CdCl_2$. The low protein diet gave considerably higher levels of cadmium in kidney, liver, and whole body, irrespective of calcium content of the diet, indicating an increase in the absorption of cadmium. Whole body retention was about 9% (range 5 to 14%) and 4.5% (range 3 to 10%) in mice fed low and high protein diets, respectively.

Lagally et al.[142] reported a decreased cadmium retention in rats fed with cadmium from scallops or cadmium sulfate mixed with scallops, compared to rats fed cadmium sulfate without scallops.

Diets containing 20.9 mg Cd per kilogram wet weight and consisting of lobster digestive glands (hepatopancreas tomalley) together with a mixture of porcine liver/kidney or a casein and high protein diet were fed to rats for 90 days.[270] Animals in high protein and casein groups had a cadmium concentration in the kidney twice as high as that of animals fed the lobster and porcine mixture.[270] This indicates a lower retention of cadmium bound to animal protein.

In a study by Siewicki et al.[234] mice were fed diets with $CdCl_2$ or cadmium–containing oyster parts. In all tissues, oyster cadmium was retained at a much lower rate than was $CdCl_2$. Oyster cadmium was preferentially retained in the kidney compared to the liver. Fox et al.[90] also studied uptake of oyster cadmium from the diet. They used Japanese quail as experimental animals and did not report any marked differences in uptake in comparison to $CdCl_2$.

Revis and Osborne[219] provided data from rats fed food containing different protein levels and exposed to cadmium in drinking water. It appeared that cadmium transferred into the body may depend on the cysteine concentration in the diet. Cadmium bound to metallothionein is absorbed and distributed differently from $CdCl_2$ (Section II.B.5). In a long-term study, Cherian[40] found similar differences.

d. Zinc

There is limited specific evidence available concerning an influence of dietary zinc on cadmium absorption,[260] even though a low zinc level in the diet has been shown to increase the susceptibility of animals to cadmium-induced hypertension (Volume II, Chapter 11). Some studies have shown increased levels of cadmium in animal tissues when the animals were fed cadmium in combination with low dietary zinc. Jacobs et al.[117] gave young Japanese quail 0.15 mg Cd per kilogram in the diet plus 60 or 30 mg Zn per kilogram. The latter zinc intake led to significantly higher amounts of cadmium retained in liver and kidney than when cadmium was given together with the former zinc intake. Lal[143] also reported increased tissue levels of cadmium in animals fed a diet containing cadmium and low zinc levels.

The data by Petering et al.[212] may also be relevant to the assessment of the effect of dietary zinc on the gastrointestinal absorption of cadmium. Rats were given 100 mg Cd per liter drinking water for 30 days and then diets sufficient or deficient in zinc for 14 days. The occurrence of metallothionein-bound cadmium, zinc, and copper in renal cortex cells was measured. There were large differences in the zinc-metallothionein concentrations, but no differences in the cadmium-metallothionein concentrations.

Differences in metallothionein concentrations in tissues, depending on metal exposure, were detected by Onosaka et al.[207] While cadmium exposure gave the highest metallothionein concentration in liver, zinc gave the highest concentration in pancreas.

e. Iron

Hamilton and Valberg[107] reported that mice with enhanced iron absorption absorbed more ^{109}Cd than those with normal iron absorption. In a more detailed study, Valberg et al.[272] showed that the average gastrointestinal uptake in iron-deficient mice 4 hr after an intragastric dose of 400 nmol of cadmium chloride (24 μg Cd) labeled with $^{109}CdCl_2$ was 25%. This was significantly greater than the 16% uptake in iron-normal animals. More cadmium was absorbed into the body of the iron-deficient mice, 3.8%, than the iron-normal mice, 2%. Most of the cadmium uptake after 4 hr was found in the gastrointestinal mucosa; about 8% in the ileum, 5% in the jejunum and colon, and 2% in stomach and duodenum. After 72 hr about 1.5% was still in duodenum and less than 0.5% in each of the other tissues. The radiocadmium in the duodenum was bound to a protein with a molecular weight of about 12,000.

After subcutaneous injection of radiocadmium, the rate of excretion of radioactivity from the body was similar in iron-normal and iron-deficient mice. However, a greater proportion of the injected dose accumulated in the duodenum of the iron-deficient animals than in the duodenum of iron-normal animals.[272]

Flanagan et al.[81] studied the influence of low dietary iron and iron deficiency induced by bleeding in mice. They found that low dietary iron increased cadmium absorption. Iron deficiency induced by bleeding also increased cadmium absorption, but had a less pronounced effect.

f. Other Metals

Mahaffey et al.[157] exposed male rats to dietary lead (200 mg/kg diet), cadmium (50 mg/kg diet), or arsenic (50 mg/kg diet) as arsenate either alone or in combination for 10 weeks. Cadmium exposure reduced both bone and kidney lead and iron levels and liver iron and bone zinc, but increased kidney zinc levels. Neither lead nor arsenic had any effect on the absorption or distribution of cadmium.

In a study of Japanese quail subjected to a variety of dietary deficiencies and excesses of trace elements,[90] it was found that a copper deficiency caused an increased cadmium concentration in the kidney but not in the other tissues studied. It was not clear whether this was due to an increased gastrointestinal cadmium absorption or due to a changed intertissue distribution of cadmium.

4. In Humans

The gastrointestinal absorption in humans has been reported to be 1 to 7%, with much higher values among people with low body iron stores. Rahola et al.[218] studied the fate of ^{115m}Cd given orally to five human male volunteers, aged 19 to 50 years. They received single doses of 4.8 to 6.1 μCi ^{115m}Cd, mixed with a calf kidney suspension. The total ingestion of cadmium was about 100 μg. During the first 3 to 5 days after administration, about 70% of the activity was eliminated, primarily in the feces. A rapid elimination continued until about 6% (4.7 to 7%) of the dose remained in the body. This indicates an average absorption of at least 6%.

Kitamura[129] reported on two balance studies performed on a 55-year-old man. Cadmium absorption was calculated as the difference between ingested amount of cadmium and amount recovered in feces. Absorption values of 8 to 10% can be calculated from the data.[97]

Yamagata et al.[284] studied gastrointestinal absorption of ^{115m}Cd which had been taken up by rice plants from a labeled nutrient solution. A human volunteer ate 100 g of this rice (total dose 150 μg), and the amount of ^{115m}Cd in feces and the amount retained in the body was measured daily for 7 days. Approximately 93% was recovered in feces within 3 days and 4.4% was retained in the body after 6 days. The authors also studied ^{115m}Cd given in a similar way in an acid solution (lemon juice) to two volunteers. For the two subjects, 71 and 76% were recovered in feces after 3 days and approximately 7% remained in the body after 6 days indicating a tendency for higher absorption with this form of administration. A whole body scanning device was used to measure body retention of cadmium. The authors pointed out that the accuracy of these measurements was uncertain.

McLellan et al.[161] prepared a breakfast of oat-porridge (oat and milk powder) and a drinkable solution of $^{115m}CdCl_2$ in water. This was given to 14 healthy volunteers. The total amount of Cd ingested was 22 to 29 μg. The meal also contained a ^{51}Cr stool marker. There was a rapid initial disappearance of radiocadmium from the body. On the average, 86% of the dose was excreted after 1 week.[166]

Cadmium absorption was calculated from whole body measurements of ^{115m}Cd 1 to 2 weeks after complete (>99%) elimination of ^{51}Cr from the body. The time for (complete) elimination of the ^{51}Cr marker varied from 1 to 5 weeks. The absorption value of cadmium was thus based on the data 2 to 6 weeks after the time of ingestion. The average absorption was 4.6 ± 4.0% (SD).[161]

Flanagan et al.[80] studied the gastrointestinal uptake of cadmium measured as whole body retention of ^{115m}Cd after the initial clearance of unabsorbed cadmium from the intestinal tract (1 week after the radiochromium marker was no longer detectable) in healthy human volunteers (the same method as used by McLellan et al.[161]). A single dose of 25 μg of ^{115m}Cd was given and whole body counting was performed at about weekly intervals. Body iron stores were evaluated by determinations of serum ferritin levels. It is evident from Figure 7 that persons with low iron stores (mostly females) often have considerably higher absorption of cadmium than persons whose iron stores are high. The highest gastrointestinal absorption values were 17 and 22%. The average cadmium absorption in males was 2.6 ± 0.6% (SEM, n = 10) and in females it was 7.5 ± 1.8% (SEM, n = 12).

Shaikh and Smith[233] administered 50 μg of ^{109}Cd in beef kidney homogenate in the way described by Rahola et al.[218] and the subjects subsequently drank 450 mℓ of milk. The study was performed on volunteers. Subjects were observed for 56 to 808 days after cadmium ingestion. Fractional absorption was calculated by extrapolation of the final component of the whole body retention curve to zero time. Absorption values were 1.1 to 3.0% in six males followed up for 100 days and 1.4 to 7.0% in five females followed up for more than 183 days. The person with the 7% absorption figure had a low serum ferritin value indicating depleted iron stores. The findings fit approximately into the same range as the other studies even though there may be methodological difficulties involved in measuring accurately the low energy radiation emitted from ^{109}Cd decay.

There are some differences among the reported absorption figures. In addition to the evident influence of iron intake, the quality of protein may also be of importance (Section II.B.3). There is evidence also in humans that the type of protein binding of cadmium also influences gastrointestinal absorption. In a group of New Zealand oyster fishermen with very high cadmium intakes (up to 500 μg/day) from oysters,[160] the blood cadmium levels were disproportionately low compared to those of Japanese farmers with similar cadmium intakes from polluted rice.[133] This may be explained alternatively by a lower gastrointestinal absorption of cadmium bound to the protein in these oysters (Section II.B.3.c) than of cadmium in polluted rice or by a different distribution of "oyster cadmium" between blood and tissues[203] (Section II.B.5).

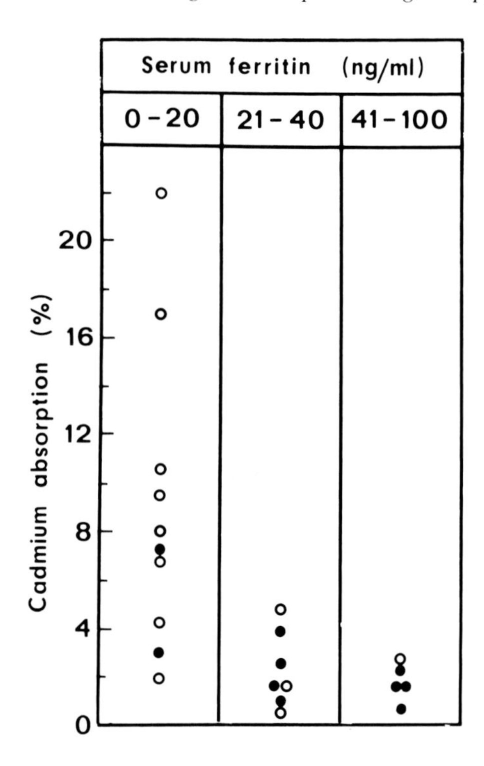

FIGURE 7. Cadmium absorption in human volunteers in relation to iron status. Females — open circles; males — closed circles. (From Flanagan, P. R., McLellan, J. S., Haist, J., Cherian, M. G., Chamberlain, M. J., and Valberg, L. S., *Gastroenterology*, 74, 841—846, 1978. With permission.)

5. Mechanism of Gastrointestinal Absorption; Uptake of Metallothionein-Bound Cadmium

It has been generally believed that gastrointestinal uptake of cadmium would be preceded by ionization of the various cadmium compounds ingested and that the uptake through the gastrointestinal mucosa would be the same regardless of cadmium compound as long as such compounds are ionized in the gastrointestinal fluids.[27,259]

In a study by Foulkes,[84] the short-term kinetics of cadmium uptake by the intestinal mucosa were studied after perfusion of intestinal segments of living rats with glucose-$CdCl_2$. Mucosal uptake was a saturable process in this experimental system. Once taken up by gastrointestinal mucosa, cadmium is partly bound to metallothionein in the mucosa cells.[272]

The data on a limited number of soluble cadmium salts, reviewed in Section II.B.1, give a certain amount of support for the idea that different soluble cadmium compounds will be absorbed to a similar extent. However, other data indicate that different mechanisms may sometimes be operating. Cherian et al.[44] gave $^{109}CdCl_2$ and ^{109}Cd-metallothionein to mice (60 μg Cd per mouse) and found approximately the same systemic absorption. However, the tissue distribution of cadmium differed greatly in the two groups. A significantly greater deposition of cadmium was observed in the kidney of the mice fed cadmium-metallothionein than in the $CdCl_2$ group.

Since it is known that metallothionein-bound cadmium given parenterally is rapidly and almost completely taken up by the kidney,[201,256] while cadmium bound to other ligands in plasma is mainly taken up by the liver, the observations by Cherian et al.[44] indicate that cadmium bound to metallothionein may be partly taken up undigested into the circulation.

Further support for this hypothesis has been provided by Cherian[36] who showed that 4 hr after gavage of cadmium-metallothionein, a major portion of the ingested protein was taken up intact by gastrointestinal mucosa. A limited proportion was taken up systemically and was subsequently distributed to the kidney where it was degraded.

Kello et al.[125] studied the binding of cadmium to metallothionein in intestinal mucosa after gavage of $CdCl_2$. Pretreatment of animals with cadmium increased the proportion of cadmium bound to metallothionein in the mucosa. An increased total systemic absorption after pretreatment was also reported by Engström and Nordberg.[72] However, the rate of transfer was slightly lower in pretreated animals according to Kello et al.[125] An involvement of metallothionein in intestinal turnover of cadmium had thus been demonstrated, however, the details of this involvement have not yet been elucidated.

Hietanen[112] reviewed the mechanism of gastrointestinal cadmium absorption and suggested that cadmium is stored in intestinal mucosal cells before absorption and "lost" from absorption at the same rate as the desquamation of mucosal epithelium. Most of the intestinal cadmium is found in the upper and middle parts of the small intestine[247,250] and Hietanen[112] suggests that these are the sites of absorption.

It may be noted that cadmium in intestinal mucosal cells may be on its way to being either excreted (Section V.B) or absorbed. The mere existence of cadmium in the mucosa does not prove that it is in the process of being absorbed into the blood or lymph.

Another aspect of importance when discussing the mechanisms of cadmium absorption is the marked increase of cadmium absorption when the body iron stores are low[80] and the decreased gastrointestinal iron absorption when cadmium intake is high.[22] This indicates that Cd^{2+} and Fe^{2+} compete for a common binding site in the intestinal mucosa. A similar relationship may exist for Ca^{2+}. Another possible mechanism for the decreased iron absorption is the intestinal mucosal damage caused by cadmium.[88]

It should be pointed out that Flanagan et al.,[80] in the studies described above, measured whole body retention of cadmium, and the reported increased absorption of cadmium might be related to an increased level in gastrointestinal mucosa, since there were no data on increased levels in other tissues.

6. *Conclusions*

The data reviewed above indicate that when no specific modifying factors occur (calcium deficiency, iron deficiency or protein deficiency), the average gastrointestinal absorption in humans is about 3 to 7%. In individuals with low body iron stores, absorption rates as high as 20% have been recorded. In small experimental animals, absorption in the range of 1 to 2% is commonly reported, but these values are affected by the other dietary factors mentioned above. The dose and age of the animal also influence absorption rates. The absorption appears to be lower for insoluble than for soluble cadmium compounds. The mechanism for absorption is not yet fully elucidated and the half-time of cadmium in intestinal mucosa may vary and affect the absorption rate estimates.

III. TRANSPORT AND DISTRIBUTION IN BLOOD

Many of the effects of cadmium are systemic. Therefore, the transport of absorbed cadmium in blood is of crucial importance for the toxicology of cadmium. The main storage tissues for cadmium (kidneys, liver, and muscles) also receive the cadmium through blood transport. With time the stored cadmium may return to the blood for redistribution within the body or for excretion.

Cadmium in whole blood of humans with any level of cadmium intake can be measured using sensitive methods (Chapter 5), but the cadmium level in plasma of people with low cadmium intake has not yet been reliably measured.

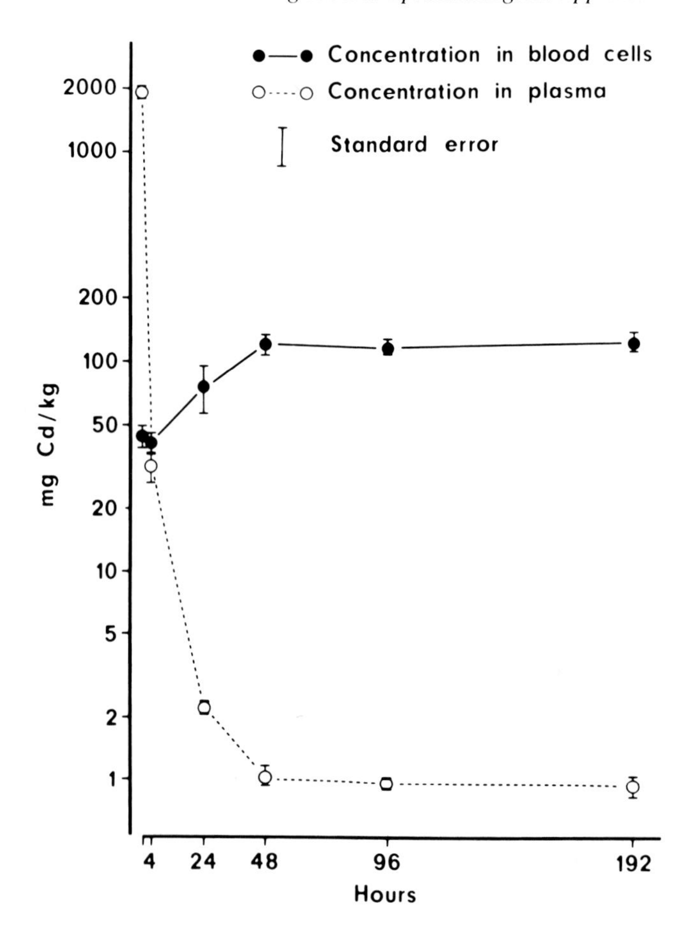

FIGURE 8. Concentrations of cadmium in plasma and blood cells, respectively, in mice given a single subcutaneous injection of ^{109}Cd (1 mg Cd/kg body weight) and killed at various times after injection. Vertical bars indicate the standard error and the circles indicate mean values. (From Nordberg, M., *Environ. Res.*, 15, 381—404, 1978. With permission.)

The chemical form of the cadmium entering plasma after absorption from the lungs or the intestines is not known, but it is likely to be bound or become bound to a great extent, to proteins soon after absorption (Chapter 4). In the first hours after parenteral exposure, cadmium in plasma is partly dialyzable.[209,210]

A. In Animals

1. Distribution between Plasma and Blood Cells after Single Exposure

This type of study can reveal the short-term kinetics of cadmium entering the blood stream. During the first hours after parenteral exposure to inorganic soluble cadmium compounds, the plasma level is higher than the blood cell level, but within a day after exposure, the plasma level decreases to below the blood cell level, as was first shown in dogs.[276]

Numerous studies have confirmed this finding for different routes of exposure (intravenous, intraperitoneal, subcutaneous), different species (mice,[181,199] rats,[74,99,211,285]rabbits [126]), and different cadmium compounds. Detailed short-term kinetics of i.v.-injected cadmium in plasma of rats have also been described.[91] The time-related changes of cadmium levels in plasma and blood cells of mice are shown in Figure 8.

Some studies[99,119,210] report that the maximum concentration in whole blood occurs already

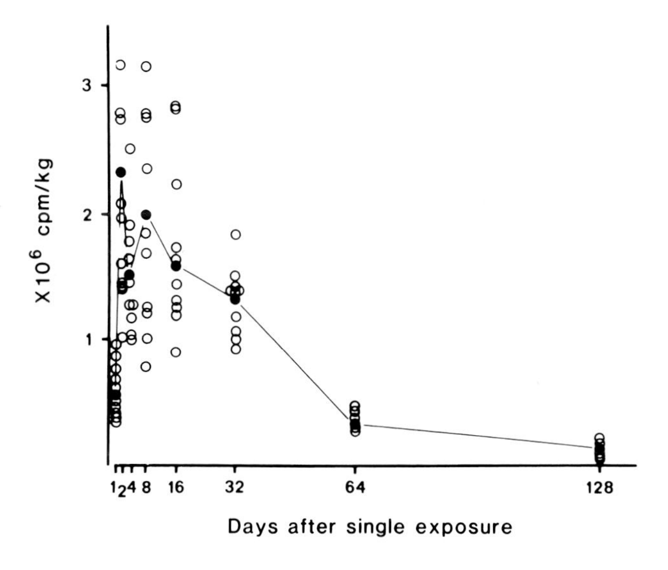

FIGURE 9. Blood levels of ^{109}Cd after a single subcutaneous injection. (Modified from Matsubara-Khan, J., *Environ. Res.*, 7, 54—67, 1974.)

a few minutes after injection. At that time most of the cadmium is in plasma (Figure 8). A part of the rapid initial fall of the plasma cadmium concentration (Figure 8) may be due to the distribution of cadmium in the extracellular space. The concentration decreases continuously during the first days.

With blood cells the situation is different. A first peak is seen within 1 hr, followed by a decrease and after 24 hr an increase to a second peak at 50 to 100 hr.[74,99,154,181,199] The reasons for these redistribution phenomena in blood are not fully known, but the complex protein binding of cadmium in blood cells and plasma may furnish an explanation (Sections III.A.2 and 3).

If the animals are followed up for a long time it can be seen that the blood cadmium levels eventually reach a peak and then start falling with a half-time of about 1 month, as was shown in a study[158] of mice given single subcutaneous injections of $^{109}CdCl_2$ (6.5 ng per mouse) (Figure 9).

2. Distribution between Plasma and Blood Cells after Repeated Exposure

These studies are more likely to reflect the steady-state situation of cadmium in plasma and blood and may provide information about the relationship between cadmium in blood and in other tissues, which is of particular importance for the use of cadmium in blood as an index for tissue burden (Section VI.).

Friberg[94] gave rabbits repeated subcutaneous injections of $^{115}CdSO_4$ (0.65 mg Cd per kilogram body weight and day). After 10 weeks' exposure, the whole blood cadmium concentration was 1 mg/kg, and no cadmium could be measured in plasma. A study of two rabbits exposed to subcutaneous injections of cadmium found blood cell levels 10 to 18 times greater than in plasma.

Carlson and Friberg[31] gave 1 mg ^{115}Cd per kilogram body weight to two rabbits daily for 1 week and 0.65 mg ^{115}Cd per kilogram to two rabbits for 3 weeks. Three weeks after the end of exposure, most of the cadmium in the blood was in the cells and only 1 to 6% was in plasma. With a packed cell volume of 0.40, 3% in plasma would indicate that the cadmium concentration in the cells was 45 times higher than in plasma.

Red cells from these rabbits were hemolyzed and centrifuged, and all the cadmium was recovered in the supernatant. About 40% of the cadmium in the hemolysate was dialyzable through polyvinyl tubing.[31] The nondialyzable cadmium migrated with hemoglobin in column type zone electrophoresis in starch.

Nordberg et al.[193] exposed mice to ^{109}Cd (subcutaneous 0.25 to 0.5 mg/kg body weight) for 6 months. They found a blood cell plasma concentration ratio of 60 in the low dose group and a ratio of 20 in the high dose group. Some hemolysis was observed in the blood samples from the high dose group, but the animals without observable hemolysis had similar ratios. In addition, the data demonstrate that while blood cell concentrations changed in proportion to increase in dose (from 1.2 to 2.4 mg/kg, a factor of 2), plasma cadmium increased by a factor of 6.

In rats repeatedly injected subcutaneously with $CdCl_2$, 0.5 mg/kg (5 μmol/kg) for 14 weeks, Shaikh and Hirayama[231] found plasma levels of cadmium which were relatively low after 4 and 6 weeks, but which increased after 10 and 14 weeks, i.e., at the time when liver and renal levels started to decrease and proteinuria appeared. Part of plasma cadmium was bound to metallothionein (see Section III.A.3). Shaikh and Hirayama[231] discussed the possibility that the increased plasma concentrations of cadmium reflected a release from the liver.

Suzuki[252] gave similar subcutaneous injections (0.5 mg/kg) to rats for 6 months and observed similar relationships between hepatic cadmium and blood cadmium as those observed by Shaikh and Hirayama[231] for plasma cadmium. Blood cadmium increased rapidly after 10 weeks to a concentration of 1.6 mg/ℓ. Plasma concentrations also rose and it was suggested that this was partly a result of hemolysis which is a well-documented effect after high cadmium exposure by the injection route (Volume II, Chapter 11, Section III.). An increase in spleen concentrations of cadmium which started at the same time[252] is consistent with the observations of hemolysis. Rats exposed subcutaneously to 3 mg Cd per kilogram body weight for up to 6 weeks[245] displayed a peak liver cadmium level already after 3 to 5 weeks. The spleen cadmium level increased continuously,[246] and the increase accelerated after 5 weeks of exposure. This finding agrees with the suggestion of hemolysis occurring at the time when liver levels decrease.

Animals exposed subcutaneously can also be expected to suffer liver damage (Volume II, Chapter 11, Section II.). This liver damage may be related to the increased plasma values. The data are in accordance with those reported by Nordberg et al.[193] and Shaikh and Hirayama.[231]

Cadmium concentrations in plasma following the repeated exposure of young and adult rats were studied by Wong et al.[282] These authors found no difference in plasma concentrations, while blood cadmium was higher in newborn than adult animals.

3. Protein Binding of Cadmium in Plasma

In plasma, cadmium is predominantly bound to proteins of high molecular weight (albumin or larger) a short time after exposure. To a large extent cadmium bound in this form will be taken up by the liver (Section IV.A). When plasma concentrations decrease and reach a low level a few days after a single exposure, cadmium is also bound to metallothionein which will be selectively taken up in the kidney (Chapter 4). The latter type of protein binding in plasma is also found in long-term exposure to cadmium.

After a single subcutaneous injection of $CdCl_2$ (1 mg/kg) in mice, Nordberg[181] and Nordberg et al.[199] demonstrated, using gel chromatography, that cadmium is initially bound almost exclusively to high molecular weight proteins in blood plasma (molecular weight of albumin or larger). This is also the case a short time (1 day) after gastrointestinal uptake of cadmium

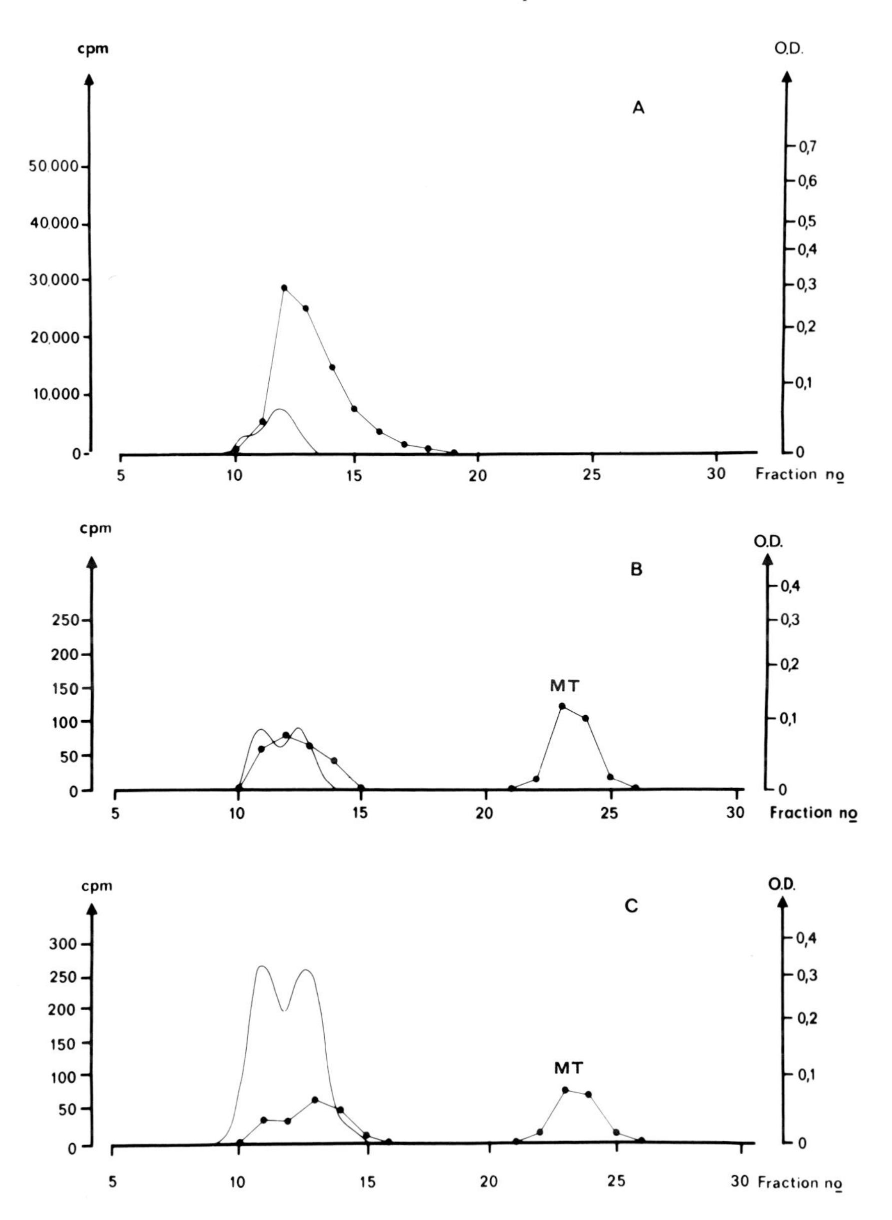

FIGURE 10. Gel chromatography on Sephadex G-75 of plasma from mice given a single subcutaneous injection of $^{109}CdCl_2$ (1 mg Cd/kg). Column dimensions were 355 × 26 mm. Elution with 0.01 *M* Tris buffer in 0.05 *M* NaCl, pH 8, at a flow rate of 14 mℓ/hr. Volume of fractions: 5 mℓ OD at 254 nm was continuously monitored: (A) 20 min after injection: 20,000 cpm = 45.4 ng Cd; (B) 96 hr after injection: 100 cpm = 0.028 ng Cd; (C) 192 hr after injection: 100 cpm = 0.028 ng Cd. (From Nordberg, M., *Environ. Res.*, 15, 381—404, 1978. With permission.)

as shown by Engström.[67,68] Four to eight days after a single subcutaneous injection, Nordberg[199] found that plasma concentrations were low, but part of plasma cadmium was bound to a low molecular weight protein of the same molecular weight as metallothionein (Figure 10).

After long-term exposure of mice to cadmium by repeated subcutaneous injections, Nordberg et al.[193] found, using gel chromatography, that the main part of plasma cadmium was bound to proteins of a molecular weight corresponding to albumin and larger proteins. A small but significant proportion of plasma cadmium was also found at a molecular weight corresponding to metallothionein. Since proteins of the size of metallothionein are ultrafiltrable, the binding of cadmium to metallothionein in plasma is of special interest in relation to uptake by the kidney. A specific transport mechanism to the renal tubule was first suggested by Piscator[213] (Section V. this chapter and Chapter 4). This mechanism involves glomerular filtration and tubular reabsorption of metallothionein-bound cadmium.

It was calculated by Nordberg et al.[193] that the low concentrations of cadmium bound to the metallothionein-like protein in blood plasma of mice could still explain the high concentrations that accumulated in the kidney after long-term exposure. An initial uptake in the liver of albumin-bound cadmium and a subsequent release of metallothionein-bound cadmium from that organ would thus explain the redistribution phenomena which occur with time after a single exposure to cadmium (Section IV.A).

Further support for such a mechanism involving metallothionein has been provided by Cherian and Shaikh[41] and Shaikh and Hirayama[231] who found metallothionein-bound cadmium in plasma of rats during long-term exposure via injection. An idea expressed by Cherian and Shaikh[41] was that metallothionein-bound cadmium would not appear in plasma until toxic organ concentrations were reached. However, as discussed above, metallothionein-bound cadmium does occur in plasma even after a single dose of cadmium with no measureable adverse effects. The fact that metallothionein was not detected during the early phases of long-term exposure in the study by Cherian and Shaikh[41] can be explained by the poor detection limit for metallothionein-bound cadmium in that particular study.

All of the authors mentioned used gel chromatography and identified cadmium-metallothionein by its cadmium content and molecular weight. Another method for identification of metallothionein in plasma is radioimmunoassay.[273,274] In rats exposed by injection of 0.6 mg Cd per week for 6 to 8 weeks, metallothionein concentrations in serum increased from <2 μg/ℓ (before exposure) to 50 μg/ℓ.[274] In humans occupationally exposed to cadmium, Nordberg et al.[198] reported metallothionein concentrations in plasma ranging from <2 to 11 μg/ℓ.

There is sufficient evidence from published data to show that cadmium bound to a low molecular weight protein, in some studies immunochemically identified as metallothionein, occurs in small concentrations in plasma both during the early part of the redistribution phase for cadmium (48 to 96 hr after a single injection) and during long-term exposure to cadmium. The possible dose dependence in relative distribution of cadmium in plasma between metallothionein, albumin, and other proteins after long-term exposure is not known in detail.

4. Distribution of Cadmium among and within Blood Cells

The first studies of the distribution of cadmium in blood were those of Friberg[94] and Truhaut and Boudene.[267] Both of these studies demonstrated that after long-term cadmium exposure, cadmium in blood is predominantly located in the blood cells. As discussed previously in this chapter, cadmium concentrations in blood cells are higher than concentrations of cadmium in plasma a short time after a single administration of cadmium (Figure 8). Although cadmium concentrations in blood cells reach a peak within a few hours and thereafter decrease, a redistribution phase starts after approximately 24 hr. At this time the concentrations in blood cells increase again to reach a peak at 50 to 100 hr.[74,99,154,181,193,199]

Hildebrand and Cram[113] studied blood cells in culture incubated with cadmium and showed that metallothionein was induced only in lymphocytes and not in mature enucleated erythrocytes. Garty et al.[99] reported that while the concentration of cadmium in white blood cells was greater than in red blood cells, most of the cadmium in blood cells was recovered in red blood cells due to their considerably greater abundance in blood.

The binding of cadmium within blood cells was investigated by Carlson and Friberg.[31] They found that about 40% of blood cell cadmium was dialyzable through polyvinyl tubings and that the nondialyzable part migrated together with hemoglobin in column type zone electrophoresis in starch. It is not known to what extent the method used allowed separation of hemoglobin from low molecular weight proteins like metallothionein. Nordberg[181] studied the distribution of cadmium in the blood cells of cadmium-exposed mice by separating hemolysates on G-75 Sephadex columns. Twenty minutes after a single subcutaneous injection, most of the cadmium was in high molecular weight fractions, whereas only a minor part was in the fraction corresponding to hemoglobin and an insignificant amount in the fraction corresponding to metallothionein. After 96 hr a redistribution had taken place so that about one third of the total cadmium was found in fractions with a molecular weight corresponding to metallothionein. The main part was still in high molecular weight fractions and no cadmium was detected in the hemoglobin peak (Figure 11).

Nordberg[199] studied further the cadmium-binding protein isolated from mouse blood hemolysate. Hemolysates of blood cells from mice given a single injection of radiolabeled cadmium (1 mg/kg body weight) and killed after 96 and 192 hr were studied using gel chromatography. Fractions corresponding to the molecular weight of metallothionein were concentrated and used for further studies. Mice were given intravenous injections (approximately 2 ng Cd per mouse) and killed after 4 and 96 hr. Mice killed after 96 hr had approximately 70% of the injected dose of cadmium bound to the low molecular weight protein in the kidney. These kidney concentrations of cadmium were about 10 times higher than for the corresponding control animals injected with $CdCl_2$. Autoradiography performed on mice 4 hr after injection of ^{109}Cd bound to low molecular weight protein from hemolysate showed an accumulation of cadmium in the kidney cortex (Figure 12).

Nordberg[199,200] also reported results of isoelectric focusing studies of the low molecular weight protein obtained by gel chromatography of hemolysate. A considerable amount of zinc (analyzed by atomic absorption spectrophotometry) and ^{109}Cd was recovered in a protein with an isoelectric point (pI) of 6. This isoelectric point is different from the one recorded for metallothionein from liver,[194,202] but since the pI might change with the metal content of the protein, a high zinc content, like that in the protein from blood cell hemolysate, might change the pI of metallothionein to such a high value. However, the possibility cannot be ruled out that cadmium binding in blood cells occurs to some other zinc-containing protein.

Garty et al.[99] in their studies of blood from rats injected with ^{109}Cd found, using gel chromatography, that a considerable proportion of the cadmium in blood was bound to blood cell ghosts (membranes) and that cadmium in the hemolysate was bound to a protein similar in size to metallothionein. Ion-exchange chromatography of the cadmium-binding protein disclosed properties different from those of metallothionein from rat liver and the authors suggested that cadmium might induce a metal-binding protein (different from metallothionein) in the ghosts (membranes) of immature nucleated erythrocytes in the bone marrow and that a delay of 60 hr is needed for this protein-bound cadmium to reach the circulation.

In summary, it has been demonstrated that the main part of cadmium in blood cells is in red blood cells. In these cells cadmium is mainly bound to a protein with a molecular weight similar to metallothionein. Whereas this protein has not yet been identified, some data indicate that when injected intravenously in animals, the cadmium bound to that protein will be taken up selectively by the kidney.

B. In Humans

Due to the difficulties in accurately measuring cadmium in blood, many of the earlier studies which present such data (see review by Friberg et al.[97] and Chapter 5) cannot be used for quantitative evaluations. Plasma data are even more difficult to evaluate.

It can be seen in Chapter 5 that the median cadmium concentration in whole blood of nonsmokers in countries with "background" dietary cadmium intakes via food of 10 to 20 μg/day is about 0.2 to 1.0 μg/ℓ. Cigarette smoking adds to cadmium exposure via inhalation

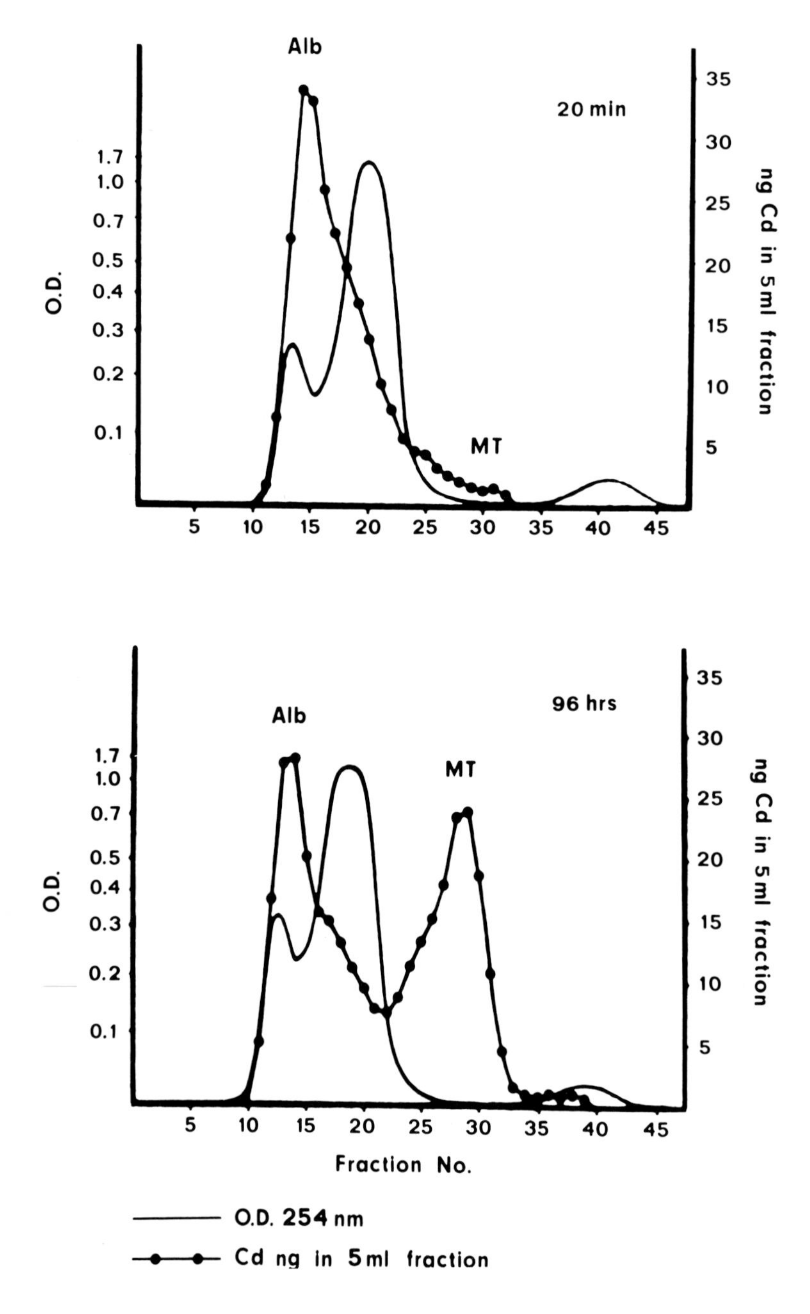

FIGURE 11. Cadmium distribution in hemolysate. (Modified from Nordberg, G. F., *Environ. Physiol. Biochem.*, 2, 7—36, 1972a.)

and this is reflected in the increase (2 to 10 times) in the blood cadmium of smokers. Studies of industrial workers[97,145,225] show even greater increases in blood cadmium, depending on exposure level.

Honda et al.[114] studied cadmium in whole blood, blood cells, and plasma of a group of 11 farmers exposed to cadmium via polluted rice and a control group of 5 persons. Cadmium was analyzed with Zeeman-effect atomic absorption spectrophotometry after APDC-MIBK extraction. In the control group, the average whole blood cadmium was 1.9 $\mu g/\ell$, which agrees with the general population sample data for Japan reported by Vahter.[271] In the

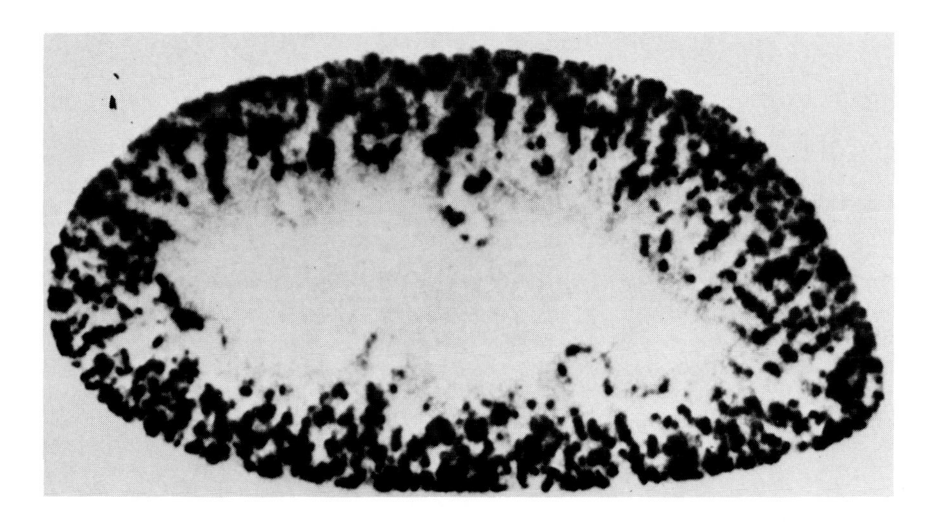

FIGURE 12. Autoradiogram of kidney from mouse injected with ^{109}Cd — low molecular weight protein obtained from hemolysate of blood cells of mice injected with $^{109}CdCl_2$. Detail of whole body autoradiogram; no other tissues displayed any radioactivity in the whole body autoradiogram. (From Nordberg, M., *Environ. Health Perspect.*, 54, 13—20, 1984. With permission.)

cadmium-exposed group, the cadmium concentrations in whole blood and plasma were 13 μg/ℓ and 0.36 μg/ℓ. The ratio whole blood to plasma concentrations was 36 for the high cadmium group. In the control group the plasma-cadmium levels were always below 0.1 μg/ℓ and too low to measure. Using a ratio of 36, the average plasma cadmium in the control group would be about 0.06 μg/ℓ.

Long-term changes in blood cadmium levels have been studied[215] in workers with short-term high exposure who showed no signs of tubular dysfunction. There is a fast component with a half-time of several months which is particularly prominent after short-term very high exposure (Figure 13).

In a 15-year longitudinal study of 5 workers exposed for several years,[118] a slow component with 7 to 16 years half-time was predominant (Figure 14). Using a two-exponential regression model, the fast component half-time in these workers was calculated to be in the range of 75 to 130 days.

Cross-sectional studies of groups of workers who had different durations of exposure and different exposure levels[108,145] have yielded such a wide scatter in results that definite conclusions about the kinetics of blood cadmium cannot be drawn. In one study[108] there was an apparent increase of the blood cadmium of 164 workers with increasing levels of cadmium in workroom air, but the time factor was not analyzed.

The increase of blood cadmium in newly employed workers during the first year follows the pattern of an accumulation curve with a half-time of a few months.[132] It was pointed out that this short-term accumulation in the blood could represent the turnover of red blood cells, which have a mean life of about 120 days (or a half-time of 83 days) (Chapter 7, Section III.B).

Both the short and long-term components of blood cadmium are likely to occur in the blood cells as well as possibly in the plasma. The long-term component is likely to reflect the accumulation of cadmium in major storage tissues (Section VI.) as their half-times are also several years (Section IV.).

Studies of plasma metallothionein (MT) in cadmium workers have been made possible by the development of the radioimmunoassay for MT.[274] About 10% of the MT weight in a cadmium-exposed person may be cadmium (Chapter 4). Thus, plasma MT may be used as an indicator of metallothionein-bound cadmium in plasma. It should be pointed out that a major part of cadmium in plasma is bound to larger proteins (Section III.A.3).

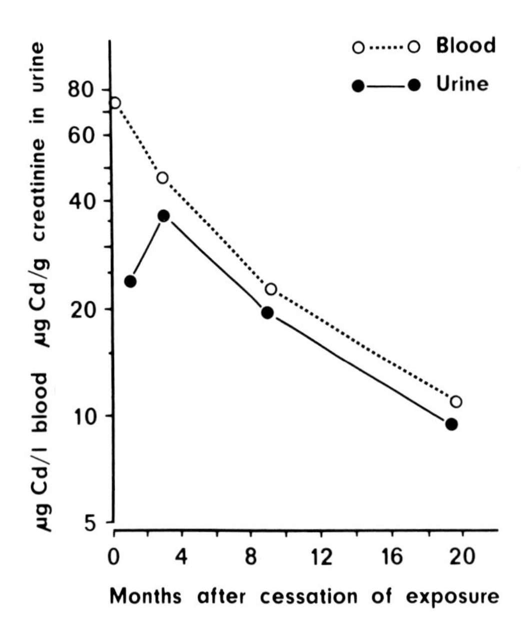

FIGURE 13. Average cadmium concentrations in blood and urine after cessation of exposure in five workers with previous high exposure to cadmium. (From Piscator, M., *Cadmium in the Environment III*, EPA-650/2-75-049, Friberg, L., Kjellström, T., Nordberg, G. F., and Piscator, M., Eds., U.S. Enviromental Protection Agency, Washington, D.C., 1975, 4—58.)

Nordberg et al.[198] reported blood cadmium and plasma MT in 29 battery factory workers, divided into two groups according to their renal tubular function status as indicated by urinary β_2-microglobulin level. In the 19 workers with low β_2-microglobulin levels, median blood cadmium was 11 $\mu g/\ell$ and median plasma MT was less than 2 $\mu g/\ell$. This was calculated to be equivalent to metallothionein-bound cadmium concentrations in plasma ranging from 0.2 to 0.5 $\mu g/\ell$.[198]

Another study of plasma MT in 33 workers[77] found an average of 2.1 $\mu g/\ell$. Blood cadmium was not measured.

C. Conclusions

In animals, the initial plasma levels of cadmium will be very high after parenteral exposure, but after some hours most of the cadmium in the blood will be found in the cells. The erythrocytes will contain the greatest amounts. The cadmium in plasma is bound to at least two different proteins, one with a molecular weight similar to albumin and one with a molecular weight similar to metallothionein. In the red blood cells, cadmium is bound to similar proteins in the cytoplasm, but a considerable portion may be bound to all membranes.

In humans with long-term high exposure, whole blood cadmium may be about 30 times higher than plasma cadmium. After cessation of long-term high exposure, the decrease of whole blood cadmium displays a slow component with a half-time of about 10 years and a fast component with a half-time of 3 to 4 months. Metallothionein concentrations have been measured in human plasma. They are associated with cadmium exposure.

IV. DISTRIBUTION TO AND RETENTION IN OTHER TISSUES

Depending on the route of uptake, a considerable proportion of the cadmium retained in

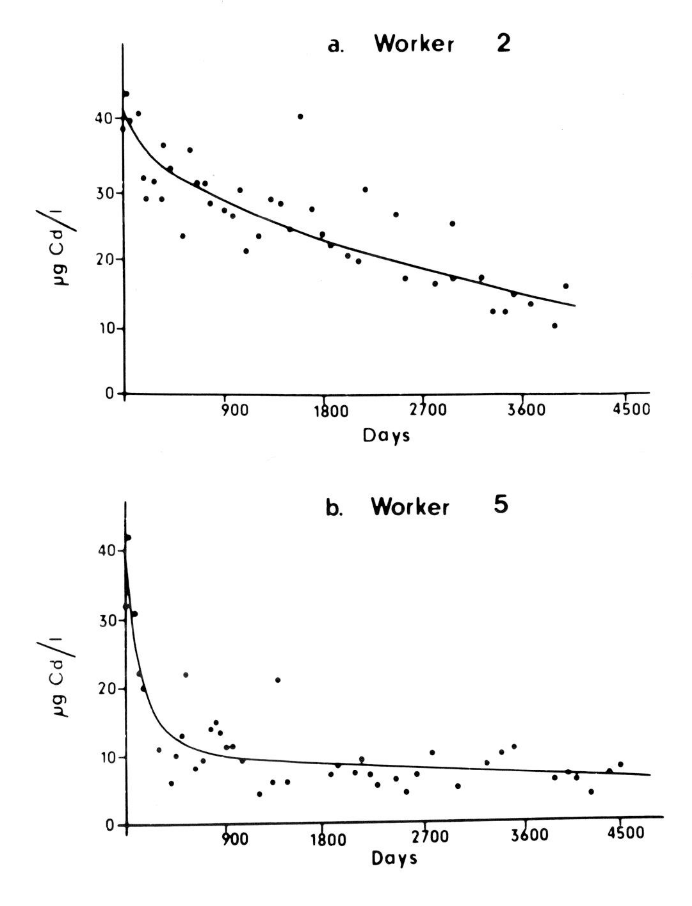

FIGURE 14. Decrease in blood cadmium in two workers after cessation of exposure (1 ng Cd/ℓ = 8.9 nmol Cd/ℓ). (From Järup, L., Rogenfelt, A., Elinder, C.-G., Nogawa, K., and Kjellström, T., *Scand. J. Work Environ. Health*, 9, 327—331, 1983. With permission.)

the body will initially be found in the lungs or the intestinal walls (see previous sections). After absorption, cadmium will be transported by the blood to other tissues and a large portion of the cadmium will be found in kidneys, liver, and muscles.

The distribution among organs is largely dependent on the time which has elapsed after a single exposure or the duration of repeated exposure, since the half-time of cadmium is quite different for the different organs. These half-times will be discussed further in Section V.D and in Chapter 7, where the whole body and interorgan kinetics of cadmium are analyzed in terms of a mathematical mode.

The distribution among organs as well as within organs is also dependent on the binding of cadmium to metallothionein (Chapter 4). The fate of the cadmium absorbed into the body is largely determined by the transport and metabolism of this protein.

A. Distribution among Organs

1. In Animals

The distribution of cadmium varies depending on length of exposure, route, and dose. At high doses and particularly after parenteral exposure, the absorbed cadmium will initially go to the liver and relatively small amounts to the kidneys. With increasing time after exposure, the kidney levels will increase due to a redistribution from the liver. The time-dependent changes in other tissues have not been studied in detail.

Gunn and Gould[104] gave a single intracardiac injection of ^{115}Cd nitrate to rats (dose not stated) and studied the animals for up to 8 months. Shortly after exposure, the highest concentration was seen in the liver (Figure 15). During the first month there was a decrease in liver levels of ^{115}Cd and an increase in kidney cortex levels, so that a month after exposure the concentration in the cortex was the same as that in the liver (Figure 15). During the following months liver levels decreased very slowly, while levels in renal cortex increased so that after 5 months the ratio between renal cortex and liver levels was about 2 and after 8 months it was about 4. The levels in kidney medulla also increased with time (Figure 15).

Several other studies using parenteral single injections of radioactive cadmium have shown similar time-dependent changes in liver and kidney levels.[24,55,154] Lucis and Lucis[153] reported that there were differences between different strains of mice in the percentage of the dose retained in the liver. Shaikh and Lucis[232] found differences in the organ distribution between mice and rats. However, the general results are similar in all these studies. Some of them report the retention of cadmium as a percentage of the dose given and others report the concentrations in the organs; this fact influences the comparability of the studies. The liver is a much larger organ than the kidneys and it may, therefore, contain more cadmium than the kidneys even though the concentration in liver is lower than that in the kidneys.

The liver and the kidneys have attracted most interest as storage tissue for cadmium. Autoradiography studies show that these organs contain the highest concentrations. Other tissues which contain considerable concentrations of cadmium are pancreas, salivary glands, and the testicles, whereas the brain has very low concentrations (Figure 16). Concentrations in testicles are of interest in relation to the acute effects occurring in this organ after injection exposure (Volume II, Chapter 11). The amount absorbed after a single oral dose is quite small (Section II.B) and it appears that initially a greater proportion of the dose will be retained in the kidneys. Decker et al.[55] gave rats ^{115}Cd nitrate (6.6 mg Cd per kilogram body weight) by gavage and determined cadmium distribution from 8 to 360 hr after the dose. After 8 hr the largest total amount of cadmium was in the liver, while the highest concentration was in the kidneys. Maximum concentrations in both organs were reached at 72 hr after exposure. Further studies in rats by Kotsonis and Klaassen[141] of various oral doses of cadmium (25 to 150 mg/kg) demonstrated approximately linear kinetics within this dose range. Miller et al.[164] gave a single oral dose of $^{115}CdCl_2$ to goats (0.04 mg Cd per goat) and found that after 14 days the concentration in the kidneys was nearly twice as high as that in the liver. A single intravenous injection (0.04 mg ^{109}Cd per goat) gave a 3 times higher concentration in the liver than in the kidneys after 14 days.[164] This agrees with other parenteral exposure studies (Figure 15).

Engström and Nordberg[71] gave cadmium at four different dose levels to groups of mice by gavage and studied whole body retention and organ distribution. The absorption fraction varied with the dose (Section II.B) and there was also a dose-dependent difference in organ distribution (Table 4). At the lower doses, a considerably smaller proportion of the dose was retained in the liver than at higher doses. It is worth noting that the lowest oral dose used (1 μg Cd per kilogram body weight) is close to the daily dose obtained from food by human beings (0.2 to 1 μg Cd per kilogram body weight; Chapter 3).

In the first weeks of a period of daily subcutaneous injections of $^{115}Cd\ SO_4$ (0.65 mg Cd per kilogram) to rabbits,[94] the highest cadmium concentration was found in the liver, as is the case following a single exposure. If the exposure is continued for several months, the liver cadmium levels will remain higher than the kidney levels.

Bonnell et al.[21] gave intraperitoneal injections of cadmium nitrate to rats (0.75 mg Cd per kilogram 3 days/week for 5 to 6 months). A linear increase of cadmium in liver was seen during the first 3 months when a peak was reached at about 350 mg/kg wet weight. After that the liver levels decreased progressively to about 200 mg/kg after 12 months in spite of continuous cadmium exposure. The kidney level also increased up to a maximum

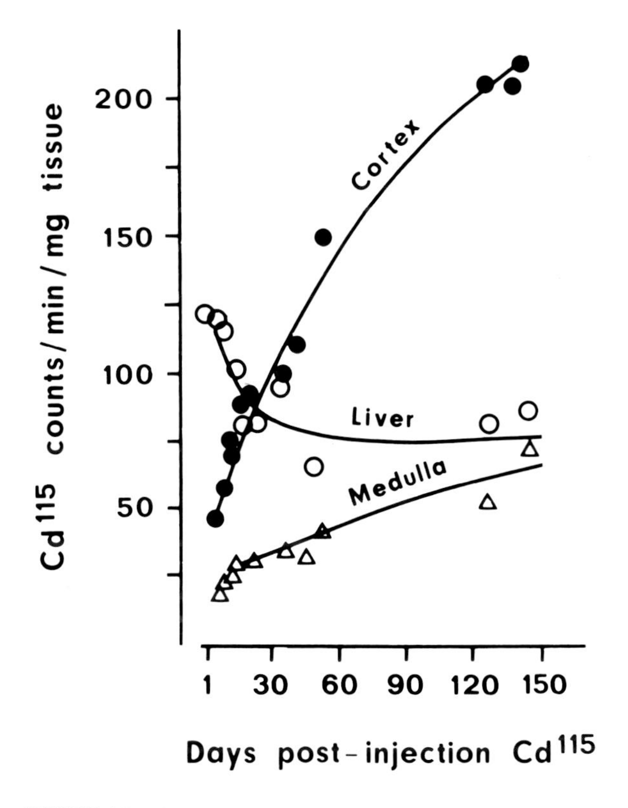

FIGURE 15. Accumulation of cadmium in kidney and liver of rats after intracardiac injection. (From Gunn, S. A. and Gould, T. C., *Proc. Soc. Exp. Biol. Med.*, 96, 820—823, 1957.)

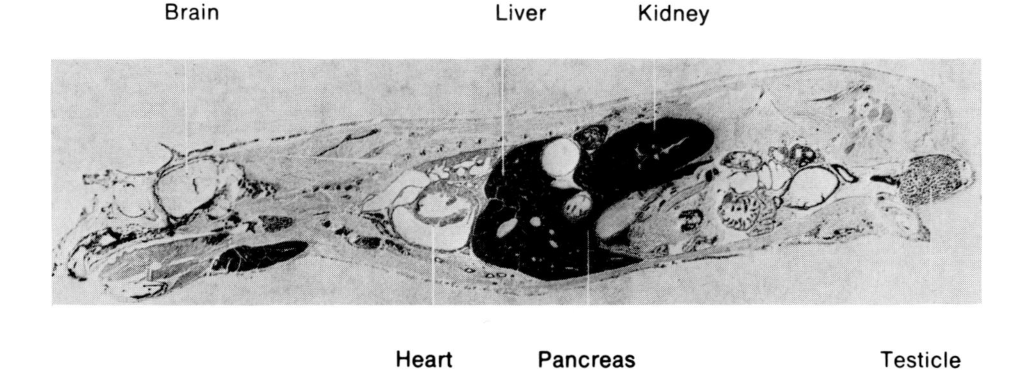

FIGURE 16. Autoradiographic distribution of ^{109}Cd in a mouse 112 days after a single intravenous injection. (From Nordberg, G. F. and Nishiyama, K., *Arch. Environ. Health*, 24, 209—214, 1972.)

of about 250 mg/kg after 4 months and it then decreased to 150 mg/kg after 12 months. At these cadmium levels in the liver and kidneys, it is likely that cadmium-induced damage occurred (Volume II, Chapters 9 and 11).

Axelsson and Piscator[9] gave rabbits cadmium chloride (0.25 mg Cd per kilogram) by subcutaneous injections 5 days a week for 11 to 29 weeks. There was an increase in liver and renal cortex levels up to 17 weeks of exposure. The levels at that time were about 450 to 400 mg/kg wet weight, respectively. Further exposure did not increase the concentrations

Table 4
CADMIUM RETENTION IN ORGANS OF MICE (% OF GIVEN DOSE IN WHOLE ORGAN) 4 MONTHS AFTER A SINGLE ORAL DOSE OF $CdCl_2$ LABELED WITH ^{109}Cd[71]

Dose (μg Cd/kg body weight)	n	% of given dose in whole organs (average)			Ratios among organs		
		Liver	Kidneys	Pancreas	Liver	Kidneys	Pancreas
37,500	10	1.338	0.375	0.051[a]	3.568	1	0.136
750	9	0.088	0.155	0.012	0.567	1	0.077
15	8	0.067	0.249	0.012	0.269	1	0.048
1	10	0.018	0.099	0.005	0.182	1	0.051

Note: n = number of surviving mice — accidental death of some animals during the whole body measurements.

[a] Value from 1 mouse excluded (0.004%).

Modified from Engström, B. and Nordberg, G. F., *Toxicology*, 13, 215—222, 1979a.

in the liver or the kidneys; on the contrary, concentrations decreased. Similar results have been reported for rats exposed subcutaneously to $CdCl_2$,[251] for rats exposed to $CdCl_2$ in drinking water,[17] and for rabbits exposed to $CdCl_2$ via daily subcutaneous injections.[178] In each study, the urinary excretion of cadmium was low until the liver and renal levels peaked but then increased considerably. The decrease in the tissue levels may therefore be due to the increased urinary excretion.

A study of 36 male rhesus monkeys[73] reported on the distribution of cadmium between different organs after feeding them cadmium in food pellets with additions of 100, 30, 10, 3, and 0 mg Cd per kilogram. The pellets themselves contained some cadmium as seen by the elevated blood, urine, and tissue levels of cadmium in some of the control animals,[177] but cadmium levels for the untreated pellets were not reported.

Between 39 and 77 weeks of exposure, seven monkeys were killed. After 2 years of exposure, five monkeys (one in each dose group) were killed and after 3 years of exposure, an additional five monkeys were killed.[177]

At autopsy, 1-gram samples were taken from 44 different tissues and these were analyzed by flame atomic absorption spectrophotometry with deuterium background correction. Some specimens were analyzed in duplicate by two different laboratories. For each animal a total dose of cadmium was calculated from the cumulated food consumption and the concentration of cadmium added to the food (range 0.02 to 22 g Cd per animal).

It can be seen in Figure 17 that the highest cadmium concentrations are found in renal cortex, liver, and renal medulla. The cadmium levels in all tissues reported in Figure 17 increase with the dose level. In terms of the total contribution to the body burden of cadmium in different tissues, the relative contribution of muscle, bone, and skin will be much greater than indicated by the data in Figure 17, due to their large size.

The liver and kidney contribute the largest proportions of body burdens. From the graphs in the report[177] it can be calculated that after a dose of 1 g Cd, the total body burden was about 30 mg and the amount in the liver and kidneys about 10 mg each. After a dose of 10 g Cd, the total body burden was 300 mg, the amount in liver 200 mg, and in kidneys 20 mg. There were considerable losses of cadmium from the kidneys as the doses progressed to the highest levels. Before these losses occurred, about one third of the body burden of the monkeys was found in the kidneys, one third in the liver, and one third in all other organs together.[177]

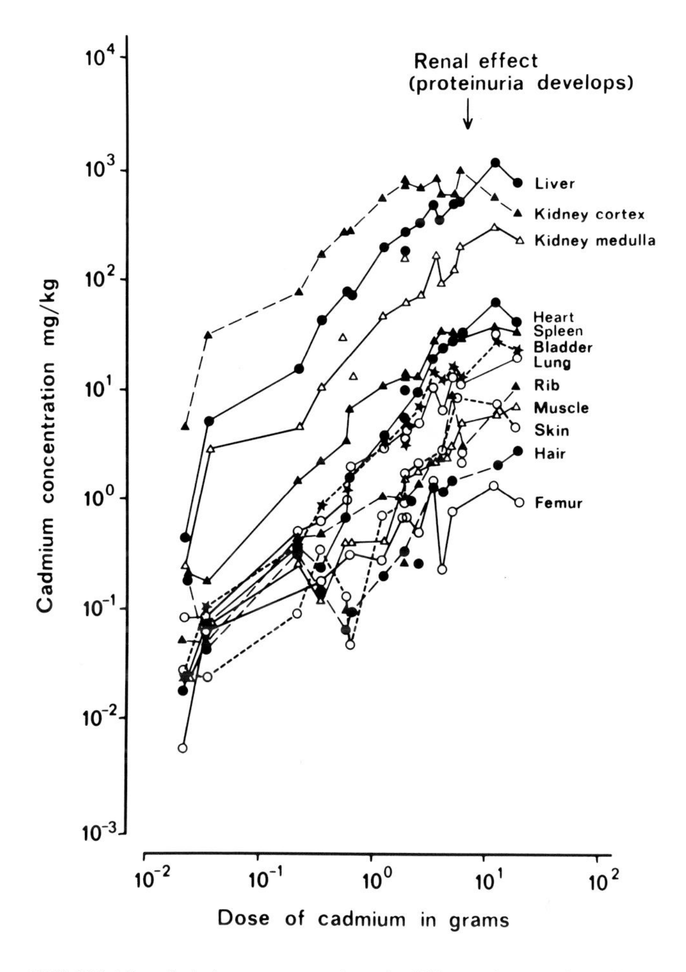

FIGURE 17. Cadmium concentrations in different tissues of monkeys fed cadmium in the diet, as a function of total ingested dose. (Modified from Nomiyama, K., Nomiyama, H., Akahori, F., and Masaoka, T., *Recent Studies on Health Effects of Cadmium in Japan*, Environment Agency, Office of Health Studies, Tokyo, 1981, 154—188.)

2. In Humans

Cadmium accumulates in a number of organs of humans with "normal" exposure via food (Chapter 5). In muscles, the accumulation continues throughout the whole life span (see Chapter 5, Figure 9). In kidneys, a maximum is reached at age 40 to 50 years (see Chapter 5, Figure 6) and in liver the accumulation levels off after 30 years of age (see Chapter 5, Figure 7). The renal cortex has higher cadmium concentrations than the renal medulla and in most studies the renal cortex level is about 10 to 30 times higher than in liver at age 50 (see Chapter 5, Table 2). We refer to Chapter 5 for detailed data on interorgan distribution of cadmium in "normal" humans. In this section we will highlight some of the more important aspects of the distribution.

As was pointed out in Chapter 5, the tissue levels increase when the exposure level increases. A study of representative populations in three countries, which incorporated careful analytical quality control procedures, showed a good correlation between cumulated daily intake and cadmium concentrations in renal cortex, liver, and muscles.[131]

Sensitive analytical techniques have revealed cadmium in all tissues and the estimated distribution in the average 45-year-old person with a daily cadmium intake from food similar

Table 5
ESTIMATED ORGAN AND BODY BURDENS OF 45-YEAR-OLD "REFERENCE" MEN[131]

		Organ and body burdens (mg) [a]		
Type of organ	**Organ weight (kg)**	**Japan (Tokyo)[b]**	**U.S. (Dallas)[c]**	**Sweden (Stockholm)[c]**
Kidneys[d]	0.26,[b] 0.30[c]	11.4 (44)	4.7 (16)	4.0 (13)
Liver	1.2,[b] 1.5[c]	3.4 (2.8)	1.7 (1.1)	1.0 (0.68)
Pancreas	0.06,[b] 0.07[c]	0.13 (2.2)	0.049 (0.70)	0.035 (0.50)
Muscles	26,[b] 30[c]	4.7 (0.18)	1.8 (0.060)	1.2 (0.040)
Blood	4.6,[b] 5.4[c]	0.005—0.028 (1—6 μg/kg)	0.006—0.032 (1—6 μg/kg)	0.006—0.032 (1—6 μg/kg)
Rest of body[e]	27.6,[b] 32.0[c]	3.3 (0.12)	1.2 (0.038)	0.85 (0.026)
Total body	60,[b] 70[c]	23	9.5	7.1

[a] Concentrations (mg/kg) used for the calculations are given in parentheses. Smokers and nonsmokers are mixed. The calculation includes also those cases for whom smoking habits were unknown. Among the cases for whom smoking habits were known, present or former smokers constituted 77% in the American group and 82% in the Swedish group.
[b] Data from Sumino et al.[243]
[c] Data from "reference" man (ICRP).[115]
[d] It was assumed that concentration in kidney cortex was 25% higher than in whole kidney.
[e] Concentrations are 64% of concentration in muscles, based on assessment of data by Sumino et al.[243]

to that found in Japan, U.S., and Sweden (40, 20, and 16 μg/day, respectively) is given in Table 5. It can be seen that the kidneys contain about 50 to 56% of the estimated body burden, the liver 14 to 18%, the muscles 17 to 20%, and all other tissues combined contain less than the liver.

It is interesting to note that the ratio of the cadmium concentrations in liver and kidney cortex increases with increasing cadmium concentration in the liver (Figure 18). Figure 18 shows mainly individuals without any signs of renal tubular damage. When such damage occurs, the kidney cortex cadmium concentration decreases (Section IV.A.1 and Volume II, Chapter 9) and the ratios will be even greater. In the 12 workers with varying degrees of renal tubular damage for whom data on liver cadmium and renal cortex cadmium are available the ratio was in the range 0.4 to 10 (Volume II, Chapter 9, Table 4), which is much greater than the ratio of 0.03 to 0.17 shown in Figure 18. The increasing ratio with increasing liver cadmium concentration, even before renal damage occurs, may reflect the greater proportion of body burden in the liver at higher exposure levels (Section IV.A.1).

The monkeys studied by Nomiyama et al.[177] had, at a dose of 1 g Cd, a liver to kidney cortex cadmium ratio of about 2.5. The liver to muscle cadmium ratio was about 1000 for the monkeys and 16 to 18 for the low-exposed humans (Table 5).

Concentration ratios for liver, kidneys, and muscles similar to those reported in Table 5 have been found in other autopsy studies of groups in general populations (Chapter 5). The higher the total body burden, the lower the proportion found in kidneys. A detailed study by Sumino et al.[243] of a group of Japanese adults with an average body burden of about 33 mg Cd found 36% in the kidneys, 26% in the liver, and 21% in the muscles (Table 6). The cadmium dose, which is likely to cause the critical effect (renal damage, see Volume II,

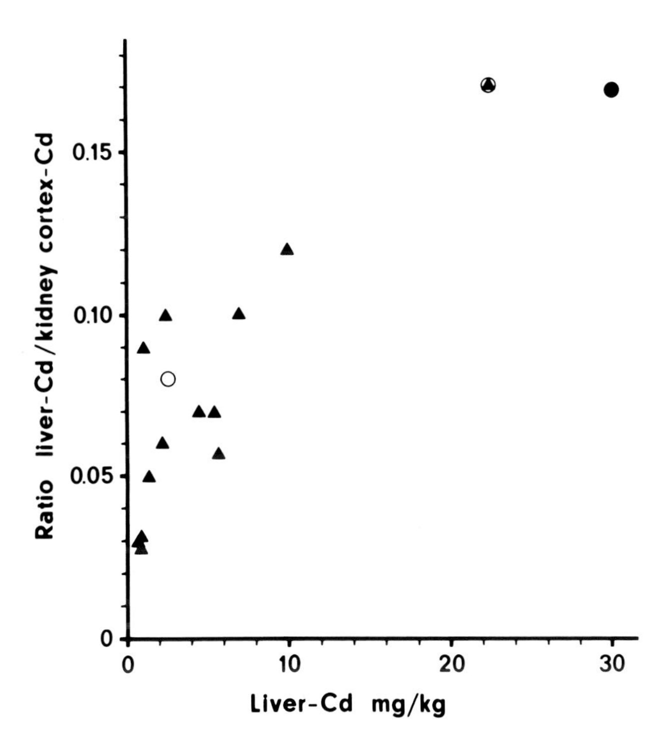

FIGURE 18. Ratio between human liver cadmium and kidney cortex cadmium (averages from several studies) at age 40 to 59 as a function of liver cadmium after long-term exposure. (▲) From Chapter 5, Table 2. (○) Controls.[65,66] (●)Midpoint workers without effects.[224] (▲) All workers.[65,66]

Chapter 9), is higher than the dose experienced by the Japanese group studied by Sumino et al.[243] Therefore, for toxicity evaluations, the best estimate of the proportion of whole body burden that is in the kidneys appears to be about one third. This agrees with the study on monkeys[177] and the data on humans with higher than "normal" exposure levels.

It should be pointed out that the body burden calculations in Tables 5 and 6 are approximate as not all body tissues were measured and individual body burdens were not measured. The "rest of body" in Table 5 contributes a substantial amount to the body burden, but was not measured in that study. There are also differences in the cadmium concentrations of different muscles, e.g., the heart has five times higher levels than the other muscles sampled in the monkey study (Figure 17), and recent autopsy data[150] have shown differences in cadmium concentrations of different bones. However, the data referred to here are the best available estimates of body and tissue burdens for humans.

Very limited autopsy data on cadmium concentrations in tissues of cadmium workers are available. Most of the workers examined had cadmium-induced renal damage and therefore disproportionately low renal cadmium levels. These data will be discussed in detail in Volume II, Chapter 9. Data from two cross-sectional studies of renal and liver cadmium of cadmium workers, based on in vivo neutron activation analysis[65,223,224] have been used in Figure 18. As yet, the whole body burden cannot be measured with this technique.

The degree of accumulation with age in a specific tissue depends on the half-time in that tissue. The different tissues are interrelated via blood. The cadmium that leaves the liver will to some degree be secreted to blood bound to metallothionein (Chapter 4) and will then be available for accumulation in other tissues. The turnover of cadmium in most tissues will mean that the cadmium is returned to blood. The liver and kidneys are special cases as they have specific excretion routes for cadmium (bile and urine).

Table 6
CONTENTS OF CADMIUM AND AVERAGE WEIGHT IN ORGANS AND WHOLE BODY OF JAPANESE ADULTS[243]

Tissues	Average weight (g)	Cadmium (mg)
Muscles	24,000	7.0
Bone	8,500	0.82
Fat	6,600	0.45
Blood	4,500	0.76
Skin	4,200	1.3
Connective tissue	1,800	—[a]
Liver	1,500	8.5
Brain	1,300	0.16
Digestive tract	1,000	0.75
Lung	900	0.65
Heart	300	0.048
Kidneys	250	12
Spleen	150	0.12
Pancreas	100	0.27
Total	55,000	>33

Note: Correct total values from data in table: 55,100 and 32.828.

[a] Not calculated because there were less than five samples available.

The apparent half-time in a tissue as measured from accumulation curves in normally exposed people will be longer than the "true" half-times, because of the influx of cadmium via blood from the other tissues. The half-times calculated for each organ in the kinetic model for cadmium (Chapter 7) are thus shorter than those estimated earlier[97] from accumulation curves.

Longitudinal studies of people with particularly high exposure levels, using in vivo neutron activation to measure liver and kidney levels, may be of particular value in measuring half-times. One study[82] reports on liver cadmium levels in cadmium workers measured using in vivo neutron activation analysis with an interval of 3 to 4 years. After cadmium exposure ceased, the liver levels decreased with an average half-time of 6.4 years in workers without kidney damage.[82] The in vivo technique could possible also be used to study the transfer of cadmium from one tissue to another, during the "recovery phase" after a high exposure. No such studies have yet been carried out.

B. Distribution within Organs

1. Uptake of Cadmium by the Kidneys and Subcellular Distribution

As was discussed in Section III., cadmium does occur in blood plasma partly bound to metallothionein. Such cadmium will be efficiently taken up to the kidneys by glomerular filtration and subsequent tubular reabsorption (Section III). As will be discussed in more detail in Section V., metallothionein-bound cadmium that is infused intraarterially will remain in a low molecular weight and freely filtrable form[83,174] so that the plasma clearance of metallothionein-bound cadmium will be equal to the glomerular filtration rate.

Subsequent to glomerular filtration, cadmium bound to metallothionein will be reabsorbed by the proximal renal tubules.[201] When the filtered load of cadmium-metallothionein is excessive, reabsorption is saturated (see Section V.) and when the cadmium concentration in the renal cortex exceeds approximately 200 mg/kg wet weight in a course of long-term accumulation, reabsorption is also depressed (Volume II Chapter 9). When renal tubular damage is induced by a large single injection of cadmium-metallothionein,[195] this type of deficient reabsorption of cadmium-metallothionein can probably occur at a considerably lower concentration of cadmium in the renal cortex. Observations of cadmium in renal lysosomes[86,87] and degradation of metallothionein[280] in the renal tissue as well as the observation[241] of a large proportion of renal cadmium in nonmetallothionein fractions of the renal cells at observation times less than 4 hr after injection of Cd-MT favor the role of nonmetallothionein-bound Cd for elicitation of renal toxicity. For example, Squibb et al.[241] found 94% of cytoplasmic Cd to be nonmetallothionein bound at 1 hr after injection of Cd-MT. The differences in subcellular distribution in the kidney cortex after injection of cadmium-metallothionein and cadmium chloride, respectively, have not been clearly elucidated. Norberg et al.[189,192] reported time-related differences after Cd-MT or $CdCl_2$ injection. Cherian et al.[43] considered that the subcellular distribution was the same regardless of whether cadmium had been given as $CdCl_2$ or Cd-MT. In view of the differences in organ distribution (Section IV.A) between systemically administered Cd-MT and $CdCl_2$, it may be expected that some differences may occur at some specific time interval after injection. Waku[275] reported 90 % of renal cadmium to be cytosolic and metallothionein bound after long-term oral exposure in rats. In this type of exposure situation, sufficient time has been available for internal renal metallothionein synthesis to take place. Orally exposed rats on a low zinc intake[212] had a lower proportion metallothionein-bound cadmium in noncytosolic cellular fractions than did animals that had normal zinc intake. Metabolism studies after injection of ^{3}H and ^{109}Cd-labeled cadmium-metallothionein[240] indicate that cadmium-metallothionein is rapidly degraded by proximal tubule cell lysosomes with subsequent release of Cd^{2+} into nonmetallothionein compartments where the sensitive sites for cadmium-induced damage occurs (Figure 19). After entry into other compartments of the renal tubular cell, cadmium induces the internal synthesis of metallothionein and while cadmium has a very long biological half-time, the metallothionein moiety has no similarly long half-time. Renal metallothionein has been shown to undergo constant turnover[221] (Chapter 4), thereby releasing some quantities of nonmetallothionein-bound cadmium, which will be able to damage the sensitive site in the renal tubular cell (Figure 19).

When sufficient concentrations of nonmetallothionein-bound cadmium are reached at the sensitive site, low molecular weight proteinuria develops (Volume II, Chapter 9). The mechanism underlying the development of such proteinuria might involve an inhibition of tubular protein reabsorption and degradation due to an inhibition of normal lysosome formation and the fusion of apical pinocytotic vesicles with primary lysosomes — events observed after injection of a single large dose of cadmium-metallothionein (0.6 mg Cd per kilogram body weight) in rats.[240]

Johnson and Foulkes,[121] in a study on the proposed role of metallothionein in the transport of cadmium from liver to kidney (Figure 19), were able to confirm such a role by observing a much more pronounced renal uptake of cadmium when rats were infused over a 3-day period with cadmium-metallothionein than when they were infused with $CdCl_2$. Pretreatment with cadmium, zinc, or mercury did not alter the tissue distribution. Selenke and Foulkes[230] demonstrated binding of cadmium-metallothionein to isolated renal brush border membranes providing further support that renal tubular reabsorption takes place via the brush border of the proximal renal tubule.

The uptake of injected cadmium-metallothionein by the kidney was also studied by Suzuki.[244] Suzuki considered that the liberated cadmium ions from degraded cadmium-metallotionein were related to observed necrosis of renal tubular lining cells.

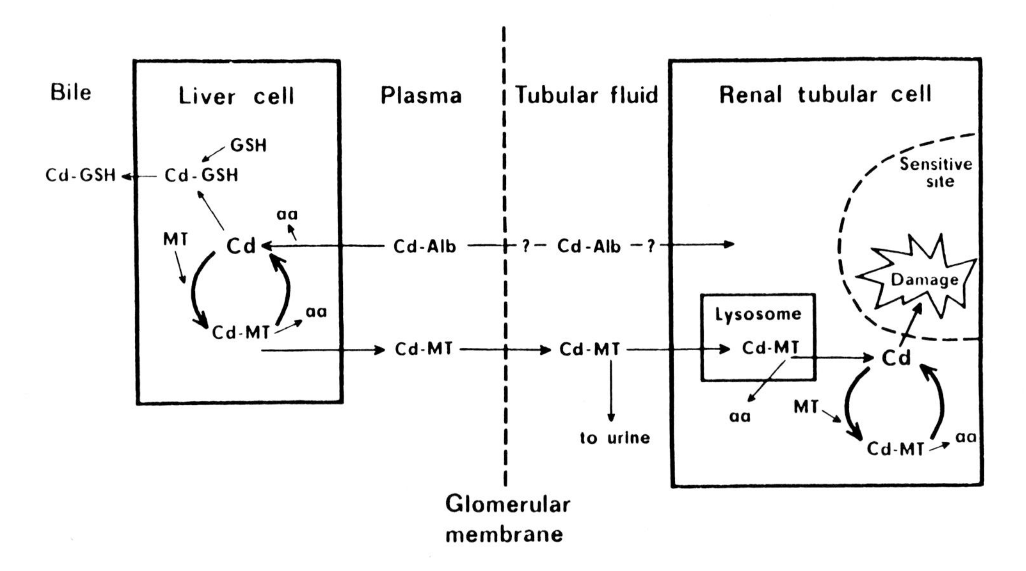

FIGURE 19. Basic flow scheme of cadmium in the body demonstrating the role of binding forms in blood and metallothionein synthesis and degradation. (Modified from Nordberg, G. F., *Environ. Health Perspect.*, 54, 213—218, 1984.

2. Distribution within Kidneys

The distribution of cadmium within various histological parts of the kidneys has special significance for the evaluation of the toxicity of cadmium, because the renal damage of cadmium initially occurs in the proximal tubules located in the renal cortex (Volume II, Chapter 9). The renal cortex concentration of cadmium gives, therefore, an important estimate of the critical organ concentration. In view of the mechanisms for renal uptake of cadmium from blood plasma discussed in the foregoing Section IV.B.1, it can be expected that cadmium would accumulate selectively in the renal cortex. It was shown already in Section IV.A.1 that the renal cortex has a higher concentration of cadmium than the renal medulla after single parenteral exposure (Figure 15). A number of studies have confirmed this to be true for different exposure routes, exposure types, and species of animals.

Autoradiographic studies of mice given tracer dose of ^{109}Cd[16] have shown that within the renal cortex the greatest accumulation was found in the outer cortex in the area of the proximal tubules. Especially, the first segment of the proximal tubules showed high cadmium levels, whereas the glomeruli had lower levels. Similar results were found when immunohistological techniques were used to locate the metallothionein in horse kidneys.[140]

When ^{109}Cd-metallothionein was injected intravenously in mice, a selective uptake by the renal cortex was demonstrated by autoradiography.[201] Mice exposed to ^{109}Cd bound to low molecular weight protein from hemolysate[199] (Section III.A.4) showed by autoradiography an accumulation of ^{109}Cd in the renal cortex (Figure 12). Ninety-six hours after a single exposure, the kidney cadmium levels were 10 times higher in these mice than in mice exposed to the same dose of $^{109}CdCl_2$. Thus, cadmium bound to low molecular weight protein is particularly prone to accumulate in the kidney. Within the renal tubular cells the location of cadmium is closely related to the location of metallothionein (Chapter 4).

The subcellular distribution of cadmium was discussed in Section IV.B.1. Initially, after injection of Cd-metallothionein, cadmium is partly found in the mitochondrial-lysosomal fraction. X-ray microanalysis also found a part of the cadmium in the lysosomes of rabbit kidney tubular cells.[87]

The distribution within human kidneys was discussed in Chapter 5, Section IV. In most autopsy studies, a ratio of 2 to 3 between the cadmium concentration in renal cortex and medulla was found. This ratio can be strongly affected by the technique used when taking specimens from the kidneys. Livingston[151] found a continuously increasing cadmium concentration from the center of one kidney cut into eight layers.

A recent study[254] of kidneys from 20 autopsied men in the age range of 30 to 59 years showed that on average 65% of the kidney weight was cortex, 27% was medulla, and 8% was renal pelvis ("the remaining part"). The arithmetic average cadmium concentration in renal cortex was 18 mg/kg, in renal medulla 7 mg/kg, and in the remaining part 3 to 4 mg/kg. The average ratio of renal cortex cadmium concentration to whole kidney cadmium concentration was 1.30. The earlier estimate of 1.5[97] appears to be in error. This is due to the fact that the estimate was based on the data by Geldmacher-v. Mallinckrodt and Opitz[100] and involved the assumption that the kidney consisted of 50% cortex and 50% medulla. No measurement of the relative size of these tissues had been made. A recent reevaluation of these data[135] indicated that the ratio should have been 1.15 instead of 1.5. A ratio of 1.5 has been widely used, but it is now apparent that the true ratio for humans is more likely to be lower. We suggest that a factor of 1.25 may be used for future calculations. However, this figure may vary with species, age, and condition of the kidneys. It is clear, for instance, that the monkeys studied by Nomiyama et al.[177] have a different intrarenal cadmium distribution than humans. The renal cortex cadmium concentration was, at each dose level, about ten times higher than the concentration in renal medulla (Figure 17).

3. Liver

Autoradiographic studies on the distribution of cadmium in different parts of the liver using single injection exposure[14,96] and repeated exposure[94,181] have revealed that cadmium is equally distributed in the different parts of the liver immediately after a single exposure, but is concentrated in the periphery of the liver lobules 8 or more days after a single exposure or after repeated exposure.

Within the liver cells, cadmium is found in the nuclei, mitochondria, microsomes, and supernatant fractions after ultracentrifugation.[122] Rats were given a single subcutaneous injection of $CdCl_2$ (10 mg/kg) and 6 to 12 hr after injection approximately 30% of the liver cadmium was found in the supernatant, whereas, after 24, 48 and 168 hr about 60% was in that fraction.[122]

Piscator[213] reported that a large part of cadmium in the liver of cadmium-exposed rabbits was found in the supernatant and consisted of metallothionein. Nordberg et al.[192] studied the distribution of cadmium after subcutaneous injection in mice. There was a gradual increase of the cadmium concentration in the supernatant (centrifugation at 18,000 g) from 43% of total liver cadmium at 20 min to more than 80% at 6 and 18 days after injection. Distribution of cadmium among proteins in this supernatant showed that there was a redistribution of cadmium from high molecular weight proteins to low molecular weight proteins (Chapter 4), which is related to variations in the acute toxicity of cadmium[7] (Volume II, Chapter 10).

After long-term exposure, liver cadmium is mainly bound to metallothionein as shown in a number of studies (Chapter 4). Waku[275] studied, e.g., the subcellular distribution of cadmium in tissues from rats exposed for 12 months to cadmium via drinking water. Approximately 90% of tissue cadmium was found in cytosolic metallothionein in liver. The mitochondrial fraction contained 3 to 5% and the microsomal fraction 5 to 7% of tissue cadmium. After solubilization of these particulate fractions, approximately 90% of the cadmium in these fractions was also found to be bound to metallothionein.

4. Other Organs

Limited data are available regarding the distribution of cadmium within other organs. The testes and the epididymis have attracted particular attention due to the acute effects of cadmium on these organs (Volume II Chapter 11). Whole body autoradiography of mice has shown that, after intravenous exposure, cadmium is localized in the interstitial tissue of the testicles.[15,181] Microautoradiographic methods showed that results were identical irrespective of type of administration or dosage and that cadmium was localized in the capillaries and in the immediately adjacent interstitial tissue.

More detailed knowledge has been contributed by Nordberg.[181] A group of eight mice were given subcutaneous injections of 1.5 mg Cd per kilogram body weight as ^{109}Cd and two mice each were killed by cervical dislocation 4 hr, 1 day, 6 days, and 18 days after injection. A second group of eight animals were given intravenous injection of 30 μCi per mouse of carrier-free $^{109}CdCl_2$ and killed 2 hr, 4 hr, 8 hr, and 1 day after injection. A third group of three mice was given 0.5 mg Cd per kilogram subcutaneously 5 days/week for 5 months. Testicles with attached epididymis were removed from the mice and taken for autoradiography according to the method by Ullberg.[269]

In mice given repeated injections, cadmium was found in the interstitial tissue of the caput epididymis, and it also seems to be present in the epithelium of the duct in the cauda epididymis.[181] A similar distribution in the epididymis was found in mice given a single injection of cadmium. Using histochemical methods, cadmium was found in the walls of testicular capillaries of hamsters 1.5 hr after injection.[236] Johnson and Sigman[120] reported the same localization 2.5 to 40 min after injection in rats. However, they used a technique which, to a great extent, removes cadmium from the specimen by ordinary histological technique and, therefore, cannot be considered reliable.

Nordberg[181] showed that for testicles from mice subcutaneously injected with 1 mg Cd per kilogram body weight, 50% of testicular cadmium was in the 105,000-g supernatant and only 26% of the cadmium supernatant was in a small protein corresponding to metallothionein. Corresponding values 4 days after injection were 74% in the supernatant and 55% in the metallothionein. Animals pretreated with 0.25 mg Cd per kilogram for 12 days prior to the 1 mg/kg dose had more cadmium bound to metallothionein than nonpretreated animals.

C. Mechanisms of Transport and Distribution

Immediately after exposure, cadmium is present as cadmium-albumin in blood plasma (Section III.A.3). Such cadmium will be taken up mainly by the liver. During the first hours after uptake, cadmium in the liver will not be bound to metallothionein.[7,148,192] The synthesis of metallothionein, however, will be induced and after 24 hr cadmium will mainly be bound to metallothionein, which to a large extent remains in the liver[7] (Chapter 4). A small proportion will be released to the blood (Section III.), while the major proportion will take part in a constant turnover in the liver, whereby nonmetallothionein-bound cadmium will be released which will again induce synthesis of new metallothionein and so on (Chapter 4, Section V; Chapter 7). Figure 19 displays these relationships.

As cadmium itself stimulates metallothionein synthesis in the liver, the possibility exists that the transport of cadmium from plasma to liver will be more rapid and more complete during continuous exposure than after a single exposure. This may explain the prominence of slow half-time components in the experiments with continuous exposure as compared to those with single exposure (Section V.D.1).

Cadmium-metallothionein that occurs in blood plasma will be quickly transported to the kidneys where it is filtered through the glomeruli and reabsorbed by the proximal renal tubule (Section IV.). The metallothionein moiety will be broken down by the lysosomes and nonmetallothionein-bound cadmium will be released in the kidney cells (Figure 19). Such cadmium will stimulate renal metallothionein synthesis and there will be a constant turnover of metallothionein (Chapter 4 and Section IV. in this chapter). A majority of renal cadmium will be bound to metallothionein in a long-term exposure situation. However, a certain amount of nonmetallothionein-bound cadmium will occur and when this concentration exceeds a certain level at the sensitive sites in the cell, toxicity to the renal tubule will occur (Chapter 4 and Volume II, Chapter 9).

D. Placental Transfer

Cadmium accumulates in the placenta to a considerable degree and large single doses can give rise to necrosis of the placenta. The passage of cadmium across the placenta is dependent on dose, gestational age, and probably also animal species. Data from both experimental animals and humans are available and have been reviewed by Miller and Shaikh.[162]

1. In Experimental Animals

Several studies are available on the acute effects of cadmium on the placenta and conceptus (Volume II, Chapter 11, Section VI.). The placenta in rats is destroyed by parenteral doses greater than 2 mg Cd per kilogram body weight.[208,209] At similar doses, teratogenic effects have been induced (Volume II, Chapter 11, Section VI.).

The placental accumulation and placental transfer of cadmium during the latter part of gestation was studied by Berlin and Ullberg.[15] They gave pregnant mice single intravenous $^{109}CdCl_2$ injections (carrier-free ^{109}Cd) on the 18th day after conception. No cadmium was detected in the fetus, but there was an accumulation of cadmium in the placenta. Tanaka et al.[257] gave pregnant mice single intravenous injections of ^{115m}Cd as chloride (about 15 μg Cd per mouse) 24 to 36 hr before delivery. The mean uptake in newborns was 0.09% of the dose.

Sonawane et al.[239] studied the placental transfer of intravenously injected $CdCl_2$ in rats and the influence of dose and gestational age. Doses were 0.1, 0.4, and 1.6 mg Cd per kilogram and the gestational ages 12, 15, and 20 days. Higher percentages of administered cadmium accumulated in the fetus with increasing dose and increasing gestational age. For example, after pregnant rats were injected with low, middle, and high doses of cadmium on day 12 of gestation, fetuses accumulated 0.0001, 0.0028, and 0.0095% of the three injected doses, respectively.

The percentage of administered cadmium detected in placental tissue did not change consistently with dose, but cadmium levels did increase with gestational age. Placental to maternal blood cadmium concentration ratios increased with gestational age but not with dose. Maternal liver to fetal liver concentration ratios were 295, 137, and 27 for the low, middle, and high doses, respectively, 21 hr after pregnant rats were treated on day 20 of gestation.[239]

There are also some studies which describe the transfer of cadmium to the embryo which takes place during earlier stages of embryonic and fetal development. After single intravenous doses of 0.5 to 0.85 mg Cd per kilogram to pregnant hamsters on the 8th day of gestation,[78] cadmium was found both in the placenta and in the fetus. The cadmium was given during the period in which the major organ systems are being established in the hamster embryo.

A more detailed localization of cadmium in the embryo was reported by Dencker[56] using autoradiographic techniques. Cadmium was localized in the primitive gut wall and vitelline duct. The pattern of embryonic uptake was followed up over time. Uptake was found to stop abruptly at day 9 of gestation in the hamster (day 9.5 in the mouse). This is the time when the vitelline duct closes. Injections of relatively high doses of cadmium to hamsters have been reported to give rise to teratogenicity when they are given during the early period of gestation (Volume II, Chapter 11, Section VI.).

The uptake of cadmium in the fetus following repeated or long-term exposure has been studied to a limited extent only. Pregnant rats exposed to cadmium oxide dust in concentrations of about 3 mg/m^3 every day from conception[52] showed an uptake of cadmium to embryonic livers.

Effects of cadmium on birth weight and possibly other fetotoxic effects may well be related to a deficient placental transfer of nutrients and essential metals (Volume II, Chapter 11, Section IV.A.1). Also of interest is the fact that chelating agents such as EDTA increase the transplacental passage of cadmium.[75]

2. In Humans

Henke et al.[111] analyzed liver and kidney samples from newborns in West Germany using neutron activation and found between 4 and 20 μg/kg wet weight in four kidneys. The concentration was less than 2 μg/kg in the liver. This means that the total content of cadmium in the newborn was less than 1 μg. It also means that the fetal liver and kidney concentrations are about 1000 times lower than those that would be expected in the mother (see Section IV.A.2).

Atomic absorption analysis of tissues from human stillborn fetuses (Kyoto, Japan) of 85 to 185 days gestational age, revealed cadmium concentrations in the range 0.02 to 0.22 mg/kg (liver), 0.02 to 0.062 mg/kg (kidneys), and 0.02 to 0.144 mg/kg (brain) wet weight.[34] These values are higher than those reported from West Germany[111] and may be due to the higher level of exposure in Japan. Since concentrations were low and atomic absorption analysis was carried out without elimination of possible interference, analytical errors cannot be excluded.

Concentrations in the placentas of 44 Swedish women were found to be less than 10 μg/kg wet weight using a spectrographic method.[214] Even though this method is also somewhat uncertain, the indication is that not more than about 5 μg cadmium will accumulate in the placenta, which weighs about 500 g. In comparison, a human kidney will usually accumulate about 100 μg over a 9-month period (see Section IV.A.2).

In a study by Roels et al.,[222] the concentrations of lead, mercury, and cadmium were determined in placenta from 474 European women and were compared with the level found in maternal and newborn blood. The influence of some individual and environmental factors (area of residence, smoking habits, drinking habits, age, occupation, previous pregnancies) on heavy metal accumulation in the placenta was also investigated. The median values of the three heavy metals in placenta, wet weight, were 75 μg Pb per kilogram, 11 μg Hg per kilogram, and 11 μg Cd per kilogram. In comparison with maternal blood, the placenta does not concentrate lead or mercury, but concentrates cadmium about tenfold. Cadmium concentration in placenta was significantly correlated with that in maternal blood ($r = +0.38$). Of these three heavy metals, only cadmium showed increased accumulation in placenta of smokers. Current residence, maternal age, and occupation had no significant effect on the observed accumulation of the heavy metals in placenta.

Additional data on "normal" levels of cadmium in the placenta are given in Chapter 5. The occurrence of cadmium in the placenta may be of importance for the health of the fetus (Volume II, Chapter 11, Section VI.).

E. Conclusions

Most absorbed cadmium will be found in the liver and the kidneys following the types of human exposure that occur in the general or occupational environment. All organs will contain some cadmium, and in many organs, particularly the muscles, kidneys, and liver, cadmium accumulates over the whole lifetime of the individual. The accumulation in the kidneys is of particular importance for the toxicity of cadmium.

After long-term low level exposure, about one third to one half of the body burden of cadmium is found in the kidneys, one sixth in the liver, and one fifth in the muscles. The higher the exposure level, the greater will be the proportion located in the liver, with a correspondingly lower proportion in the kidneys.

Within the kidney, the highest concentrations are found in the cortex. Particularly high concentrations are found in the proximal tubular cells. The average ratio of cadmium concentrations in renal cortex to whole kidney was estimated to be 1.25. This ratio is of importance for the recalculation of whole kidney cadmium concentrations to renal cortex concentrations.

Some cadmium is transferred via the placenta to the fetus, but due to the short period of cadmium accumulation in the fetus, the tissue cadmium levels at birth are very low.

V. EXCRETION

The main excretion routes for cadmium are urine and feces. Urinary cadmium is influenced by daily intake, body burden, and damage to the kidney. Different animal studies may have focused on one or more of these variables and there may be difficulties in comparing studies. Fecal cadmium is also a function of several factors, such as unabsorbed daily intake, biliary excretion, and "true" intestinal excretion (Figure 4). An additional problem is the difficulty in separating urine and feces in metabolic studies of animals.

A. Urinary Excretion

1. Single Exposure: Animals

After single exposures the daily urinary excretion is only a small fraction of the dose given, in some studies as low as 0.002%, but during the first days after exposure the excretion may be as high as several percent of the dose. This is dose–dependent and may be the result of renal "overflow" of plasma cadmium.

In rats, Lucis et al.[154] found that 1% of a single subcutaneous injection of ^{109}Cd solution (carrier free) had been excreted with the urine after 1 week. In rats given a single subcutaneous injection of cadmium, Shaikh and Lucis[232] found a daily urinary excretion over 4 days of between only 0.003 to 0.007% of the dose.

Nordberg[181] gave mice a single subcutaneous dose (0.25 mg Cd per kilogram body weight) of radioactive cadmium and after that daily doses (0.25 mg Cd per kilogram body weight) of nonradioactive cadmium. There was an initial combined urinary and fecal excretion of about 5% of the single dose given during the first week. During subsequent weeks, less than 1% of the dose was excreted.

The selective uptake of cadmium in the kidney cortex of mice when cadmium-metallothionein was given as an intravenous injection[201] was referred to in Section III.A. In this study, single intravenous injections of metallothionein, or $CdCl_2$ (0.3 or 0.8 mg Cd per kilogram body weight) were given. Four hours after exposure, 0.03 to 1.23% of the lower dose cadmium-metallothionein dose was excreted in urine. The corresponding values for the higher dose group were 0.03 to 2.18%. Clearly the direct transport of cadmium-metallothionein to the kidneys facilitates its urinary excretion.[201]

2. Repeated Exposure: Animals

After repeated exposure, excretion increases over time. There is an association between body burden and urinary excretion. When the kidney has accumulated so much cadmium that renal damage occurs, urinary excretion increases dramatically.

Friberg[94] gave subcutaneous injections of cadmium sulfate containing ^{115}Cd in daily doses of 0.65 mg Cd per kilogram to rabbits for 10 weeks. For a period varying between 6 and 7 weeks, the daily cadmium excretion was very small, less than 1% of the daily injected dose. During the last weeks of the experiment there was a sharp rise in cadmium excretion, up to about 100 times the amount excreted during the first weeks. This rise in cadmium excretion coincided with the appearance of proteinuria in the rabbits (Figure 20).

Similar observations concerning an increase in urinary excretion have later been made in rats[17,231,251] and rabbits.[9,175]

Nordberg and Piscator[190] and Nordberg[181] found that during repeated subcutaneous ex-

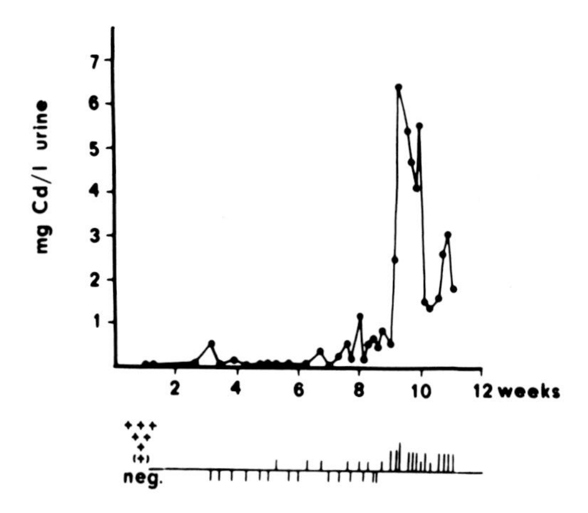

FIGURE 20. Cadmium concentrations and occurrence of protein in urine of one cadmium-exposed rabbit selected from the classical study by Friberg.[94] (+) Proteinuria demonstrated only with trichloroacetic acid; (+ to + + +) Proteinuria demonstrated also with nitric acid. (Modified from a figure by Friberg, L., *Arch. Ind. Hyg. Occup. Med.*, 5, 30—36, 1952.)

posure to 0.25 or 0.5 mg Cd per kilogram body weight for up to 6 months, the urinary excretion was very low (0.007 to 0.009% of body burden per day). At the time that pathological urinary proteins were observed on electrophoresis (after 21 weeks) in the group given the highest exposure, there was a sharp increase in urinary cadmium excretion. It was shown by Nordberg and Piscator[190] by means of gel filtration, that a large part of the cadmium in urine, during this phase of pronounced excretion, was bound to proteins with a molecular weight corresponding to that of metallothionein.

Nordberg,[181,182] in the studies referred to above, also showed that the urinary excretion on a group basis was significantly related to total body burden during the period of low urinary excretion, before pathological proteinuria appeared. At the highest dose level 0.02% was excreted daily, whereas at the lower dose levels about 0.01% was excreted (Figure 21).

Elinder and Pannone[60] gave repeated subcutaneous injections up to a total dose of 6 mg Cd per kilogram body weight. The 24-hr urinary excretion 4 weeks after the last injection was 0.001% of body burden of cadmium. This is the proportion of body burden excreted via urine that seems to decrease with time after the last injection.

Nomiyama and Foulkes[174] collected urine from cadmium-poisoned rabbits (subcutaneous injections of 0.5 mg Cd /kg/day for 12 weeks). Mean renal cortex cadmium was 262 mg/kg. Urine was collected with urethral catheters while infusing inulin or cadmium-metallothionein (labeled with ^{115m}Cd) through a femoral artery catheter at a level above the renal arteries. Blood samples were collected via a catheter in the other femoral artery. There was no significant uptake of cadmium-metallothionein by erythrocytes and at all observation times up to 20 min all plasma ^{115M}Cd remained in the low molecular weight fraction as evaluated by gel chromatography. The filtered load of cadmium-metallothionein (Cd-MT) was calculated as the product of the plasma concentration and inulin clearance.

The excretion of metallothionein-bound cadmium was compared with the "filtered load" in control animals (Figure 22) and in the poisoned animals.[174] The range of filtered loads was 2.2 to 5.0 μg Cd per minute and the reabsorptions were calculated as the difference between excretion and filtered load.

Figure 22 shows that the tubular reabsorption was complete at filtered loads below 1.2

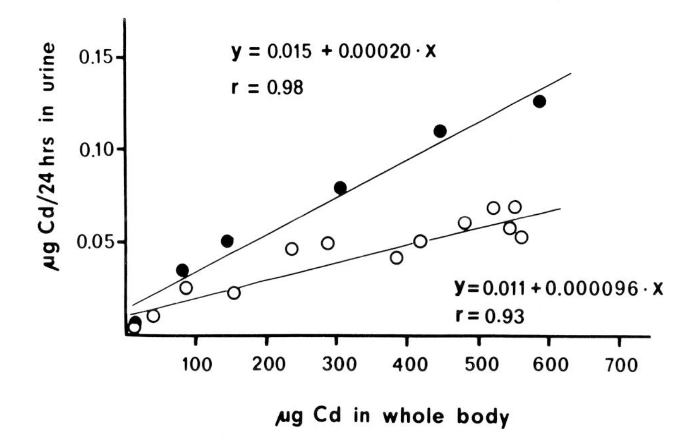

FIGURE 21. Relation between urinary excretion and whole body content of cadmium in mice given daily injections of ^{109}Cd (0.25 or 0.5 mg Cd/kg) and followed by whole body and excretion measurements. (From Nordberg, G. F., *Environ. Physiol. Biochem.*, 2, 7—36, 1972a. With permission.)

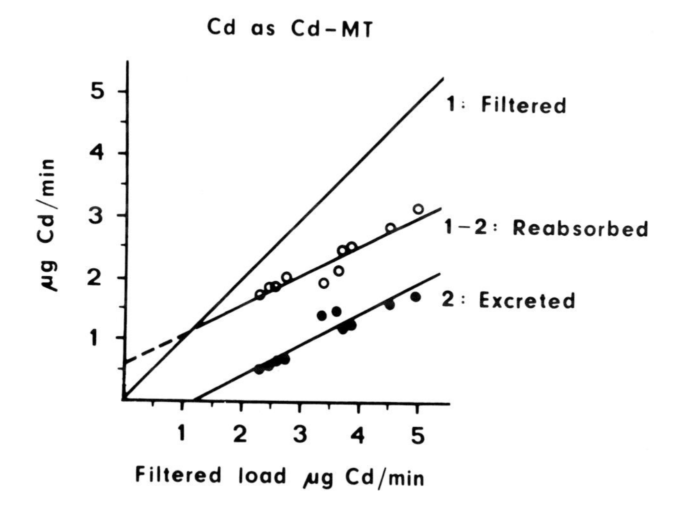

FIGURE 22. Excretion of cadmium-metallothionein at different filtered loads. Control animals; five successive clearance periods were obtained, and each point represents an observed (●) or calculated (○) value from one kidney. Lines were calculated by least-squares analysis. (From Nomiyama, K. and Foulkes, E. C., *Proc. Soc. Exp. Biol. Med.*, 156, 97—99, 1977. With permission.)

μg Cd per minute. Above this level an increasing amount was excreted (the slope is 0.5).

The average inulin clearance of the four control rabbits was 20 mℓ/min.[174] Thus, a filtered load of 1.2 μg/min[174] corresponds to a plasma concentration of 0.06 μg Cd per milliliter

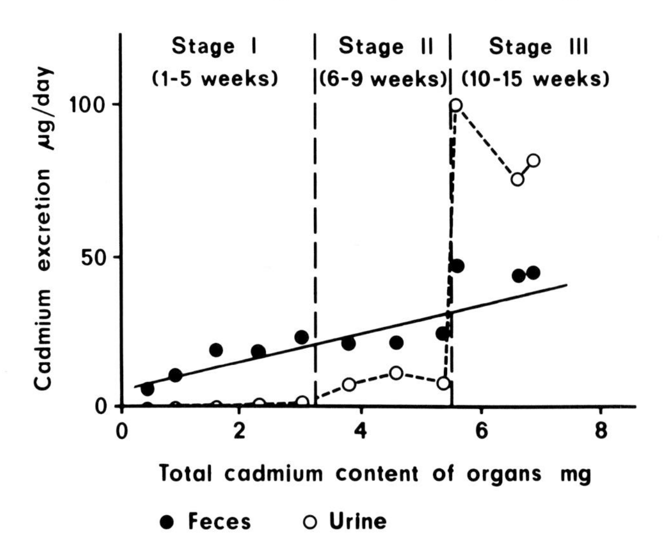

FIGURE 23. Relationships between total cadmium content of the organs (liver, kidneys, lung, spleen, and testes) and daily cadmium excretion in urine or feces of rats given a subcutaneous injection of 0.5 mg Cd/kg, 6 days/week. Each point represents the mean value of three to five animals. (From Suzuki, Y. and Yoshikawa, H., *Ind. Health*, 21, 43—50, 1983. With permission.)

(as Cd-MT) or 60 μg Cd per liter which is much higher than what has been reported for people with high exposures to cadmium in the general environment (less than 0.36 μg/ℓ[114]) and for cadmium workers (0.2 to 0.5 μg/ℓ[198]) (Section III.C). It appears unlikely that human cadmium exposures could lead to such high plasma concentrations that tubular reabsorption capacity is exceeded.

Foulkes[83] made further observations on rabbits and found no evidence of tubular secretion of injected Cd-MT. Further evidence was reported demonstrating that Cd-MT is freely filtrable and that its tubular reabsorption can be saturated (as discussed above). Of several proteins tested, only myoglobin interfered with Cd-MT reabsorption. Reabsorption was also depressed following cadmium exposure at the time when renal critical concentrations of cadmium exceeded approximately 200 mg/kg wet weight (accumulated in a course of chronic cadmium poisoning). These observations furnish experimental confirmation of the mechanisms for urinary excretion of cadmium in poisoned and nonpoisoned animals suggested by Friberg,[94] Piscator,[213] Nordberg and Piscator,[190] and Friberg et al.[97]

More recently, further evidence in accordance with the mentioned mechanisms has been reported. In a subchronic study of rats[253] given subcutaneous injections of $CdCl_2$ (0.5 mg Cd per kilogram body weight, 6 days/week for 15 weeks), the relationships between total body burden, urinary cadmium, and fecal cadmium were clearly seen (Figure 23). There is a steady but small increase of urinary cadmium until week 6, when renal cadmium stops increasing and proteinuria develops.[253] For 3 weeks, urinary cadmium increases above its trend line, which may be an indication of the increased excretion due to renal damage. Then at 10 weeks, urinary cadmium increases dramatically (Figure 23), which coincides with a dramatic decrease in liver cadmium levels. Probably liver damage has developed, large amounts of Cd-MT are released into the blood, and much of this is excreted in the urine (see also discussion in Section III.A.2).

Several studies of rhesus monkeys exposed to cadmium via the diet have been published by Nomiyama and the Japanese Monkey Experiment Research Team. In one study, 10 male rhesus monkeys were divided into groups exposed to 0, 3, 30, or 300 mg Cd per kilogram in food pellets.[176] Every 2 weeks, urine and feces samples were collected (method not stated). There was a great individual and week-to-week variation in the urinary cadmium concentration. In the two higher dose groups the concentration increased dramatically during the first 10 to 15 weeks and then leveled off at a level of 20 μg/ℓ for the 30 mg/kg group and at 500 μg/ℓ for the 300 mg/kg group.

In another study of 36 male rhesus monkeys divided into groups exposed to 0, 3, 10, 30, or 100 mg Cd per kilogram in food pellets,[177] the urinary cadmium concentrations again increased rapidly during the first weeks of exposure. The 100 mg/kg dose group had 1000 μg Cd per liter in urine after 10 weeks of exposure and the 30 mg/kg group had 100 to 500 μg Cd per liter in urine. These results do not agree with the results referred to above from the first experiment.[176] In the second experiment,[177] the control group of monkeys had urinary cadmium in the range 10 to 100 μg/ℓ over a period of several months. These results indicate that some analytical problems occurred and the data are, therefore, difficult to interpret. The urines were collected while the monkeys were exposed to high amounts of cadmium in pellets. Contamination from pellets or from feces is very difficult to avoid. No quality control data for the analyses were reported. The control group had a mean blood cadmium of 20 μg/ℓ, which is very high (Section III.B).

In a third monkey experiment,[128] the results were quite different even though the exposure conditions were the same. Four groups of four to ten monkeys were given 3 mg Cd per kilogram in food pellets for 1 year and after that 30 mg Cd per kilogram food for more than 2 years. The data were not reported in such detail that exact comparisons with the earlier two monkey studies could be made, but it was clear that a significant increase of urinary cadmium occurred only after 24 months of feeding (Figure 24). From that time onwards, there appeared to be an increase of urinary cadmium with increasing body burden.

The results of the third monkey experiment agree with the results from other animal experiments, in which only a small fraction of urinary excretion is related to recent intake and the major part is related to body burden.

3. In Humans

The "normal" cadmium excretion in urine is in the order of about 1 μg/day (Chapter 5), with a good correlation between the average urinary cadmium level and the average body burden. Urinary cadmium increases with age in a similar way to kidney cadmium (Chapter 5). At very high inhalation exposure, urinary cadmium may increase considerably as a function of recent intake.

A large number of reports on urinary cadmium in cadmium workers have shown that in isolated cases their excretions are elevated up to 1 mg/24 hr. The highest cadmium excretions are found in people with renal damage. However, in cadmium-exposed workers without renal damage,[223] the urinary cadmium increased with the estimated body burden.

In five workers who had experienced high cadmium exposure for 1 to 2 years, Piscator[215] found the average urinary cadmium levels decreased after cessation of exposure with a half-time of a few months (Figure 13). The decrease in average urinary cadmium corresponds to the decrease in average blood cadmium during 20 months of follow up.[215]

Buchet et al.[23] studied a group of 148 cadmium-exposed workers as well as a control group and groups exposed to lead or inorganic mercury. On an average, the workers with renal tubular damage as assessed by urinary β_2-microglobulin, had a higher level of urinary cadmium. This agrees with a study by Lauwerys et al.[145,146] in which 25 cadmium workers with renal damage had an average urinary cadmium of 48 μg/g creatinine and 16 cadmium workers in a similar exposure situation, but with no renal damage, had an average of 16 μg/g creatinine. The average blood cadmium levels in the groups were 29 and 3.9 μg/ℓ, respectively.

Within these two groups there was no correlation between duration of employment with

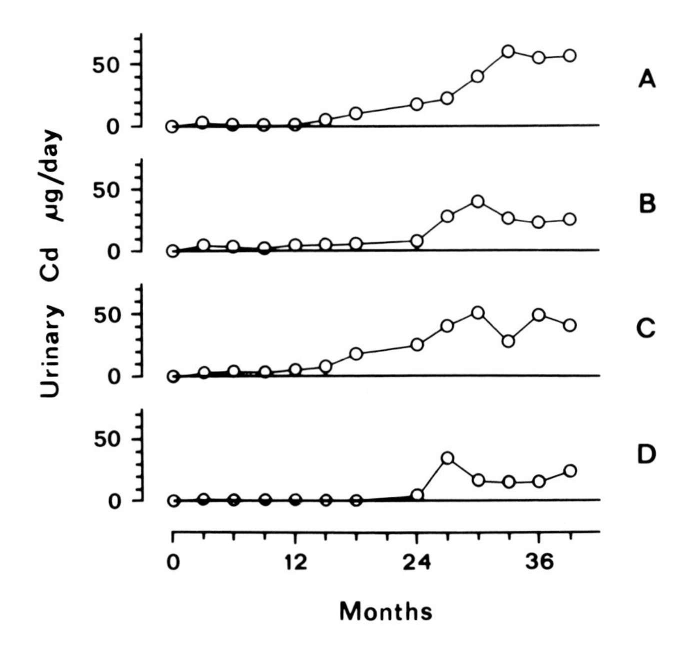

FIGURE 24. Average urinary cadmium during the first 39 months of the Japanese Tertiary Monkey Experiment in four groups of monkeys. (A) Ten monkeys fed 50 mg Cd/kg food, with concurrent vitamin D and calcium deficiency. (B) Four monkeys fed 50 mg Cd/kg food, with concurrent calcium deficiency. (C) Four monkeys fed 50 mg Cd/kg food, with concurrent vitamin D deficiency. (D) Four monkeys fed 50 mg Cd/kg food, with sufficient calcium and vitamin D in the diet. (Modified from Kimura, M., Watanabe, M., and Otaki, N., *Effects of Nutritional Factors on Cadmium-Administered Monkeys*, The Tertiary Monkey Experiment Team, Japan Public Health Association, Tokyo, 1983, 66—82.)

cadmium (an indication of body burden) and urinary cadmium. However, in a group of 113 workers with lower cadmium exposures and without renal damage, there was a significant correlation.[145] The scatter of individual values was considerable: 0.1 to 4 µg/g creatinine in the low exposure group; 0.1 to 80 µg/g creatinine in the exposed group without renal damage; and 8 to 195 µg/g creatinine in the group with renal damage.

In the early studies of urinary cadmium excretion of cadmium workers (reviewed by Friberg et al.[97]), values of up to more than 400 µg/ℓ were reported.[237,267] The analytical methods were not validated by quality control. After exposure to cadmium ceases, cadmium excretion will decrease as mentioned earlier, whereas cadmium-induced proteinuria will persist (Volume II, Chapter 9). Thus, groups of retired workers may have lower urinary cadmium than active workers.[1] Evaluations of urinary cadmium in workers need to take several factors into consideration: current exposure, total dose (or body burden), and presence of renal damage.

In general populations exposed to high cadmium doses via food,[169] the urinary cadmium levels were on an average about five times higher in exposed groups than in control groups. On a group basis there was a close correlation between urinary cadmium and the prevalence of signs of renal dysfunction. This may be an indication of the higher urinary excretion which occurs after the development of proteinuria as has been seen in animal studies (Section V.A.2) or of a higher average body burden in the people affected. Detailed longitudinal studies may be required to resolve this issue. Also in the general population the individual variations in urinary cadmium are great[133] and it is therefore not possible to use individual urinary cadmium values to obtain anything but a crude and uncertain estimate, e.g., of body burden.

There was a good correlation between urinary metallothionein and cadmium in studies of 67 people exposed in the general environment[265] and 94 cadmium workers.[224] Within groups of retired and active workers as well as workers with or without signs of renal tubular dysfunction,[224] the relationship between urinary cadmium and metallothionein was similar. There appears to be a constant ratio of about 10 between the urinary metallothionein and cadmium concentrations. This coincides with the maximum binding capacity of metallothionein for cadmium, and in the cadmium–exposed people studied these relationships may be interpreted as indicating that cadmium in urine is mainly bound to metallothionein.

B. Fecal Excretion

1. Total Fecal Excretion after Single Exposure: Animals

Animal experiments have shown that injected cadmium will be partially excreted with the feces. Both direct excretion from the mucosa and biliary excretion contribute, but the former is likely to contribute most. Decker et al.[55] gave rats a single intravenous dose of ^{115}Cd (0.63 mg/kg). After the first 24 hr, 7.3% of the dose was found in the feces and after 72 hr, 18.5% had been excreted via feces. Similar results have been reported for rats,[232] dogs,[24] and goats.[163]

Berlin and Ullberg,[15] using whole body autoradiography after a tracer dose of ^{109}Cd intravenously injected into mice, found that cadmium rapidly accumulated in the mucous membrane of the intestinal tract during the first 24 hr. It accumulated mainly in the secretory part of the gastric mucosa and in the colonic mucosa. There was also some cadmium in the contents of the stomach and colon 20 min after injection.

In a study of rats given a tracer dose of ^{109}Cd in a single subcutaneous injection, cadmium accumulated rapidly in the wall of the stomach.[154] The wall of the small intestine contained the largest concentration. Radioactivity in the intestinal contents increased during the first 24 hr. Thus, there appears to be some direct secretion of cadmium from the intestinal mucosa, but it is difficult to quantify.

Elinder and Pannone[60] gave two groups of five rats subcutaneous injections of $^{109}CdCl_2$. One group received 12 injections of 0.5 mg Cd per kilogram body weight (total dose 6 mg Cd per kilogram body weight) and the other group received 8 injections of 0.125 mg Cd per kilogram body weight (total dose 1 mg Cd per kilogram body weight). About 4 weeks after the last injection, bile, feces, and urine levels of cadmium were measured. The average 24–hr fecal excretion was 0.038% of body burden.

2. Total Fecal Excretion after Repeated Exposure: Animals

Ceresa[33] determined the daily excretion of cadmium in rabbits given high subcutaneous doses of $CdSO_4$ (5.5 mg Cd per kilogram). The mean fecal excretion was 1.8% of the injected amount and slightly higher than the urinary excretion.

A long-term experiments on rabbits given 0.25 mg Cd per kilogram as chloride 5 days/week for 29 weeks by subcutaneous injections, Axelsson and Piscator[9] reported a daily fecal excretion after 11 weeks of 1.6% of the daily dose and after 17 weeks of 2.8% of the daily dose. After 23 and 29 weeks, when the urinary excretion of cadmium was greater, 11.2 and 6.6% of the daily dose were found in the feces. The body burdens of the rabbits were not measured, but these would have increased proportionately to the exposure duration. The increasing fecal excretion as a proportion of daily intake may, therefore, indicate a correlation between fecal excretion and body burden.

Nordberg[184,185] found a considerable individual scatter in the fecal cadmium of mice, but on average there was an association between total body burden and fecal cadmium.

In mice given subcutaneous injections of 0.025 mg Cd per kilogram body weight per day,

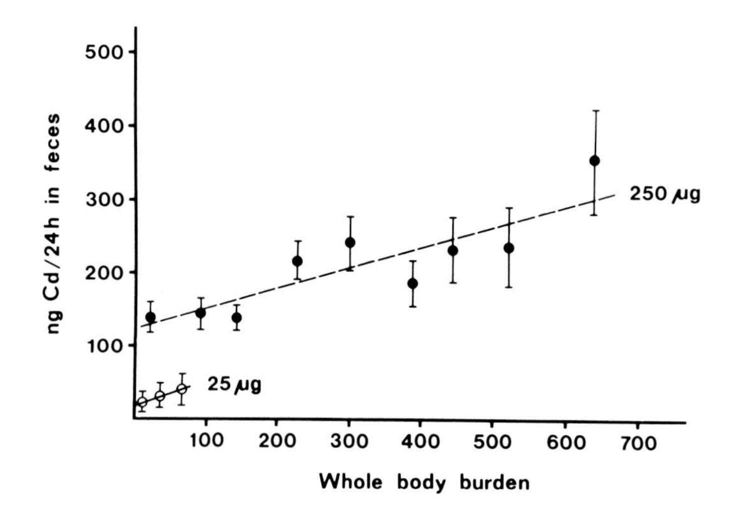

FIGURE 25. Fecal excretion of cadmium in mice given 250 μg Cd/kg body weight per day (●) or 25 μg Cd/kg/day (○) by subcutaneous injections.[184,185]

there was a good correlation between fecal excretion and body burden regression line: 13 ng Cd per day plus 0.04% of body burden (see Figure 25). In mice given a 10 times higher dose, the fecal excretion was 119 ng Cd per day plus 0.02% of body burden (Figure 25).[184,185] The part of fecal cadmium that was related to the daily intake increased tenfold as anticipated, but apparently the body burden-related part decreased with increased dose.

Suzuki and Yoshikawa[253] confirmed the good correlation between body burden and fecal cadmium excretion in rats (Figure 23).

3. Biliary and Pancreatic Excretion: Animals

Biliary excretion of cadmium is of importance for fecal excretion. Rats were given intraperitoneal injections of ^{109}Cd as sulfate[32] and a relatively constant excretion of cadmium via bile was found up to 6 hr after injection. When the bile duct of one rat was connected to the small intestine of another rat, cadmium excreted via bile was reabsorbed to a certain extent, indicating an enterohepatic circulation.

In a similar study over a longer time period,[196] rats were given subcutaneous injections of $^{109}CdCl_2$ (dose 0.25 mg Cd per kilogram body weight). Approximately 0.005% of the dose was recovered per hour in bile over 3 consecutive days. During the first few hours of the experiment, the excretion was higher and some rats excreted up to 0.15% of the dose per hour in the bile, but the level soon decreased and stabilized.[196] In all instances, the total daily biliary excretion constituted only a small fraction (on an average 5%) of the fecal excretion measured in the same animals. In the study of rats by Elinder and Pannone[60] (Section V.B.1), the 24-hr biliary cadmium excretion 4 weeks after the last injection was 0.013 to 0.015% of body burden.

Klaassen and Kotsonis[137] noticed that shortly after cadmium administration the plasma concentration of cadmium was lower than that in bile. This was interpreted as evidence of an active secretion of cadmium into the bile. Elinder and Pannone,[60] however, found that the plasma concentration of cadmium was higher than that in bile. One possible explanation for this discrepancy in results is that after long-term exposure only a minor portion of the plasma-borne cadmium is available for excretion in bile. Shortly after parenteral doses most of the cadmium will be found in high molecular weight proteins, and cadmium bound to these proteins is probably available to the liver cells to be actively secreted into the bile.

Nordberg et al.[196] in their studies discussed above, also collected pancreatic fluid and found that very little additional cadmium (between 1 to 10% of the biliary excretion of cadmium) was found in pancreatic fluid under the same experimental conditions.

The cumulative biliary excretion 24 hr after intravenous administration of 66, 90, or 120 μg Cd per rat as $CdCl_2$ was 0.83, 1.18, and 5.68% of the dose.[47] The highest excretion rate of Cd^{2+} was detected between 15 and 30 min after administration. The mean amount of cadmium found in the contents of the entire gastrointestinal tract and feces was 5.5% of the administered dose.

In further studies by Cikrt and Havrdova,[46] effects of dosage and cadmium pretreatment on the binding of cadmium in rat bile were investigated. With increasing dose (from 0.6 to 2.6 mg Cd per kilogram body weight), a higher cumulative biliary excretion of cadmium was observed (4.8 to 12.1%) and a higher percentage of the cadmium was excreted in a low molecular weight form (44% as compared to 97%). On the other hand, after cadmium pretreatment, a drastic decrease in the cumulative biliary excretion of cadmium was observed (0.03 to 0.07%) and a greater percentage of that excreted into the bile was bound to high molecular weight compounds.

Stowe[242] found that pretreatment with zinc also decreased the biliary excretion of cadmium in rats. It is thus likely that the stimulation of metallothionein synthesis by zinc or cadmium influences the transport of cadmium via liver to the bile.

The effect on bile excretion by pretreatment with cadmium was also studied by Cherian.[35] Intravenous injection of ^{109}Cd-labeled $CdCl_2$ (1 mg/kg) to control rats and rats that had been injected with $CdCl_2$ (0.25 mg/kg) 24 hr earlier revealed a decrease in the biliary excretion of ^{109}Cd in the latter group. The pretreatment resulted in induced synthesis of metallothionein in rat liver and kidneys. The binding of cadmium to liver tissue increased and the renal accumulation of cadmium was unaltered by this cadmium pretreatment.

In the same study,[35] rat liver metallothionein was isolated from rats injected repeatedly with $CdCl_2$. Intravenous injection of this type of cadmium-metallothionein caused a different distribution of cadmium in the tissues and biological fluids when compared to injection of $CdCl_2$. After injection of cadmium-metallothionein, a major proportion of cadmium was deposited in the kidneys and the biliary excretion of cadmium was minimal. In animals injected with $CdCl_2$, hepatic uptake was larger and urinary excretion smaller than for the other animals.

Biliary excretion of cadmium also increases with increasing dose of $CdCl_2$.[42] Cumulative biliary excretion of cadmium during 5 hr was 0.065% of the the dose for rats injected with 0.1 mg Cd per kilogram compared to 16.9% of the dose at 2 mg Cd per kilogram. During the 5–hr experimental period, most of the cadmium in liver cytosol was bound to high molecular weight proteins and less than 10% was bound to the metallothionein fraction. The biliary cadmium was recovered as a low molecular weight compound (less than 4000) in experiments with various doses of cadmium. No cadmium was attached to high molecular weight proteins or metallothionein in the bile. The low molecular weight protein cadmium complex in bile was partially characterized as Cd-glutathione by thin-layer chromatography and amino acid analysis.[42]

It should be pointed out that biliary cadmium excretion is about 100% higher 10 to 20 min after intravenous exposure, if the exposure is via the portal vein rather than the femoral vein.[263] This indicates that there may be differences in the distribution and excretion pathways depending on exposure route. Ingested cadmium and inhaled cadmium may be handled differently, but these possible differences need to be much more extensively studied before any firm conclusions can be drawn.

4. Fecal Excretion in Humans

Cadmium in feces mainly reflects the unabsorbed part of ingested cadmium (Section II.B.4) and the actual fecal excretion is difficult to measure. There are a number of studies on cadmium in feces of general population groups (Chapter 5), and the aim of these studies has often been to measure daily cadmium intake via food. Rahola et al.[218] found that less than 0.1% of an ingested and absorbed dose of radioactive cadmium was excreted in the feces.

Some estimates[130] indicate that in the general population with long-term low level exposure, the fecal excretion may be as much as six times greater than the urinary excretion. The bile may contribute a considerable part of the fecal cadmium excretion of humans. Cadmium levels in bile of 0.1 to 11.4 μg/kg were measured in patients undergoing gallstone operations[62] (average about 3.2 μg/kg). With a daily bile flow of 500 mℓ[57] this would contribute 1.6 μg Cd to feces each day. Some of the biliary cadmium may be reabsorbed in the intestines as suggested by Caujolle et al.[32] and Tsuchiya et al.,[268] but this mechanism has not been studied.

In one study,[2] cadmium in feces was used as an indicator of cadmium dust swallowed from industrial air. In workers with past high exposures, fecal cadmium may be a valid indicator of "true" fecal excretion after the unabsorbed cadmium from food is deducted (Chapter 7, Section VI.C).

5. *Conclusions*

There is evidence that absorbed cadmium is excreted via some parts of the alimentary tract. Cadmium excretion is mainly through the mucosa of the alimentary tract with a certain contribution from biliary excretion. Pancreatic excretion appears to be quantitatively insignificant.

The daily excretion is partly related to the daily uptake and partly to the body burden. How much is contributed to each of these by biliary cadmium is not known, but it appears that biliary excretion is a rapid process which is influenced by the binding of cadmium to metallothionein in liver. There are indications that enterohepatic circulation of cadmium may exist.

C. Other Excretion Routes

Truhaut and Boudene[267] reported that cadmium was excreted in hair of rats and rabbits injected with cadmium. This was confirmed in mice by Berlin and Ullberg.[15] Miller et al.[163] found that the concentration in goat hair was about 20% of the concentration in the liver 15 days after a single oral dose of ^{109}Cd as the chloride. After an intravenous dose, the corresponding figure was about 1%. When this experiment was repeated by Miller et al.,[164] the concentrations in hair after oral exposure were only 3% of the concentrations in the liver. External contamination may explain the earlier high results.

Nordberg and Nishiyama[188] found that in mice given a single intravenous injection of $^{109}CdCl_2$ (4.5 μg Cd per kilogram), the decrease in cadmium levels in hair corresponded to the decrease in whole body retention. Less than 0.5% of the total excretion between days 41 and 105 was via the hair. The main analytical problem with hair is the possibility of external contamination[188] and there is no effective way to separate exogenous and endogenous cadmium in hair.[168] In an industrial environment hair cadmium is most likely heavily contaminated by exogenous cadmium, and within a group of cadmium workers no correlations were, therefore, found between hair cadmium and urine, blood, kidney, or liver cadmium.[66]

Other excretion routes may be worth comment. From the data of Lucis et al.[155] it can be calculated that excretion was less than 0.05%/g milk per day after a single subcutaneous post-partum injection given to female rats. Tanaka et al.[257] gave pregnant mice ^{115m}Cd (15 μg Cd per mouse) in a single intravenous injection 24 to 36 hr before delivery. Only 0.09% of the dose was transferred via the placenta and found in the newborns (see also Section IV.C). The cadmium concentration in the rat milk was not measured. Sucklings born to female mice not given radioactive cadmium were suckled by the female mice that had been given radioactive cadmium before delivery. Whole body measurements of the sucklings showed that after 14 days, 0.3% of the dose given the sucklers was found in the sucklings. As some cadmium was eliminated in the feces of the sucklings, the total amount excreted via the sucklers' milk must have been more than 0.3%.

Cadmium may be excreted through saliva. Using multielement emission spectrometry, Dreizen et al.[58] studied the concentration of cadmium in marmoset saliva after stimulation with pilocarpine and mecholyl. The average concentrations were 0.044 and 0.078 mg/kg, respectively, with a maximum of 0.344 mg/kg. These figures are so high that they appear rather unlikely. The data were not supported by analytical quality assurance studies and need to be confirmed. Another study on cadmium in rat saliva[238] concluded that the concentration in saliva was about the same as the concentration in plasma, but no quantitative data (e.g., micrograms per liter) were presented. Nevertheless, these data contradict the high concentrations reported by Dreizen et al.[58]

Some data on cadmium in "normal" human sweat (average 23 μg/ℓ) have also been reported,[48] but they cannot be considered reliable as no analytical quality control data were presented and the average urinary cadmium concentration was given as 13 μg/ℓ. This is far too high for a general population group (Chapter 5).

D. Total Excretion, Whole Body Retention, and Biological Half-Time

1. In Animals

The total excretion and whole body retention of cadmium determine to a great extent the risk of health effects after long-term exposure. The biological half-time is a measure of the retention, which can be calculated from the total excretion. As the half-time concept is based on an implied mathematical model (one- or multicompartment exponential model) of cadmium metabolism, it will be treated in more detail in Chapter 7.

Animal experiments have shown that biological half-times of cadmium vary greatly, depending on dose, administration route, single or repeated exposure, length of observation period, etc.,[173,181,259] which will be discussed further in the following text.

The half-time is often measured by giving the animal a single small dose of radioactive cadmium and following the decrease in whole body cadmium levels over a long time period. Another approach is to give repeated doses and follow the increase of the cadmium accumulation in the body.

An important aspect of these studies is the length of follow-up. It has to be long enough to detect the components with the longest half-times. In Table 7, we have summarized data from those studies that had sufficiently long follow-up periods. It is seen that the whole body half-time is often about 20% of the maximum life span of each species.

Some other experiments report shorter or uncertain half-times,[24,49,76,163,188,266] but in most cases follow-up periods were too short.

It is interesting to note that half-times as long as 50% of the maximum species life time were found in two studies[72,181] where a single radioactive dose was followed by continued exposure to nonradioactive cadmium. The reason for this may be the stimulation of hepatic metallothionein production that the continuous cadmium exposure leads to.

In four male CBA mice, the elimination rate of a single subcutaneous dose of $^{109}CdCl_2$ (0.25 mg Cd per kilogram) was measured from whole body measurements.[181] Repeated doses of nonradioactive cadmium (0.25 mg/kg 5 days/week) were given during the 25 weeks following the subcutaneous administration of radioactive cadmium. The decrease in the whole body value during that time (corrected for radioactive decay) was only about 10% which corresponds to a biological half-time of a couple of years.

Other groups of mice were exposed continuously to ^{109}Cd, 5 days/week for 25 weeks, and the whole body retention and the excretion in urine and feces were determined at weekly intervals.[183] One group of four mice received 0.25 mg/kg doses and another group of four mice 0.5 mg/kg doses. In the group receiving 0.25 mg/kg, a linear increase in whole body

Table 7
BIOLOGICAL HALF-TIMES REPORTED IN DIFFERENT SPECIES OF ANIMALS, BASED ON LONG-TERM WHOLE BODY MEASUREMENTS AFTER SINGLE EXPOSURES OR REPEATED EXPOSURES

				Biological half-time components			
					Slowest phase		
Species	Maximum life span[a] (days)	Exposure route	Follow-up duration (days)	Fast phases (days)	(days)	(% of life span)	Ref.
Mice	1275	Oral single	360	—	200	16	220
		i.p. single	400	40	200—300	16—24	220
	Age 7 weeks	i.p. single	131	7, 178	261	20	255
	Age 16 weeks			43, 159	254	20	255
	Age 50 weeks			91, 375	630	49	255
	Male			42, 135	220	17	255
	Female			46, 178	349	27	255
		s.c. single + background	180		≈ 730	≈ 57	181
		s.c. repeated	180		≈ 730	≈ 57	181
		i.p. single	60		60	5	139
		oral single	540		238	19	72
		oral single + background	540		612	48	72
Rat	1700	i.m. single	64		300	18	59
		s.c. single	570		Very long		279
		i.p. several	750		Liver 475	28	139
					Kidney 150	9	139
Rabbit	2500	s.c. several	730		Liver 357	14	175
					Kidney 497	20	175
Monkey	7700	Oral single	135		>730	>9	191

[a] According to Altman and Dittmer.[5]

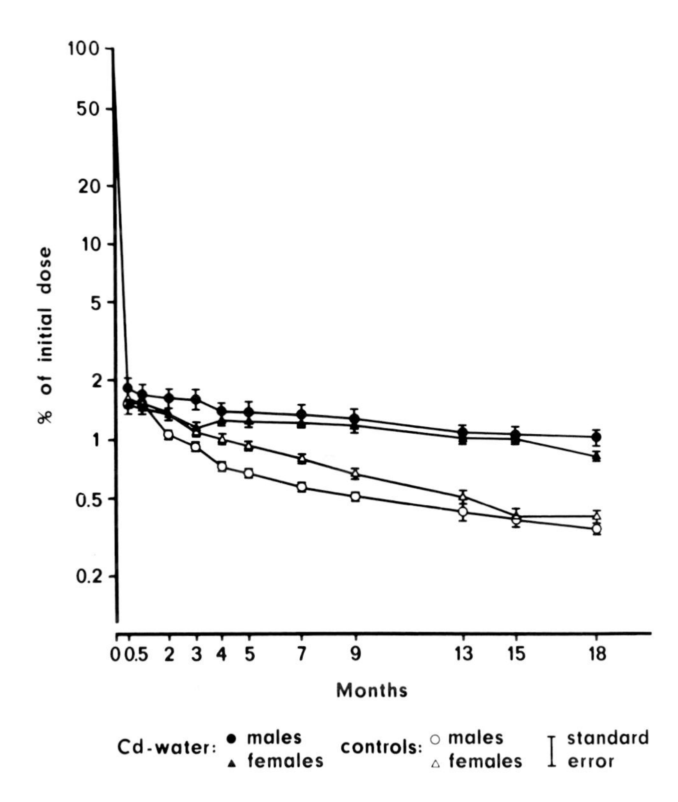

FIGURE 26. Whole body retention of a single oral dose of radiolabeled cadmium (^{109}Cd) in mice with or without subsequent exposure to nonradioactive cadmium during 18 months. The values given are corrected for radioactive decay. (Modified from Engström, B. and Nordberg, G. F., *Acta Pharmacol. Toxicol.*, 45, 315—324, 1979b. With permission.)

retention was seen throughout the experiment, which indicates a half-time of several years.

Engström and Nordberg[70] exposed four groups of eight mice to a single oral dose of $^{109}CdCl_2$ and subsequently two of the groups (one male, one female) were exposed for 18 months to 50 mg Cd per liter in drinking water. Whole body retention was measured at monthly or bimonthly intervals and it was seen that the gastrointestinal absorption was similar in all groups, but the biological half-time was over twice as long in the groups given subsequent cadmium exposure (Table 3). It appears that the early short-term half-time components become less prominent (Figure 26) and after 18 months the retention is more than twice that of the groups given only the single cadmium dose.

It is possible that the continued cadmium exposure maintains a high synthesis rate of metallothionein in the liver which may increase the rate of uptake of cadmium from plasma to liver and may increase the half-time of cadmium in the liver by enlarging the metallothionein pool that the cadmium can attach to.

The above-mentioned values for biological half-times of cadmium will only be valid before renal damage has occurred. If there is renal tubular dysfunction, the excretion of cadmium will increase (Section V.A) and this will decrease considerably the biological half-time.

The biological half-times of cadmium in different tissues is discussed further in Chapter 7.

2. *In Humans*

The total excretion in humans is not known, but all data favor a retention of almost all cadmium that is absorbed, as the tissue accumulation continues for the whole life span (Section IV.A.2). The accumulation curve for body burden is almost straight from age 0 to age 40 or 59 as seen in autopsy studies from several countries (Chapter 5). As the human body has several compartments with very different half-times for cadmium (Chapter 7), a simple half-time calculation based on total excretion may give misleading results. Of particular importance are any components of the total excretion that are directly related to daily uptake.

Many studies on urinary cadmium after low level long-term exposure (Section II) have shown an increase with age similar to that of body burden. In areas with daily cadmium intakes via food of 10 to 20 μg Cd per day, nonsmoking adults excrete 0.5 to 1 μg/day (Chapter 5). This constitutes about 0.005 to 0.01%/day of a body burden of 10 mg (Table 5). In addition, cadmium excretion in feces contributes about the same amount to total excretion (Section V.B.4). This implies a whole body half-time for humans of about 10 to 20 years, which is about 12 to 25% of the maximum life span, similar to the findings for other species (Table 7).

The whole body half-time of ^{115m}Cd or ^{109}Cd given as a single oral dose to human volunteers has been measured by whole body counting (Section II.8.4). In one study,[233] one person was followed up for more than 2 years after a single dose of ^{109}Cd and the half-time was calculated to be 26 years. In three other similar studies,[80,161,218] ^{115m}Cd-exposed people were followed up for about 100 days only and half-times between 93 and 202 days were reported. Only one of these studies took confidence limits into account[218] and it was concluded that the half-time could be 130 days to infinity. In addition to the consideration of follow-up time, methodological difficulties in the accurate measurement of whole body ^{109}Cd should be borne in mind when drawing conclusions from these studies.

VI. INDEXES OF EXPOSURE AND RETENTION

A. Fecal Cadmium

As was mentioned in Section II.B, the gastrointestinal absorption of cadmium is only a few percent with a maximum of 20% in people with low iron stores. The fecal excretion is also only a fraction of the daily intake via food (Section V.B), so in the main, the amount of cadmium in feces will reflect the amount swallowed during the last few days. Kjellström et al.[134] showed that within 3 days after the consumption of a large dose of cadmium-polluted rice, almost 100% of the cadmium was recovered in the feces. The average fecal cadmium of adults in Sweden (16 μg/day) closely agreed with an estimate of average daily intake from food (17 μg) based on a calculation from food intake habits and average cadmium concentrations in food.[130]

The daily cadmium intake in some other countries has been measured by both feces analysis and food analysis (Chapter 3, Table 7), and the intake estimates are similar using either method in spite of the fact that considerable geographical variations in intake may occur within a country. In Japan, the feces method gave values of 24 to 81 μg Cd per day and the food analysis method gave values of 31 to 113 μg Cd per day, for areas not considered to be cadmium-polluted (Chapter 3, Table 7). The corresponding values for polluted areas were 149 to 255 and 180 to 391 μg Cd per day (Chapter 3, Table 8).

In people with high exposure to cadmium via inhalation, some of the inhaled cadmium will be deposited in the upper respiratory tract, cleared to the throat, and swallowed (Section I.A.2). This leads to a high gastrointestinal exposure from the inhaled cadmium. A high fecal elimination of cadmium was shown in 15 cadmium workers who had fecal cadmium of up to 2577 μg/day.[2] An important source of cadmium ingestion in factory workers may be the contamination of fingers, sandwiches, cigarettes, etc.

Kjellström et al.[134] found a difference in fecal cadmium elimination between adult nonsmokers (16 μg/day) and adult smokers (19 μg/day). The 3 μg/day difference is likely to represent mainly the deposited inhaled cadmium that has been cleared from the lungs and the excreted cadmium from the gastrointestinal tract.

In conclusion, the fecal cadmium of people with cadmium exposure via food only can be used as a measure of the most recent intake. In people with additional cadmium intake from air, fecal cadmium may be used as an indicator of the cadmium amount cleared from the respiratory tract or as an indicator of contamination of foodstuffs by fingers.

B. Blood Cadmium

Cadmium in blood increases rapidly during the first 3 to 4 months after a sudden increase in exposure[132] and then reaches an apparent steady state level, which is likely to reflect the average exposure during those months. Lauwerys et al.[147] found a similar increase of blood cadmium among cadmium workers.

In a cross-sectional study of a group of cadmium workers, kidney and liver cadmium levels were measured with in vivo neutron activation.[223] There was no increase of blood cadmium with body burden, but the elevated levels (5 to 30 μg/ℓ) reflected the high daily intakes via air.

On the other hand, when high inhalation exposure ceases, blood cadmium decreases with two distinct half-time components.[118] One component has a half-time of a few months and probably reflects the turnover rate of red blood cells (Section III.C). The average half-time in blood of workers who have ceased exposure has been reported as 41 days[281] assuming a one-compartment exponential component (Chapter 7). The short half-time component may explain the gradual increase in blood cadmium which occurs after a sudden increase in exposure. The increase of blood cadmium takes place over 3 to 4 months.[132] The other half-time component is several years long[118] and is likely to reflect body burden (Section III.C). The relative contribution to blood cadmium of body burden and recent intake depends very much on the history of exposure.

There is a component of the blood cadmium that is associated with the long-term accumulation of cadmium in tissues such as liver, kidneys, and muscles. This can be used as a body burden indicator after the short half-time component has disappeared in situations when the exposure to cadmium has been high and subsequently ceased or dramatically decreased (Section III.C).

C. Urinary Cadmium

As shown in Section V.A, the urinary excretion is closely correlated with body burden or kidney burden in both animals and humans before renal effects occur. Subsequent to the development of renal damage (sometimes combined with liver damage), there is a dramatic increase in the urinary cadmium excretion, which makes it impossible to use it as an indicator of retention after such damage has occurred.

In addition, high urinary cadmium excretion has been seen after short-term very high exposures, even without renal damage (Figure 13). Under such circumstances a significant proportion of urinary cadmium may be associated with recent intake. However, after long-term low level exposure and exposure before renal effects occur, there appears to be very good agreement between average kidney burden or body burden and the average daily urinary cadmium excretion (Figure 27).

It is important to stress the difficulties in interpreting cross-sectional data such as those in Figure 27. Cohort effects in the data may confuse the picture. No sufficient longitudinal data for urinary cadmium in humans are available. However, the animal studies strongly support a body burden relationship after long-term low level exposure. Taking all the data together, it is estimated that for humans about 0.005 to 0.007% of body burden is excreted daily in the urine (Section V.A.3). A proportion of the urinary cadmium is likely to be a

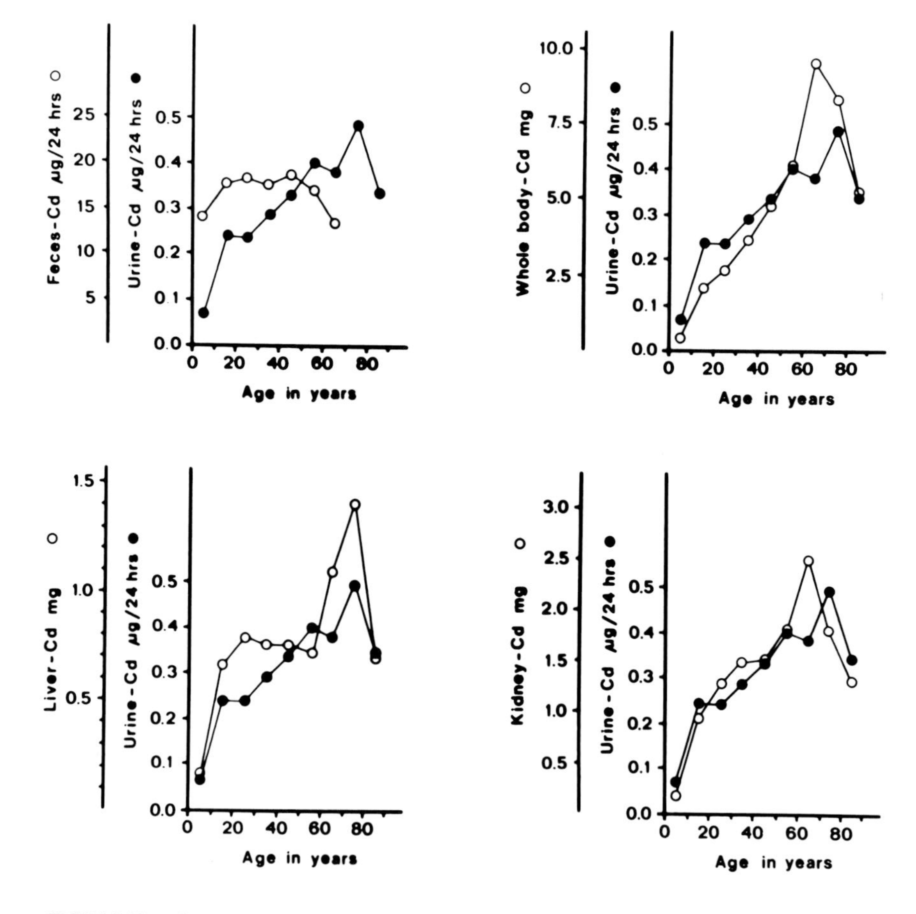

FIGURE 27. Comparison of daily intake, daily urinary excretion, and body liver and kidney burdens of cadmium for nonsmoking Swedes. (From Kjellström, T., Accumulation and Renal Effects of Cadmium in Man. A Dose-Response Study, Doctoral thesis, Karolinska Institute, Stockholm, 1977.)

function of daily intake (Section V.A.), but this may be quite small at long-term low level exposure.

D. Other Indicators

The most accurate way to measure body burden is to analyze the major storage tissues (kidneys, liver, and muscles) directly at autopsy. As this can only be done after death, the in vivo neutron activation method is of great value, but its validity is yet to be proven (Chapter 2). It has been used to record the increased levels in cadmium workers,[65,223] but the detection limit is too high to measure "normal levels". Muscle biopsies could also possibly be used to collect material for analysis, but no such data are available.

VII. INFLUENCE OF OTHER COMPOUNDS ON THE METABOLISM OF CADMIUM

A. Dietary Factors

A number of dietary factors influence the gastrointestinal absorption of cadmium (Section II.B.3) or directly affect the toxicity of cadmium. The latter are referred to in Volume II of this monograph, i.e., in the chapters dealing with specific effects. Some dietary factors such as zinc, selenium, and cysteine also affect other aspects of cadmium metabolism.

The interactions between different trace elements were discussed by the Task Group on Metal Interaction,[260] and a number of reports on specific interactions are included in the same issue of the journal *Environmental Health Perspectives*.

Interactions of dietary factors with cadmium metabolism will not be discussed further here. Interactions with chelating agents are discussed in some detail, however, as this has a great bearing on the treatment of cadmium poisoning.

B. Chelating Agents

A number of chelating agents can influence the distribution and toxicity of cadmium. A discussion of renal changes resulting from the combined administration of cadmium and chelating agents will be given in Volume II, Chapter 9. Distribution changes occurring as a result of this type of combined exposure frequently involve a change in critical organ. In several instances the critical organ is not the kidney. It is, therefore, useful to discuss distribution and toxicity jointly regardless of the organ involved. A discussion of this follows, in which an attempt is made to discover explanations for the various experimental findings reported in the literature.

One important aspect when discussing the influence of various externally administered chelating agents is the chelation by metallothionein, which takes place in the body after exposure to cadmium, and its influence on cadmium toxicity. Chelation by metallothionein has been discussed in detail in Chapter 4 and only a few aspects will be taken up here.

The relationship between cadmium chelation by metallothionein and cadmium toxicity was originally suggested by Piscator,[213] Nordberg,[180] and Nordberg et al.[191] and has been repeatedly confirmed by other authors (Chapter 4). The deleterious effects of cadmium are due to binding of "free" cadmium to specific sensitive sites in the cell and cadmium-metallothionein or other types of cadmium chelates may be considered as inactive (Figure 19). Cadmium-metallothionein and chelated cadmium can, however, undergo breakdown in the tissues and nonchelated cadmium may be released. It is therefore assumed that toxicity develops when the tissue concentration of nonchelated cadmium exceeds a certain level.

In this section, available evidence concerning the influence of various chelating agents on cadmium distribution and toxicity will be summarized. There are differences in how each chelating agent interacts with cadmium. Explanatory schemes in relation to present knowledge about mechanisms of cadmium toxicity for each chelating agent will be set forth. More detailed reviews of the original data up to 1974 have been given by Friberg et al.[97] and in the later literature by Klaassen et al.[136] and Nordberg.[186]

1. Influence on Acute Effects

Metallothionein induction is important for the long-term effects of cadmium. However, in the acute exposure situation, induced metallothionein synthesis has not yet occurred and the lethal effects of cadmium injection in this situation are related to liver toxicity[191] (Volume II, Chapter 11). The increased mortality when cadmium is injected subcutaneously in combination with STPP (sodium tripolyphosphate) or NTA (nitrilotriacetic acid)[45,70] is related to a more rapid uptake in blood and a higher level of cadmium-albumin in plasma which effectuates a more rapid uptake by the liver.[7] Thus a high concentration of nonmetallothionein-bound cadmium is obtained in the liver before metallothionein synthesis has occurred. This leads to liver damage at lower cadmium doses than when cadmium is given without STPP or NTA. A similar explanation may be valid for the increased reproductive toxicity observed after injection of NTA in combination with cadmium.[45] Neither type of increase in toxicity is observed after repeated oral exposure.[170,171]

The increased renal toxicity of cadmium when given in combination with BAL (British antilewisite, 2,3-dimercaptopropanol) and the beneficial influence of this combination on the acute effects in other organs (e.g., the lungs[156,264]) may be explained by the removal of cadmium from the cadmium-albumin complex in blood plasma and in tissues and the formation of a cadmium-BAL complex instead. This complex may displace cadmium to tissues (mainly liver and kidneys) other than those at the site of uptake (e.g., lungs) and protection against effects on the uptake organ may be observed, whereas renal toxicity could possibly be increased by this type of chelation therapy. BAL increased the concentration of cadmium in the kidney in animal experiments[167,262] and an increased renal toxicity of cadmium-BAL has been reported after repeated exposure.[53]

Similar to the situation for the BAL-Cd complex, the complex with cadmium and EDTA (ethylene diaminetetra-acetic acid) is stronger than the cadmium-albumin complex at physiological pH, and a removal of cadmium from the albumin site to the EDTA complex will probably take place in blood plasma and tissues. The cadmium-EDTA complex will probably pass through the kidneys into the urine, since urinary cadmium excretion is increased more than 100-fold after EDTA injection, whereas the kidney concentration of cadmium is decreased.[95] A small amount of cadmium will be released from the EDTA complex inside the kidneys, but toxic concentrations will not occur after acute cadmium exposure when relatively small amounts of cadmium are available compared to the amounts available after chronic cadmium exposure. This explains the favorable effects of EDTA treatment for acute cadmium intoxication induced by injection in experimental animals.[29,30,74,95] In these animal studies, the effectiveness of such treatment was shown to decrease with the time interval between injection of cadmium and chelation therapy. As discussed by Cantilena and Klaassen[30] and Klaassen et al.,[136] this phenomenon is probably related to the binding of cadmium to metallothionein and other ligands in the tissues and body fluids, which takes some time.

Combined treatment with two or more chelating agents has been tried in acute cadmium poisoning in animals. Although an advantage of such combined treatment has been reported, this report has later been shown to be erroneous and there is no advantage of, e.g., a combination of EDTA and salicylic acid compared to EDTA alone.[28]

In addition to the chelating agents just discussed, hydroxyethylene diamine triacetic acid (HEDTA), diethylenetriamine pentaacetic acid (DTPA), and 2.3, dimercaptosuccinic acid (DMSA), also have been shown to increase survival after a single injection of cadmium in animals.[29,74] Decreased cadmium concentrations in cells and tissues were also observed.[10,11,29]

The influence of chelation therapy on long-term cadmium exposure is different from the acute situation due to the involvement of metallothionein, which forms very stable complexes with cadmium.

2. Lack of Influence of STPP or NTA on Chronic Cadmium Toxicity

When cadmium is combined with sodium tripolyphosphate, cadmium tripolyphosphate is formed which quickly releases its cadmium to cadmium-albumin in blood plasma.[6,7,68] An increased concentration of cadmium-albumin will momentarily be formed and a more rapid uptake will occur in the liver.[7] These events are illustrated in Figure 28. The total amount transferred will be approximately the same as that without the chelating agent and since the liver is not the critical organ in long-term exposure, when relatively small amounts of nonmetallothionein-bound cadmium are absorbed daily, toxic liver concentrations of nonmetallothionein-bound cadmium will not arise. Cadmium will be released in a similar way from the liver to the kidneys as occurs without the chelating agent and under such circumstances a change in the renal toxicity cannot be expected.

Similar conditions can be expected for NTA. In animals orally exposed to cadmium and NTA for various time periods,[70,170,171,180,183,227] slight differences in kidney concentrations were sometimes noted, but these concentrations were not consistently higher or lower than in animals given cadmium only. In the long-term studies by, Engström and Nordberg[70] when renal toxicity was evaluated, no influence of NTA or STPP in combination with cadmium was noted.

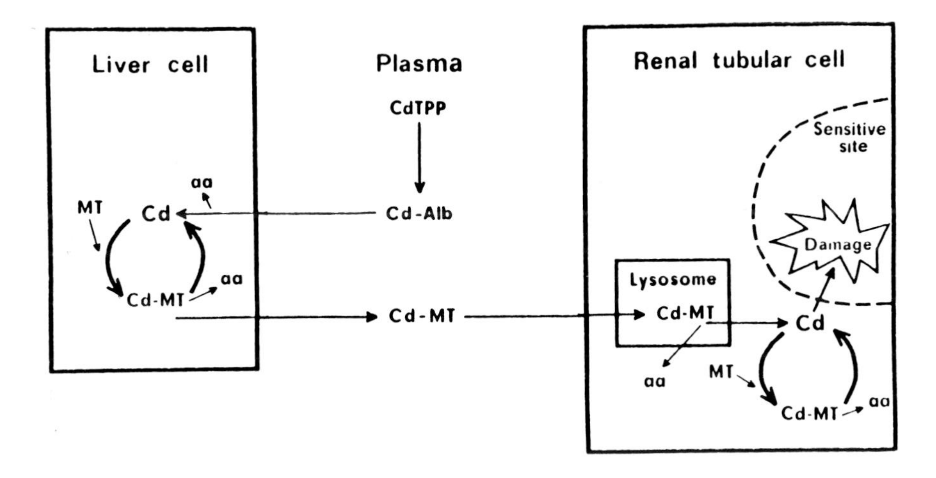

FIGURE 28. Scheme illustrating events taking place when cadmium and STPP are given in a long-term exposure situation. CdTPP: cadmium-tripolyphosphate is formed from Cd and STPP immediately after administration. (Modified from Nordberg, G. F., *Environ. Health Perspect.*, 54, 213—218, 1984.)

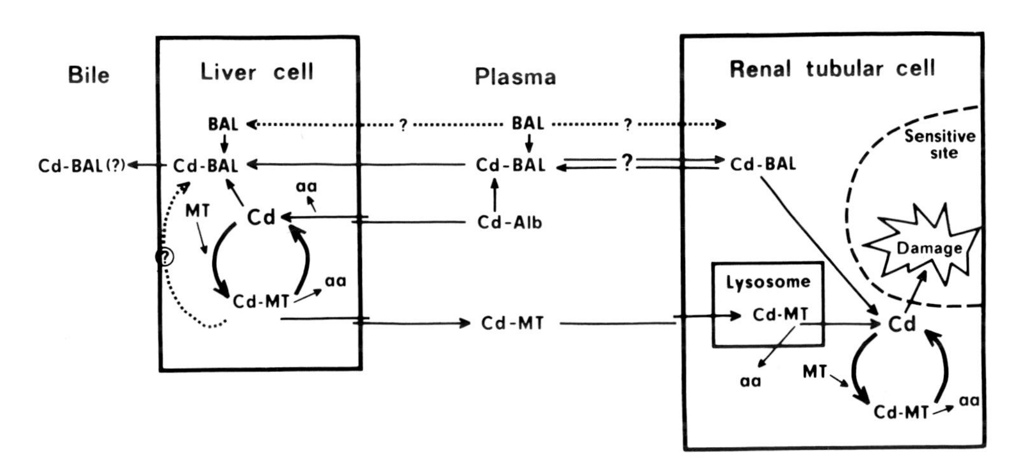

FIGURE 29. Scheme illustrating the flow of cadmium when BAL is given to animals or humans repeatedly exposed to cadmium. Modified from Nordberg, G. F., *Environ. Health Perspect.*, 54, 13—20, 1984.

3. Flow Scheme of Cadmium-BAL

The cadmium-BAL complex, like BAL, is more lipid soluble and appears to have the ability to move more freely across cell membranes than cadmium-albumin. The cadmium-BAL complex has also been shown to be more readily excreted in bile than non-BAL-bound cadmium.[37-39] Cadmium-BAL is probably more stable and more readily formed than cadmium-albumin. BAL will thus probably displace cadmium from cadmium-albumin and thereby decrease the influx of cadmium into the liver, but may increase the influx into the kidneys (Figure 29).

The liver is the critical organ for acute lethality from injected cadmium (Volume II, Chapter 11). Therefore, a decreased influx into the liver may decrease the acute mortality. This mechanism may serve as an explanation for the decreased acute mortality in experimental animals injected with a single large dose of cadmium and BAL.[101]

The increased influx into the kidneys, on the other hand, can increase renal toxicity. If the kidneys do not receive sufficient amounts of BAL constantly, the cadmium-BAL complex may be broken down and added to the nonmetallothionein-bound cadmium in the kidneys.

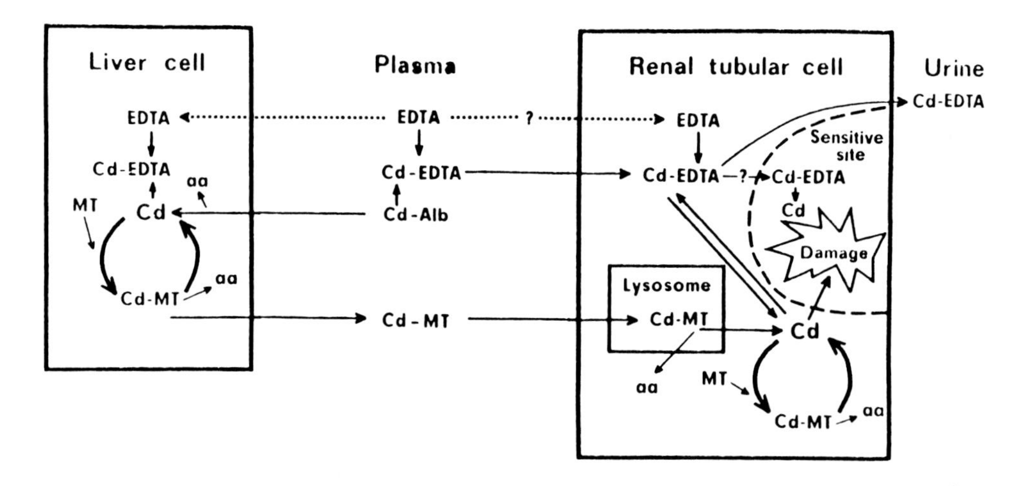

FIGURE 30. Scheme illustrating the flow of cadmium when EDTA is given to animals (or humans) repeatedly exposed to cadmium. (Modified from Nordberg, G. F., *Environ. Health Perspect.*, 54, 213—218, 1984.)

If a borderline amount of this type of cadmium is already present in the kidneys, the cadmium released from BAL may elevate the concentration above the critical concentration (for non-metallothionein-bound cadmium) and kidney damage may occur. This mechanism, which is somewhat hypothetical, would correspond to the observations of an increased renal toxicity when cadmium is given together with BAL in repeated injections.[53] However, if cadmium-albumin is not present in plasma in any appreciable amount, BAL may have a beneficial effect in providing a means for increased excretion of cadmium from the liver, mainly via bile.[37-39]

The effects of BAL would thus, to a large extent, be dependent on whether ongoing exposure to cadmium or mobilization of cadmium from absorption sites, like the lungs, is taking place. When this is not the case, beneficial effects of BAL might be expected. In experiments on metallothionein-containing cells in vitro, BAL was able to mobilize cadmium from the cells more efficiently than a number of other chelating agents. There were, however, signs of toxicity to cells in culture when exposure was to Cd-BAL.[11] The side effects that may occur as a result of BAL treatment should be kept in mind in addition to the risk of increased renal toxicity from the breakdown of the Cd-BAL[11] complex. Thus, additional evidence should be obtained from experiments where kidney damage has been induced by long-term cadmium exposure. No such experiments have been published as yet.

4. Flow Scheme of Cadmium-EDTA

The cadmium-EDTA complex, like the one with DTPA, may be excreted more readily in urine compared to cadmium-albumin or cadmium-metallothionein,[29,95] and thus EDTA injection may decrease renal concentrations of cadmium (Figure 30). In the studies by Cherian,[37-39] a combined treatment with DTPA and BAL decreased both renal and liver concentrations of cadmium in rats. However, increased renal toxicity may still be possible if EDTA is not provided in sufficient amounts at all times to maintain the cadmium-EDTA complex. A release of nonchelated cadmium in the kidney from the cadmium-EDTA complex may cause kidney damage. If such a release occurs, the borderline toxic concentration of nonmetallothionein-bound cadmium that may already be present in long-term cadmium exposure may be increased. The critical concentration for nonmetallothionein-bound cadmium may thus be exceeded and renal toxicity may occur.

An alternative explanation for such renal toxicity may be that cadmium-EDTA is more readily transferred to the sensitive sites where cadmium may be released from EDTA and bound to the sensitive receptor (Figure 30). The mechanisms mentioned are suggested as possible explanations for the increased renal toxicity found in animals treated with EDTA after repeated exposure to cadmium observed by Friberg.[95]

VIII. SUMMARY AND CONCLUSIONS

After inhalation, cadmium is absorbed and retained to a considerable degree in the body. In addition, dust particles retained in the bronchial mucus and cleared to the gastrointestinal tract may be absorbed.

Observations on human beings are scarce, but data on smokers indicate that absorption of cadmium fumes may well be 25 to 50%. Animal experiments are in favor of an absorption of between 10 to 60% of inhaled cadmium. Considerable differences appear to exist for different cadmium compounds and different particle sizes.

There is a need for further studies on the absorption and retention of different cadmium compounds after inhalation. Most animal data, even though there are large individual variations, indicate a gastrointestinal absorption of about 2% of ingested cadmium. Absorption is increased considerably, with a factor of two or more, by a low iron, calcium, or a low protein intake. In humans, the absorption is in the range of 3 to 7%, with values of up to 20% for people with low iron stores.

It is evident, both from human and animal data, that nutritional factors are of great importance for the absorption of cadmium. Another observation that is of interest in this context is the possible influence of the protein composition of the diet and the binding form of cadmium in the diet. Additional studies of these factors are necessary before any firm conclusions can be drawn. The possibility certainly exists, however, that absorption may vary considerably depending on type of foodstuff.

After absorption, cadmium is mainly transported via the blood plasma, although the main part of blood cadmium is bound to blood cells. In plasma, cadmium may be bound partly to albumin (and other high molecular weight proteins) and partly to metallothionein. After being transported, cadmium is distributed to various organs in the body. Two important organs of cadmium uptake are the liver and the kidneys. Depending on type and amount of uptake a smaller or larger proportion of absorbed cadmium is taken up by the liver. A larger absorbed dose seems to give rise to a larger proportion in the liver. Animal experiments also indicate that the whole body retention and biological half-time are greater if a continued cadmium exposure is maintained than if only a single cadmium dose is given.

A redistribution takes place over time, so that the longer the time after a single exposure or after long-term repeated exposure, a larger proportion of whole body cadmium is found in the kidneys. In people with long-term low level exposure, as much as one half of the body burden may be present in the kidneys. Approximately one sixth is in the liver and one fifth in the muscles. At higher exposure levels that may be toxic, about one third of the body burden accumulates in the kidneys.

The kidney is the critical organ (Volume II, Chapter 9). Within the kidneys, the cadmium is primarily located in the proximal tubules in the cortex. In humans, the cortex cadmium concentration can be estimated to be on average 1.25 times higher than the whole kidney cadmium concentration.

Cadmium is excreated via urine, bile, and intestinal mucosal sloughing. Urinary excretion increases with body burden and dose. In addition, there is a dramatic increase of urinary cadmium when renal damage develops.

Fecal cadmium is more difficult to measure, but several animal studies show that both biliary excretion and excretion across the mucosa do occur and that this excretion may be at least as great as the urinary excretion.

The whole body half-time of cadmium is very long, in the order of 20 to 50% of the maximum lifetime for most animal species and humans. Human data indicate that the muscles have one of the longest half-times, more than 30 years. Kidneys have a half-time of 10 to 30 years and liver has a half-time of 5 to 15 years.

Fecal cadmium is a good indicator of the daily intake of cadmium via food. Blood cadmium, particularly during high exposure, is a good indicator of the average exposure level during the last months, but after the cessation of high exposure, a large proportion of blood cadmium is also related to body burden. Urinary cadmium is usually a good indicator of body burden, before renal damage develops. Direct measurement of tissue levels with in vivo neutron activation is a new way of measuring body burden.

Several explanatory schemes concerning the interactions of cadmium with acute or chronic exposure to chelating agents have been presented. It is important for the future that additional experimental research be made to confirm or reject the hypothetical schemes proposed here. In particular, the possibility of a beneficial effect of chelation therapy on renal toxicity needs experimental confirmation. At present, no recommendation for chelation therapy in human cases of chronic cadmium poisoning can be made. In cases of acute poisoning, however, this type of therapy may be beneficial. A prerequisite for a favorable effect in acute cases is that chelation therapy be given as early as possible after exposure.

REFERENCES

1. **Adams, R. G., Harrison, J. F., and Scott, P.,** The development of cadmium-induced proteinuria, impaired renal function and osteomalacia in alkaline battery workers, *Q. J. Med.*, 38, 425—443, 1969.
2. **Adamsson, E., Piscator, M., and Nogawa, K.,** Pulmonary and gastrointestinal exposure to cadmium oxide dust in battery workers, *Environ. Health Perspect.*, 28, 219—222, 1979.
3. **Albert, R. E., Lippmann, M., Spiegelman, J., Liuzzin, A., and Nelson, N.,** The deposition and clearance of radioactive particles in the human lung, *Arch. Environ. Health*, 14, 10—15, 1967.
4. **Albert, R. E., Lippmann, M., and Briscoe, W.,** The characteristics of bronchial clearance in humans and the effects of cigarette smoking, *Arch. Environ. Health*, 18, 738—755, 1969.
5. **Altman, P. L. and Dittmer, D. S., Eds.,** *Biology Data Book*, 2nd ed., Federation of American Societies for Experimental Biology, Bethesda, Md., 1972.
6. **Andersen, O.,** Chelation of cadmium, *Environ. Health Perspect.*, 54, 249—266, 1984.
7. **Andersen, O., Hägerstrand, I., and Nordberg, G. F.,** Effect of the chelating agent sodium tripolyphosphate, *Environ. Res.*, 29, 54—61, 1982.
8. **Ando, M., Sayato, Y., Tonomura, M., and Osawa, T.,** Studies on excretion and uptake of calcium by rats after continuous oral administration of cadmium, *Toxicol. Appl. Pharmacol.*, 39, 321—327, 1977.
9. **Axelsson, B. and Piscator, M.,** Renal damage after prolonged exposure to cadmium. An experimental study, *Arch. Environ. Health*, 12, 360—373, 1966.
10. **Bakka, A.,** Studies on Possible Functions of Metallothionein, Doctoral thesis, University of Oslo, Oslo, Norway, 1983.
11. **Bakka, A., Aaseth, J., and Rugstad, H. E.,** Influence of certain chelating agents on egress of cadmium from cultured epithelial cells containing high amounts of metallothionein: a screening of Cd-releasing and toxic effects, *Acta Pharmacol. Toxicol.*, 49, 432—437, 1981.
12. **Barltrop, D. and Khoo, H. E.,** The influence of nutritional factors on lead absorption, *Postgrad. Med. J.*, 51, 795—800, 1975.
13. **Barrett, H. M., Irwin, D. A., and Semmons, E.,** Studies on the toxicity of inhaled cadmium. I. The acute toxicity of cadmium oxide by inhalation, *J. Ind. Hyg. Toxicol.*, 29, 279—285, 1947.
14. **Berlin, M. and Friberg, L.,** Bone-marrow-activity and erythrocyte destruction in chronic cadmium poisoning, *Arch. Environ. Health*, 1, 478—486, 1960.
15. **Berlin, M. and Ullberg, S.,** The fate of ^{109}Cd in the mouse. An autoradiographic study after a single intravenous injection of $^{109}CdCl_2$, *Arch. Environ. Health*, 7, 686—693, 1963.
16. **Berlin, M., Hammarström, L. and Maunsbach, A. B.,** Microautoradiographic localization of water-soluble cadmium in mouse kidney, *Acta Radiol.*, 2, 345—352, 1964.
17. **Bernard, A., Lauwerys, R., and Gengoux, P.,** Characterisation of the proteinuria induced by prolonged oral administration of cadmium in female rats, *Toxicology*, 20, 345—357, 1981.

18. **Bernstein, M.,** Inhalation of cadmium pigments and other cadmium compounds, in *Cadmium 81, Proc. 3rd Int. Cadmium Conf., Miami*, Cadmium Association, London, 1982, 179.
19. **Boisset, M. and Boudene, C.,** Effect of a single exposure to cadmium oxide fumes on rat lung microsomal enzymes, *Toxicol. Appl. Pharmacol.*, 57, 335—345, 1981.
20. **Boisset, M., Girard, F., Godin, J., and Boudene, C.,** Cadmium content of lung, liver and kidney in rats exposed to cadmium oxide fumes, *Int. Arch. Occup. Environ. Health*, 41, 41—53, 1978.
21. **Bonnell, J. A., Ross, J. H., and King, E.,** Renal lesions in experimental cadmium poisoning, *Br. J. Ind. Med.*, 17, 69—80, 1960.
22. **Bremner, I.,** Heavy metal toxicities, *Q. Rev. Biophys.*, 7, 75—124, 1974.
23. **Buchet, J. P., Roels, H., Lauwerys, R., Bruaux, P., Clays-Thoreau, F., Lafontaine, A., and Verduyn, G.,** Repeated surveillance of exposure to cadmium, manganese and arsenic in school-age children living in rural, urban and nonferrous smelter areas in Belgium, *Environ. Res.*, 22, 95—108, 1980.
24. **Burch, G. E. and Walsh, J. J.,** The excretion and biologic decay rates of ^{115m}Cd with a consideration of space, mass, and distribution in dogs, *J. Lab. Clin. Med.*, 54, 66—72, 1959.
25. **Bus, J. S., Vinegar, A., and Brooks, S. M.,** Biochemical and physiologic changes in lungs of rats exposed to a cadmium chloride aerosol, *Am. Rev. Respir. Dis.*, 118, 573—580, 1978.
26. **Camner, P. and Philipson, K.,** Human alveolar deposition of 4 μm teflon particles, *Arch. Environ. Health*, 33, 181—185, 1978.
27. **Camner, P., Clarkson, T. W., and Nordberg, G. F.,** Routes of exposures, dose and metabolism of metals, in *Handbook on the Toxicology of Metals*, Friberg, L., Nordberg, G. F., and Vouk, V., Eds., Elsevier, Amsterdam, 1979, 65—97.
28. **Cantilena, L. R., Jr. and Klaassen, C. D.,** The effect of ethylene diaminetetraacetic acid (EDTA) and EDTA plus salicylate on acute cadmium toxicity and distribution, *Toxicol. Appl. Pharmacol.*, 53, 510—514, 1980.
29. **Cantilena, L. R., Jr. and Klaassen, C. D.,** Comparison of the effectiveness of several chelators after single administration on the toxicity, excretion, and distribution of cadmium, *Toxicol. Appl. Pharmacol.*, 58, 452—460, 1981.
30. **Cantilena, L. R., Jr. and Klaassen, C. D.,** Decreased effectiveness of chelation therapy with time after acute cadmium poisoning, *Toxicol. Appl. Pharmacol.*, 63, 173—180, 1982.
31. **Carlson, L. A. and Friberg, L.,** The distribution of cadmium in blood after repeated exposure, *Scand. J. Clin. Lab. Invest.*, 9, 1—4, 1957.
32. **Caujolle, F., Oustrin, J., and Silve-Mamy, G.,** Fixation and enterohepatic circulation of cadmium (in French), *Eur. J. Toxicol.*, 4, 310—315, 1971.
33. **Ceresa, C.,** An experimental study of cadmium intoxication (in Italian), *Med. Lav.*, 36, 71—88, 1945.
34. **Chaube, S., Nishimura, H., and Swinyard, C. A.,** Zinc and cadmium in normal human embryos and fetuses, *Arch. Environ. Health*, 26, 237—240, 1973.
35. **Cherian, M. G.,** Studies on the synthesis and metabolism of zinc-thionein in rats, *J. Nutr.*, 107, 965—972, 1977.
36. **Cherian, M. G.,** Metabolism of orally administered cadmium-metallothionein in mice, *Environ. Health Perspect.*, 28, 127—130, 1979.
37. **Cherian, M. G.,** Chelation of cadmium with BAL and DTPA in rats, *Nature (London)*, 287, 871—872, 1980.
38. **Cherian, M. G.,** Biliary excretion of cadmium in rat. III. Effects of chelating agents and change in intracellular thiol content on biliary transport and tissue distribution of cadmium, *J. Toxicol. Environ. Health*, 6, 379—391, 1980a.
39. **Cherian, M. G.,** Biliary excretion of cadmium in rat. IV. Mobilization of cadmium from metallothionein by 2,3-dimercaptopropanol, *J. Toxicol. Environ. Health*, 6, 393—401, 1980b.
40. **Cherian, M. G.,** Absorption and tissue distribution of cadmium in mice after chronic feeding with cadmium chloride and cadmium-metallothionein, *Bull. Environ. Contam. Toxicol.*, 30, 33—36, 1983.
41. **Cherian, M. G. and Shaikh, Z. A.,** Metabolism of intravenously injected cadmium-binding protein, *Biochem. Biophys. Res. Commun.*, 65, 863—869, 1975.
42. **Cherian, M. G. and Vostal, J. J.,** Biliary excretion of cadmium in rat. I. Dose-dependent biliary excretion and the form of cadmium in the bile, *J. Toxicol. Environ. Health*, 2, 945—954, 1977.
43. **Cherian, M. G., Goyer, R. A., and Delaquerrier-Richardson, L.,** Cadmium — metallothionein-induced nephropathy, *Toxicol. Appl. Pharmacol.*, 38, 399—408, 1976.
44. **Cherian, M. G., Goyer, R. A., and Valberg, L. S.,** Gastrointestinal absorption and organ distribution of oral cadmium chloride and cadmium-metallothionein in mice, *J. Toxicol. Environ. Health*, 4, 861—868, 1978.
45. **Chernoff, N. and Courtney, K. D.,** Maternal and fetal effects of NTA, NTA and cadmium, NTA and mercury, NTA and nutritional imbalance in mice and rats, *Natl. Inst. Environ. Health Sci. Prog. Rep.*, NIEH, Washington, D.C., December 1, 1970 (January 18, 1971).

46. **Cikrt, M. and Havrdova, J.,** Effects of dosage and cadmium pretreatment on the binding of cadmium in rat bile, *Experientia*, 35, 1640—1641, 1979.
47. **Cikrt, M. and Tichý, M.,** Excretion of cadmium through bile and intestinal wall in rats, *Br. J. Ind. Med.*, 31, 134—139, 1974.
48. **Cohn, J. R. and Emmett, E. A.,** The excretion of trace metals in human sweat, *Ann. Clin. Lab. Sci.*, 8, 270—275, 1978.
49. **Cotzias, G. C., Borg, D. C., and Selleck, B.,** Virtual absence of turnover in cadmium metabolism: ^{109}Cd studies in the mouse, *Am. J. Physiol.*, 201, 927—930, 1961.
50. **Cousins, R. J. and Feldman, S. L.,** Effect of cholecalciferol on cadmium uptake in the chick, *Nutr. Rep. Int.*, 8, 363—369, 1973.
51. **Cox, C. C. and Waters, M. D.,** Isolation of a soluble cadmium-binding protein from pulmonary macrophages, *Toxicol. Appl. Pharmacol.*, 46, 385—394, 1978.
52. **Cvetkova, R. P.,** Materials on the study of the influence of cadmium compounds on the generative function (in Russian with English summary), *Gig. Tr. Prof. Zabol.*, 14, 31—33, 1970.
53. **Dalhamn, T. and Friberg, L.,** Dimercaprol (2,3-dimercaptopropanol) in chronic cadmium poisoning, *Acta Pharmacol.*, 11, 68—71, 1955.
54. **Davies, N. T. and Campell, J. K.,** The effect of cadmium on intestinal copper absorption and binding in the rat, *Life Sci.*, 20, 955—960, 1977.
55. **Decker, C. F., Byerrum, R. U., and Hoppert, C. A.,** A study of the distribution and retention of cadmium-115 in the albino rat, *Arch. Biochem.*, 66, 140—145, 1957.
56. **Dencker, L.,** Possible mechanisms of cadmium toxicity in golden hamsters and mice: uptake by the embryo, placenta and ovary, *J. Reprod. Fertil.*, 44, 461—471, 1975.
57. **Diem, K. and Lentner, C., Eds.,** *Document Geigy, Scientific Tables*, 7th ed., Ciba Geigy, Basel, 1970.
58. **Dreizen, S., Levy, B. M., Niedermeier, W., and Griggs, J. H.,** Comparative concentrations of selected trace metals in human and marmoset saliva, *Arch. Oral Biol.*, 15, 179—188, 1970.
59. **Durbin, P. W., Scott, K. G., and Hamilton, J. G.,** The distribution of radioisotopes of some heavy metals in the rat, *Univ. Calif. Publ. Pharmacol.*, 3, 1, 1957.
60. **Elinder, C.-G. and Pannone, M.,** Biliary excretion of cadmium, *Environ. Health Perspect.*, 28, 123—126, 1979.
61. **Elinder, C.-G., Kjellström, T., Friberg, L., Lind, B., and Linnman, L.,** Cadmium in kidney cortex, liver and pancreas, from Swedish autopsies, *Arch. Environ. Health*, 31, 292—302, 1976.
62. **Elinder, C.-G., Kjellström, T., Lind, B., Molander, M.-L., and Silander, T.,** Cadmium concentrations in human liver, blood, and bile. Comparison with a metabolic model, *Environ. Res.*, 17, 236—241, 1978.
63. **Elinder, C.-G., Jönsson, L., Piscator, M., Linnman, L., and Stenström, T.,** Water hardness in relation to cadmium accumulation and microscopical finds of cardiovascular disease in horses, *Arch. Environ. Health*, 35, 81—84, 1980.
64. **Elinder, C.-G., Friberg, L., Lind, B., and Jawaid, M.,** Lead and cadmium levels in blood samples from the general population of Sweden, *Environ. Res.*, 30, 233—253, 1983.
65. **Ellis, K. J., Morgan, W. D., Zanai, I., Yasumura, S., Vartsky, D., and Cohn, S. H.,** Critical concentrations of cadmium in human renal cortex: dose-effect studies in cadmium smelter workers, *J. Toxicol. Environ. Health*, 7, 691—703, 1981a.
66. **Ellis, K. J., Yasumura, S., and Cohn, S. H.,** Hair cadmium content: is it a biological indicator of the body burden of cadmium for the occupationally exposed worker?, *Am. J. Ind. Med.*, 2, 323—333, 1981b.
67. **Engström, B.,** Factors Influencing Metabolism and Toxicity of Cadmium. Experimental Studies on Mice with Special Reference to Chelating Agents, Doctoral thesis, Karolinska Institute, Stockholm, Sweden, 1979.
68. **Engström, B.,** Effects of chelating agents on oral uptake and renal deposition and excretion of cadmium, *Environ. Health Perspect.*, 54, 219—232, 1984.
69. **Engström, B. and Nordberg, G. F.,** Effects of milk diet on gastrointestinal absorption of cadmium in adult mice, *Toxicology*, 9, 195—203, 1978a.
70. **Engström, B. and Nordberg, G. F.,** Effects of detergent formula chelating agents on the metabolism and toxicity of cadmium in mice, *Acta Pharmacol. Toxicol.*, 43, 387—397, 1978b.
71. **Engström, B. and Nordberg, G. F.,** Dose dependence of gastrointestinal absorption and biological half-time of cadmium in mice, *Toxicology*, 13, 215—222, 1979a.
72. **Engström, B. and Nordberg, G. F.,** Factors influencing absorption and retention of oral ^{109}Cd in mice: age, pretreatment and subsequent treatment with non-radioactive cadmium, *Acta Pharmacol. Toxicol.*, 45, 315—324, 1979b.
73. Environment Agency Japan, *Recent Studies on Health Effects of Cadmium in Japan*, Office of Health Studies, Tokyo, August 1981.
74. **Eybl, V., Sýkora, J., and Mertl, F.,** Effects of CaEDTA and CaDTPA in cadmium intoxication (in German), *Acta Biol. Med. Ger.*, 17, 178—185, 1966a.

75. **Eybl, V., Sýkora, J., and Mertl, F. K.,** Passage of Zn, Cd, Hg and chelates of these metals through the placental barrier (in Czech), *Cesk. Fysiol.*, 15, 36, 1966b.
76. **Eybl, V., Sýkora, J., and Mertl, F.,** Influence of sodium selenite, sodium tellurite and sodium sulfite on the retention and distribution of cadmium in mice. *Arch. Toxikol.*, 26, 169—175, 1970.
77. **Falck, F. Y., Jr., Fine, L. J., Smith R. G., Garvey, J., Schork, A., England, B., McClatchey, K. D., and Linton, J.,** Metallothionein and occupational exposure to cadmium, *Br. J. Ind. Med.*, 40, 305—313, 1983.
78. **Ferm, V. H., Hanlon, D. P., and Urban, J.,** The permeability of the hamster placenta to radioactive cadmium, *J. Embryol. Exp. Morphol.*, 22, 107—113, 1969.
79. **Fitzhugh, O. G. and Meiller, F. H.,** The chronic toxicity of cadmium, *J. Pharmacol. Exp. Ther.*, 72, 15, 1941.
80. **Flanagan, P. R., McLellan, J. S., Haist, J., Cherian, M. G., Chamberlain, M. J., and Valberg, L. S.,** Increased dietary cadmium absorption in mice and human subjects with iron deficiency, *Gastroenterology*, 74, 841—846, 1978.
81. **Flanagan, P. R., Haist, J., and Valberg, L. S.,** Comparative effects of iron deficiency induced by bleeding and a low-iron diet on the intestinal absorptive interactions of iron, cobalt, manganese, zinc, lead and cadmium, *J. Nutr.*, 110, 1754—1763, 1980.
82. **Fletcher, J. G., Chettle, D. R., and Al-Haddad, I. K.,** Experience with the use of cadmium measurements of liver and kidney, *J. Radioanal. Chem.*, 71, 547—560, 1982.
83. **Foulkes, E. C.,** Renal tubular transport of cadmium metallothionein, *Toxicol. Appl. Pharmacol.*, 45, 505—512, 1978.
84. **Foulkes, E. C.,** Some determinants of intestinal cadmium transport in the rat, *J. Environ. Pathol. Toxicol.*, 3, 471—481, 1980.
85. **Fowler, B. A.,** Proceedings of an International Conference on Environmental Cadmium, Bethesda, Md., June 7 to 9, 1978. Overview, *Environ. Health Perspect.*, 28, 297—300, 1979.
86. **Fowler, B. A. and Nordberg G. F.,** The renal toxicity of cadmium metallothionein, in Int. Conf. Heavy Metals in the Environment, Abstracts, Toronto, October 27 to 31, 1975.
87. **Fowler, B. A. and Nordberg, G. F.,** The renal toxicity of cadmium metallothionein: morphometric and X-ray microanalytical studies, *Toxicol. Appl. Pharmacol.*, 46, 609—623, 1978.
88. **Fox, M. R. S.,** Effect of essential minerals on cadmium toxicity. A review, *J. Food Sci.*, 39, 321—324, 1974.
89. **Fox, M. R. S., Jacobs, R. M., Jones, A. O. L., and Fry, B. E., Jr.,** Effects of nutritional factors on metabolism of dietary cadmium levels similar to those of man, *Environ. Health Perspect.*, 28, 107—114, 1979.
90. **Fox, M. R. S., Tao, S.-H., Stone, C. L., and Fry, B. E., Jr.,** Effects of zinc, iron and copper deficiencies on cadmium in tissues of Japanese quail, *Environ. Health Perspect.*, 54, 57—65, 1984.
91. **Frazier, J. M.,** Cadmium and zinc kinetics in rat plasma following intravenous injection, *J. Toxicol. Environ. Health*, 6, 503—518, 1980.
92. **Freeland, J. H. and Cousins, R. J.,** Effect of dietary cadmium on anemia, iron absorption and cadmium binding protein in the chick, *Nutr. Rep. Int.*, 8, 337—347, 1973.
93. **Friberg, L.,** Health hazards in the manufacture of alkaline accumulators with special reference to chronic cadmium poisoning (Doctoral thesis), *Acta Med. Scand.*, 138(Suppl. 240), 1—124, 1950.
94. **Friberg, L.,** Further investigations on chronic cadmium poisoning; a study on rabbits with radioactive cadmium, *Arch. Ind. Hyg. Occup. Med.*, 5, 30—36, 1952.
95. **Friberg, L.,** Edathamil calcium-disodium in cadmium poisoning, *AMA Arch. Ind. Health*, 13, 18—23, 1956.
96. **Friberg, L. and Odeblad, E.,** Localization of ^{115}Cd in different organs. An autoradiographic study, *Acta Pathol. Microbiol. Scand.*, 41, 96—98, 1957.
97. **Friberg, L., Piscator, M., Nordberg, G. F., and Kjellström, T.,** *Cadmium in the Environment*, 2nd ed., CRC Press, Boca Raton, Fla., 1974.
98. **Friberg, L., Kjellström, T., Nordberg, G., and Piscator, M.,** Cadmium, in *Handbook on the Toxicology of Metals*, Friberg, L., Nordberg, G. F., and Vouk, V., Eds., Elsevier, Amsterdam, 1979, 355—381.
99. **Garty, M., Wong, K.-L., and Klaassen, C. D.,** Redistribution of cadmium to blood of rats, *Toxicol. Appl. Pharmacol.*, 59, 548—554, 1981.
100. **Geldmacher-v. Mallinckrodt, M. and Opitz, O.,** Diagnosis of cadmium poisoning. Normal cadmium levels in human organs and body fluids (in German), *Arbeitsmed. Sozialmed. Arbeitshyg.*, 10, 276—279, 1968.
101. **Gilman, A., Philips, F. S., Allen, R. P., and Koelle, E. S.,** The treatment of acute cadmium intoxication in rabbits with 2,3-dimercaptopropanol (BAL) and other mercaptans, *J. Pharmacol. Exp. Ther.*, 87, 85—101, 1946.

102. **Goyer, R. A., Fowler, B. A., Nordberg, G. F., Shepherd, G. and Maustafa, L., Eds.,** Proceedings of the metallothionein and cadmium nephrotoxicity conference, North Carolina, 1983, *Environ. Health Perspect.*, 54, 1—295, 1984.
103. **Gruden, N.,** Influence of cadmium on calcium transfer through the duodenal wall in rats, *Toxicology*, 37, 149—154, 1977.
104. **Gunn, S. A. and Gould, T. C.,** Selective accumulation of ^{115}Cd by cortex of rat kidney, *Proc. Soc. Exp. Biol. Med.*, 96, 820—823, 1957.
105. **Hadley, J. G., Conklin, A. W., and Sanders, C. L.,** Rapid solubilization and translocation of ^{109}CdO following pulmonary deposition, *Toxicol. Appl. Pharmacol.*, 54, 156—160, 1980.
106. **Hamilton, D. L. and Smith, M. W.,** Cadmium inhibits calcium absorption by rat intestine, *Am. J. Physiol.*, 265, 54—55, 1977.
107. **Hamilton, D. L. and Valberg, L. S.,** Relationship between cadmium and iron absorption, *Am. J. Physiol.*, 227, 1033—1037, 1974.
108. **Harada, A., Hirota, M., and Kono, K.,** Surveillance of cadmium workers, in *Cadmium 79, Proc. 2nd Int. Cadmium Conf. Cannes*, Metal Bulletin Ltd., London, 1980, 183—193.
109. **Harrison, H. E., Bunting, H., Ordway, N., and Albrink, W. S.,** The effects and treatment of inhalation of cadmium chloride aerosols in the dog, *J. Ind. Hyg. Toxicol.*, 29, 302—314, 1947.
110. **Henderson, R. F., Rebar, A. H., Pickrell, J. A., and Newton, G. J.,** Early damage indicators in the lung. III. Biochemical and cytological response of the lung to inhaled metal salts, *Toxicol. Appl. Pharmacol.*, 50, 123—136, 1979.
111. **Henke, G., Sachs, H. W., and Bohn, G.,** Determination of cadmium in liver and kidney of children and adolescents using neutron activation analysis (in German), *Arch. Toxikol.*, 26, 8—16, 1970.
112. **Hietanen, E.,** Gastrointestinal absorption of cadmium, in *Cadmium in the Environment, Part II*, Nriagu, J. O., Ed., John Wiley & Sons, New York, 1981, 55—68.
113. **Hildebrand, C. E. and Cram, L. S.,** Distribution of cadmium in human blood cultured in low levels of $CdCl_2$: accumulation of Cd in lymphocytes and preferential binding to metallothionein, *Proc. Soc. Exp. Biol. Med.*, 161, 438—443, 1979.
114. **Honda, R., Yamada, Y., Kobayashi, E., Tsuritani, I., Ishizaki, M., and Nogawa, K.,** Distribution of Pb and Cd in the blood of persons exposed to lead or cadmium (in Japanese with English summary), *Hokuriku J. Public Health*, 9, 27—31, 1982.
115. International Commission on Radiological Protection, Report of the Task Group on Reference Man, Rep. No. 23, Pergamon Press, Oxford, 1975.
116. **Illing, J. and Hook, G. E. R.,** Production of low molecular weight cadmium-binding proteins in rabbit lung following exposure to cadmium chloride, *Biochem. Pharmacol.*, 31(18), 2969—2975, 1982.
117. **Jacobs, R. M., Fox, M. R. S., Jones, A. O. L., Hamilton, R. P., and Lener, J.,** Cadmium metabolism. Individual effects of Zn, Cu and Mn, *Fed. Proc.*, 36, 1152, 1977.
118. **Järup, L., Rogenfelt, A., Elinder, C.-G., Nogawa, K., and Kjellström, T.,** Biological half-time of cadmium in the blood of workers after cessation of exposure, *Scand. J. Work Environ. Health*, 9, 327—331, 1983.
119. **Johnson, A. D. and Miller, W. J.,** Early actions of cadmium in the rat and domestic fowl testis. II. Distribution of injected ^{109}Cd, *J. Reprod. Fertil.*, 21, 395—405, 1970.
120. **Johnson, A. D. and Sigman, M. B.,** Early actions of cadmium in rat and domestic fowl testis. IV. Autoradiographic location of ^{115m}Cd, *J. Reprod. Fertil.*, 24, 115—117, 1971.
121. **Johnson, D. R. and Foulkes, E. C.,** On the proposed role of metallothionein in the transport of cadmium, *Environ. Res.*, 21, 360—365, 1980.
122. **Kapoor, N. K., Agarwala, S. C., and Kar, A. B.,** The distribution and retention of cadmium in subcellular fractions of rat liver, *Ann. Biochem. Exp. Med.*, 21, 51—54, 1961.
123. **Kello, D. and Kostial, K.,** Influence of age on whole-body retention and distribution of ^{115}Cd in the rat, *Environ. Res.*, 14, 92—98, 1977a.
124. **Kello, D. and Kostial, K.,** Influence of age and milk diet on cadmium absorption from the gut, *Toxicol. Appl. Pharmacol.*, 46, 277—282, 1977b.
125. **Kello, D., Dekanic, D., and Kostial, K.,** Influence of sex and dietary calcium in intestinal cadmium absorption in rats, *Arch. Environ. Health*, 34, 30—33, 1979.
126. **Kench, J. E., Wells, A. R., and Smith, J. C.,** Some observations on the proteinuria of rabbits poisoned with cadmium, *S. Afr. Med. J.*, 36, 390—394, 1962.
127. **Kimura, M. and Otaki, N.,** Percutaneous absorption of cadmium in rabbit and hairless mouse, *Ind. Health*, 10, 7—10, 1972.
128. **Kimura, M., Watanabe, M., and Otaki, N.,** Determination of heavy metals and beta-2-microglobulin, in *Effects of Nutritional Factors on Cadmium-Administered Monkeys*, The Tertiary Monkey Experiment Team, Japan Public Health Association, Tokyo, 1983, 66—82.
129. **Kitamura, S.,** Cadmium absorption and accumulation (mainly about humans) (in Japanese), in *Kankyo Hoken Rep. No. 11*, Japanese Public Health Association, 1972, 42.

130. **Kjellström, T.,** Accumulation and Renal Effects of Cadmium in Man. A Dose-Response Study, Doctoral thesis, Karolinska Institute, Stockholm, 1977.
131. **Kjellström, T.,** Exposure and accumulation of cadmium in populations from Japan, the United States and Sweden, *Environ. Health Perspect.*, 28, 169—197, 1979.
132. **Kjellström, T. and Nordberg, G. F.,** A kinetic model of cadmium metabolism in the human being, *Environ. Res.*, 16, 248—269, 1978.
133. **Kjellström, T., Shiroishi, K., and Evrin, P.-E.,** Urinary β_2-microglobulin excretion among people exposed to cadmium in the general environment, *Environ. Res.*, 13, 318—344, 1977.
134. **Kjellström, T., Borg, K., and Lind, B.,** Cadmium in feces as an estimator of daily cadmium intake in Sweden, *Environ. Res.*, 15, 242—251, 1978.
135. **Kjellström, T., Elinder, C.-G., and Friberg, L.,** Conceptual problems in establishing the critical concentration of cadmium in human renal cortex, *Environ. Res.*, 33, 284—295, 1984.
136. **Klaassen, C. D., Waalkes, M. P., and Cantilena, L. R. J.,** Alteration of tissue disposition of cadmium by chelating agents, *Environ. Health Perspect.*, 54, 233—242, 1984.
137. **Klaassen, C. D. and Kotsonis, F. N.,** Biliary excretion of cadmium in the rat, rabbit and dog, *Toxicol. Appl. Pharmacol.*, 41, 101—112, 1977.
138. **Kobayashi, J., Nakahara, H., and Hasegawa, T.,** Accumulation of cadmium in organs of mice fed on cadmium-polluted rice (in Japanese with English summary), *Jpn. J. Hyg.*, 26, 401—407, 1971.
139. **Koizumi, N.,** Fundamental studies on the movement of cadmium in animals and human beings (in Japanese), *Jpn. J. Hyg.*, 30, 300—314, 1975.
140. **Kojima, Y. and Hamashima, Y.,** Immunohistological study of equine renal metallothionein, *Acta Histochem. Cytochem.*, 1, 205—211, 1978.
141. **Kotsonis, F. N. and Klaassen, C. D.,** Toxicity and distribution of cadmium administered to rats at sublethal doses, *Toxicol. Appl. Pharmacol.*, 41, 667—680, 1977.
142. **Lagally, H. R., Biddlé, G. N., and Siewicki, T. C.,** Cadmium retention in rats fed either bound cadmium in scallops or cadmium sulfate, *Nutr. Rep. Int.*, 21, 351—363, 1980.
143. **Lal, U. B.,** The Effects of Low and High Levels of Dietary Zinc on Pathology in Rats Exposed to Cadmium, Doctoral thesis, University of Cincinnati, Cincinnati, Ohio, 1976.
144. **Larsson, S.-E. and Piscator, M.,** Effect of cadmium on skeletal tissue in normal and calcium-deficient rats, *Isr. J. Med. Sci.*, 7, 495—497, 1971.
145. **Lauwerys, R., Buchet, J.-P., and Roels, H.,** The relationship between cadmium exposure or body burden and the concentration of cadmium in blood and urine in man, *Int. Arch. Occup. Environ. Health*, 36, 275—285, 1976.
146. **Lauwerys, R., Roels, H., Buchet, J.-P., Bernard, A., and Stanescu, D.,** Investigations on the lung and kidney functions in workers exposed to cadmium, *Environ. Health Perspect.*, 28, 137—145, 1979a.
147. **Lauwerys, R., Roels, H., Regniers, M., Buchet, J. P., Bernard, A., and Goret, A.,** Significance of cadmium concentration in blood and in urine in workers exposed to cadmium, *Environ. Res.*, 20, 375—391, 1979b.
148. **Leber, A. P. and Miya, T. S.,** A mechanism for cadmium- and zinc-induced tolerance to cadmium toxicity: involvement of metallothionein, *Toxicol. Appl. Pharmacol.*, 37, 403—414, 1976.
149. **Lewis, G. P., Coughlin, L., Jusko, W., and Hartz, S.,** Contribution of cigarette smoking to cadmium accumulation in man, *Lancet*, 1, 291—292, 1972.
150. **Lindh, U., Brune, D., Nordberg, G., and Wester, P. O.,** Levels of cadmium in bone tissue (femur) of industrially exposed workers — a reply, *Sci. Total Environ.*, 20, 3—11, 1981.
151. **Livingston, H. D.,** Measurement and distribution of zinc, cadmium and mercury in human kidney tissue, *Clin. Chem.*, 18, 67—72, 1972.
152. **Lorentzon, R. and Larsson, S.-E.,** Vitamin D metabolism in adult rats at low and normal calcium intake and the effect of cadmium exposure, *Clin. Sci. Mol. Med.*, 53, 439—446, 1977.
153. **Lucis, O. J. and Lucis, R.,** Distribution of cadmium-109 and zinc-65, in mice of inbred strains, *Arch. Environ. Health*, 19, 334—336, 1969.
154. **Lucis, O. J., Lynk, M. E., and Lucis, R.,** Turnover of cadmium-109 in rats, *Arch. Environ. Health*, 18, 307—310, 1969.
155. **Lucis, O. J., Lucis, R., and Shaikh, Z. A.,** Cadmium and zinc in pregnancy and lactation, *Arch. Environ. Health*, 25, 14—22, 1972.
156. **MacFarland, H. N.,** The use of dimercaprol (BAL) in the treatment of cadmium oxide fume poisoning, *Arch. Environ. Health*, 1, 487—496, 1960.
157. **Mahaffey, K. R., Capar, S. G., Gladen, B. C., and Fowler, B. A.,** Concurrent exposure to lead, cadmium and arsenic. Effects on toxicity and tissue metal concentrations in the rat, *J. Lab. Clin. Med.*, 98, 463—481, 1981.
158. **Matsubara-Khan, J.,** Compartmental analysis for the evaluation of biological half-lives of cadmium and mercury in mouse organs, *Environ. Res.*, 7, 54—67, 1974.

159. **Matsusaka, N., Tanaka, M., Nishimura, Y., Yuyama, A., and Kobayashi, H.,** Whole body retention and intestinal absorption of ^{115m}Cd in young and adult mice (in Japanese), *Med. Biol.*, 85, 275—279, 1972.
160. **McKenzie, J., Kjellström, T., and Sharma, R.,** Cadmium intake, metabolism and effects in people with a high intake of oysters in New Zealand, Report to the U.S. Environmental Protection Agency, Washington, D.C., 1982.
161. **McLellan, J. S., Flanagan, P. R., Chamberlain, M. J., and Valberg, L. S.,** Measurements of dietary cadmium absorption in humans, *J. Toxicol. Environ. Health*, 4, 131—138, 1978.
162. **Miller, R. and Shaikh, Z. A.,** Prenatal metabolism: metals and metallothionein, in *Reproductive and Developmental Toxicity of Metals*, Clarkson, T. W., Nordberg, G. F., and Sager, P. R., Eds., Plenum Press, New York, 1983, 151—204.
163. **Miller, W. J., Blackmon, D. M., and Martin, W. G.,** ^{109}Cd absorption, excretion and tissue distribution following single tracer oral and intravenous doses in young goats, *J. Dairy Sci.*, 51, 1836—1839, 1968.
164. **Miller, W. J., Blackmon, D. M., Gentry, R. P., and Pate, F. M.,** Effect of dietary cadmium on tissue distribution of ^{109}Cd following a single oral dose in young goats, *J. Dairy Sci.*, 52, 2029—2035, 1969.
165. **Moore, W., Jr., Stara, J. F., and Crocker, W. C.,** Gastrointestinal absorption of different compounds of ^{115}Cd and the effect of different concentrations in the rat, *Environ. Res.*, 6, 159—164, 1973a.
166. **Moore, W., Jr., Stara, J. F., Crocker, W. C., Malanchuk, M., and Iltis, R.,** Comparison of ^{115m}Cd retention in rats following different routes of administration, *Environ. Res.*, 6, 473—478, 1973b.
167. **Niemeier, B.,** The influence of chelating agents on the distribution and toxicity of cadmium (in German), *Int. Arch. Gewerbepathol. Gewerbehyg.*, 24, 160—168, 1967.
168. **Nishiyama, K. and Nordberg, G. F.,** Adsorption and elution of cadmium on hair, *Arch. Environ. Health*, 25, 92—96, 1972.
169. **Nogawa, K., Kobayashi, E., Inaoka, H., and Ishizaki, A.,** The relationship between the renal effects of cadmium and cadmium concentration in urine among the inhabitants of cadmium-polluted areas, *Environ. Res.*, 14, 391—400, 1977.
170. **Nolen, G. A., Bohne, R. L., and Buehler, E. V.,** Effects of trisodium nitrilotriacetate, trisodium citrate and a trisodium nitrilotriacetate-ferric chloride mixture on cadmium and methyl mercury toxicity and teratogenesis in rats, *Toxicol. Appl. Pharmacol.*, 23, 238—250, 1972a.
171. **Nolen, G. A., Buehler, E. V., Geil, R. G., and Goldenthal, E. I.,** Effects of trisodium nitrilotriacetate on cadmium and methyl mercury toxicity and teratogenicity in rats, *Toxicol. Appl. Pharmacol.*, 23, 222—237, 1972b.
172. **Nomiyama, K.,** *In vivo* experiments, in *Cadmium Studies in Japan — A review*, Tsuchiya, K., Ed., Elsevier, Amsterdam, 1978, 47—86.
173. **Nomiyama, K.,** Recent progress and perspectives in cadmium health effects studies, *Sci. Total Environ.*, 14, 199—232, 1980.
174. **Nomiyama, K. and Foulkes, E. C.,** Reabsorption of filtered cadmium-metallothionein in the rabbit kidney, *Proc. Soc. Exp. Biol. Med.*, 156, 97—99, 1977.
175. **Nomiyama, K. and Nomiyama, H.,** Biological half-time of cadmium in rabbits (in Japanese), in *Kankyo Hoken Rep. No. 38*, Japan Public Health Association, Tokyo, 1976, 156—158.
176. **Nomiyama, K., Nomiyama, H., Nomura, T., Taguchi, T., Matsui, J., Yotoriyama, M., Akahori, F., Iwao, S., Koizumi, N., Masaoka, T., Kitamura, S., Tsuchiya, K., Suzuki, T., and Kobayashi, K.,** Effects of dietary cadmium on rhesus monkeys, *Environ. Health Perspect.*, 28, 223—243, 1979.
177. **Nomiyama, K., Nomiyama, H., Akahori, F., and Masaoka, T.,** Further studies on effects of dietary cadmium on rhesus monkeys. X. Tissue cadmium, zinc and copper, in *Recent Studies on Health Effects of Cadmium in Japan*, Environment Agency, Office of Health Studies, Tokyo, 1981, 154—188.
178. **Nomiyama, K., Nomiyama, H., and Yotoriyama, M.,** Low molecular weight proteins in urine from rabbits given nephrotoxic compounds, *Ind. Health*, 20, 1—10, 1982.
179. **Nordberg, G. F.,** Effects of acute and chronic cadmium exposure on the testicles of mice, *Environ. Physiol.*, 1, 171—187, 1971a.
180. **Nordberg, G. F.,** Effects of NTA on the toxicity and turnover of cadmium (in Swedish), Report to the Research Council, Swedish National Environment Protection Board, Solna, Sweden, 1971b.
181. **Nordberg, G. F.,** Cadmium metabolism and toxicity. Experimental studies on mice with special reference to the use of biological materials as indices of retention and the possible role of metallothionein in transport and detoxification of cadmium, *Environ. Physiol. Biochem.*, 2, 7—36, 1972a.
182. **Nordberg, G. F.,** Models used for calculation of accumulation of toxic metals, in *Proc. 17th Int. Congr. Occup. Health 1972*, (available through the Secretariat, Av. Rogue Saenz, Pena, 110-2 piso, Officio 8, Buenos Aires), 1972b.
183. **Nordberg, G. F.,** Urinary blood and fecal cadmium concentrations as indices of exposure and accumulation, in *Proc. 17th Int. Cong. Occup. Health 1972*, (available through the Secretariat, Av. Rogue Saenz, Pena, 110-2 piso, Officio 8, Buenos Aires), 1972c.
184. **Nordberg, G. F.,** in *Cadmium in the Environment*, 2nd ed., Friberg, L., Piscator, M., Nordberg, G. F., and Kjellström, T., Eds., CRC Press, Boca Raton, Fla., 1974.

185. **Nordberg, G. F.,** Metabolism of cadmium, in *Nephrotoxic Mechanisms of Drugs and Environmental Toxins*, Porter, G. A., Ed., Plenum Medical Book Co., New York, 1982, 285—303.
186. **Nordberg, G. F.,** Chelating agents and cadmium toxicity. Problems and prospects, *Environ. Health Perspect.*, 54, 213—218, 1984.
187. **Nordberg, G. F. and Kjellström, T.,** Metabolic model for cadmium in man, *Environ. Health Perspect.*, 28, 211—217, 1979.
188. **Nordberg, G. F. and Nishiyama, K.,** Whole-body and hair retention of cadmium in mice, including an autoradiographic study on organ distribution, *Arch. Environ. Health*, 24, 209—214, 1972.
189. **Nordberg, G. F. and Nordberg, M.,** Metabolism and toxicity of metallothionein-bound cadmium, Int. Conf. Heavy Metals in the Environment, Abstract, Toronto, Ontario, Canada, October 27—31, 1975b.
190. **Nordberg, G. F. and Piscator, M.,** Influence of long-term cadmium exposure on urinary excretion of protein and cadmium in mice, *Environ. Physiol. Biochem.*, 2, 37—49, 1972.
191. **Nordberg, G. F., Friberg, L., and Piscator, M.,** in *Cadmium in the Environment*, Friberg, L., Piscator, M., and Nordberg, G. F., Eds., CRC Press, Boca Raton, Fla., 1971a, 30, 44.
192. **Nordberg, G. F., Piscator, M., and Lind, B.,** Distribution of cadmium among protein fractions of mouse liver, *Acta Pharmacol. Toxicol.*, 29, 456—470, 1971b.
193. **Nordberg, G. F., Piscator, M., and Nordberg, M.,** On the distribution of cadmium in blood, *Acta Pharmacol. Toxicol.*, 30, 289—295, 1971c.
194. **Nordberg, G. F., Nordberg, M., Piscator, M., and Vesterberg, O.,** Separation of two forms of rabbit metallothionein by isoelectric focusing, *Biochem. J.*, 126, 491—498, 1972.
195. **Nordberg, G. F., Goyer, R. A., and Nordberg, M.,** Comparative toxicity of cadmium-metallothionein and cadmium chloride on mouse kidney, *Arch. Pathol.*, 99, 192—197, 1975.
196. **Nordberg, G. F., Robért, K.-H., and Pannone, M.,** Pancreatic and biliary excretion of cadmium in the rat, *Acta Pharmacol. Toxicol.*, 41, 84—88, 1977.
197. **Nordberg, G. F., Fowler, B. A., Friberg, L., Jernelöv, A., Nelson, N., Piscator, M., Sandstead, H. H., Vostal, J., and Vouk, V. B., Eds.,** Proceedings of an International Meeting in Stockholm, 1977, *Environ. Health Perspect.*, 25, 1—157, 1978.
198. **Nordberg, G. F., Garvey, J. S., and Chang, C. C.,** Metallothionein in plasma and urine of cadmium workers, *Environ. Res.*, 28, 179—182, 1982.
199. **Nordberg, M.,** Studies on metallothionein and cadmium, *Environ. Res.*, 15, 381—404, 1978.
200. **Nordberg, M.,** General aspects of cadmium transport, uptake and metabolism by the kidney, *Environ. Health Perspect.*, 54, 13—20, 1984.
201. **Nordberg, M. and Nordberg G. F.,** Distribution of metallothionein-bound cadmium and cadmium chloride in mice: preliminary studies, *Environ. Health Perspect.*, 12, 103—108, 1975.
202. **Nordberg, M., Nordberg, G. F., and Piscator, M.,** Isolation and characterization of a hepatic metallothionein from mice, *Environ. Physiol. Biochem.*, 5, 396—403, 1975b.
203. **Nordberg, M., Cherian, M. G., and Kjellström, T.,** Defence mechanisms against metal toxicity and their potential importance for risk assessments with particular reference to the importance of various binding forms in food stuff, Rep. Ser. No. 55, Project Coal-Health-Environment, Swedish State Power Board, Vällingby, Sweden, 1983.
204. **Oberdörster, G., Baumert, H.-P., Hochrainer, D., and Stoeber, W.,** The clearance of cadmium aerosols after inhalation exposure, *Am. Ind. Hyg. Assoc. J.*, 40, 443—450, 1979.
205. **Oberdörster, G., Oldiges, H., and Zimmerman, B.,** Lung deposition and clearance of cadmium in rats exposed by inhalation or by intratracheal instillation, *Zentralbl. Bakteriol. (B)*, 170, 35—43, 1980.
206. **Ogawa, E., Suzuki, S., Tsuzuki, H., and Kawajiri, M.,** Experimental studies on the absorption of cadmium chloride, *Jpn. J. Pharmacol.*, 22, (Suppl. 63), 1972.
207. **Onosaka, S., Tanaka, K., and Cherian, M. G.,** Effects of cadmium and zinc on tissue levels of metallothionein, *Environ. Health Perspect.*, 54, 67—72, 1984.
208. **Parizek, J.,** Vascular changes at sites of oestrogen biosynthesis produced by parenteral injection of cadmium salts, the destruction of placenta by cadmium salts, *J. Reprod. Fertil.*, 7, 263—265, 1964.
209. **Parizek, J.,** The peculiar toxicity of cadmium during pregnancy — an experimental 'toxaemia of pregnancy' induced by cadmium salts, *J. Reprod. Fertil.*, 9, 111—112, 1965.
210. **Perry, H. M., Jr. and Erlanger, M. W.,** Hypertension and tissue metal levels after intraperitoneal cadmium, mercury and zinc, *Am. J. Physiol.*, 220, 808—811, 1971.
211. **Perry, H. M., Jr., Erlanger, M. W., Yunice, A., Schoepfle, E., and Perry, E. F.,** Hypertension and tissue metal levels following intravenous cadmium, mercury and zinc, *Am. J. Physiol.*, 219, 755—761, 1970.
212. **Petering, D. H., Loftsgaarden, J., Schneider, J., and Fowler, B. A.,** Metabolism of cadmium, zinc and copper in the rat kidney: the role of metallothionein and other binding sites, *Environ. Health Perspect.*, 54, 73—81, 1984.
213. **Piscator, M.,** Cadmium in the kidneys of normal human beings and the isolation of metallothionein from liver of rabbits exposed to cadmium (in Swedish), *Nord. Hyg. Tidskr.*, 45, 76—82, 1964.

214. **Piscator, M.,** in *Cadmium in the Environment*, Friberg, L., Piscator, M., and Nordberg, G. F., Eds., CRC Press, Boca Raton, Fla., 1971.
215. **Piscator, M.,** in Cadmium in the Environment III, EPA-650/2-75-049, Friberg, L., Kjellström, T., Nordberg, G. F., and Piscator, M., Eds., U.S. Environmental Protection Agency, Washington, D.C., 1975, 458.
216. **Piscator, M. and Larsson, S. E.,** Retention and toxicity of cadmium in calcium-deficient rats, in *Proc. 17th Int. Congr. Occup. Health 1972*, (available through the Secretariat, Av. Rogue Saenz, Pena, 110-2 piso, Oficio 8, Buenos Aires), 1972.
217. **Post, von, C. T., Squibb, K. S., Fowler, B. A., Gardner, D. E., Illing, J., and Hook, G. E. R.,** Production of low molecular weight cadmium-binding proteins in rabbit lung following exposure to cadmium chloride, *Biochem. Pharmacol.*, 31(18), 2969—2975, 1982.
218. **Rahola, T., Aaran, R.-K., and Miettinen, J. K.,** Half-time studies of mercury and cadmium by whole-body counting, in *Assessment of Radioactive Contamination in Man*, IAEA-SM-150/13, International Atomic Energy Agency, Unipublisher, New York, 1972, 553—562.
219. **Revis, N. W. and Osborne, T. R.,** Dietary protein effects on cadmium and metallothionein accumulation in the liver and kidney of rats, *Environ. Health Perspect.*, 54, 83—91, 1984.
220. **Richmond, C. R., Findlay, J. S., and London, J. E.,** Whole-body retention of cadmium-109 by mice following oral intraperitoneal and intravenous administration, in *Biological and Medical Research Group (H-4) of the Health Division — Annual Report, July 1965 through June 1966*, Report LA-3610-MS, Los Alamos Scientific Laboratory, Los Alamos, New Mexico, 1966, 195.
221. **Ridlington, J. W., Winge, D. R., and Fowler, B. A.,** Long-term turnover of cadmium metallothionein in liver and kidney following a single low dose of cadmium in rats, *Biochem. Biophys. Acta*, 673, 177—183, 1981.
222. **Roels, H., Hubermont, G., Buchet, J. P., and Lauwerys, R.,** Placental transfer of lead, mercury, cadmium and carbon monoxide in women. III. Factors influencing the accumulation of heavy metals in the placenta and the relationship between metal concentration in the placenta and in maternal and cord blood, *Environ. Res.*, 16, 236—247, 1978.
223. **Roels, H. A., Lauwerys, R. R., Buchet, J.-P., Bernard, A., Chettle, D. R., Harvey, T. C., and Al-Haddad, I. K.,** *In vivo* measurement of liver and kidney cadmium in workers exposed to this metal: its significance with respect to cadmium in blood and urine, *Environ. Res.*, 26, 217—240, 1981.
224. **Roels, H. A., Lauwerys, R. R., Buchet, J.-P., Bernard, A., Garvey, J. S., and Linton, H. J.,** Significance of urinary metallothionein in workers exposed to cadmium, *Int. Arch. Occup. Environ. Health*, 52, 159—166, 1983.
225. **Sakurai, H.,** Epidemiological studies, in *Cadmium Studies in Japan — A Review*, Tsuchiya, K., Ed., Elsevier, Amsterdam, 1978, 133—267.
226. **Sasser, L. B. and Jarboe, G. E.,** Intestinal absorption and retention of cadmium in neonatal rat, *Toxicol. Appl. Pharmacol.*, 41, 423—431, 1977.
227. **Scharpf, L. G., Jr., Ramos, F. J., and Hill, I. D.,** Influence of nitrilotriacetate (NTA) on the toxicity, excretion and distribution of cadmium in female rats, *Toxicol. Appl. Pharmacol.*, 22, 186—192, 1972.
228. **Schmidt, P. and Gohlke, R.,** Animal experiments on the toxicity of cadmium stearates (in German), *Z. Gesamte Hyg.* 17, 827—831, 1971.
229. **Schroeder, H. A., Nason, A. P., and Balassa, J. J.,** Trace metals in rat tissues as influenced by calcium in water, *J. Nutr.*, 93, 331—336, 1967.
230. **Selenke, W. and Foulkes, E. C.,** The binding of cadmium-metallothionein to isolated renal brush border membranes, *Proc. Soc. Exp. Biol. Med.*, 167, 40—44, 1981.
231. **Shaikh, Z. A. and Hirayama, K.,** Methallothionein in the extra cellular fluids as an index of human toxicity, *Environ. Health Perspect.*, 28, 267—271, 1979b.
232. **Shaikh, Z. A. and Lucis, O. J.,** Biological differences in cadmium and zinc turnover, *Arch. Environ. Health*, 24, 410—418, 1972.
233. **Shaikh, Z. A. and Smith, J. C.,** Metabolism of orally ingested cadmium in humans, in *Mechanisms of Toxicity and Hazard Evaluation*, Holmstedt, B., Lauwerys, R., Mercier, M., and Roberfroid, M., Eds., Elsevier, Amsterdam, 1980, 569—574.
234. **Siewicki, T. C., Balthrop, J. E., and Sydlowski, J. S.,** Iron metabolism of mice fed low levels of physiologically-bound cadmium in oysters of cadmium chloride, *J. Nutr.*, 113, 1140—1149, 1983.
235. **Skog, E. and Wahlberg, J. E.,** A comparative investigation of the percutaneous absorption of metal compounds in the guinea pig by means of the radioactive isotopes: ^{51}Cr, ^{58}Co, ^{65}Zn, ^{110m}Ag, ^{115m}Cd, ^{203}Hg, *J. Invest. Dermatol.*, 43, 187—192, 1964.
236. **Sköld, G.,** Effect of cadmium poisoning on testes, *Acta Pathol. Microbiol. Scand.*, 53, 440, 1961.
237. **Smith, J. C. and Kench, J. E.,** Observations on urinary cadmium and protein excretion in men exposed to cadmium oxide dust and fume, *Br. J. Ind. Med.*, 14, 240—249, 1957.
238. **Smith, L. W. and Borzelleca, J. F.,** Excretion of cadmium and mercury in rat saliva, *Toxicol. Appl. Pharmacol.*, 54, 134—140, 1980.

239. **Sonawane, B. R., Nordberg, M., Nordberg, G. F., and Lucier, G. W.,** Placental transfer of cadmium in rats: influence of dose and gestational age, *Environ. Health Perspect.*, 12, 97—102, 1975.
240. **Squibb, K. S. and Fowler, B. A.,** Intracellular metabolism and effects of circulating cadmium metallothionein in the kidney, *Environ. Health Perspect.*, 54, 31—35, 1984.
241. **Squibb, K. S., Pritchard, J. B., and Fowler, B. A.,** Cadmium-metallothionein nephropathy: relationships between ultrastructural/biochemical alterations and intracellular cadmium binding, *J. Pharmacol. Exp. Ther.*, (in press).
242. **Stowe, H. D.,** Biliary excretion of cadmium by rats: effects of zinc, cadmium and selenium pretreatments, *J. Toxicol. Environ. Health*, 2, 45—53, 1976.
243. **Sumino, K., Hayakawa, K., Shibata, T., and Kitamura, S.,** Heavy metals in normal Japanese tissues, *Arch. Environ. Health*, 30, 487—494, 1975.
244. **Suzuki, K. T.,** Studies of cadmium uptake and metabolism by the kidney, *Environ. Health Perspect.*, 54, 21—30, 1984.
245. **Suzuki, K. T., Yaguchi, K., Ohnuki, R., Nishikawa, M., and Yamada, Y. K.,** Extent of cadmium accumulation and its effect on essential metals in liver, kidney, and body fluids, *J. Toxicol. Environ. Health*, 11, 713—726, 1983a.
246. **Suzuki, K. T., Ohnuki, R., Yaguchi, K., and Yamada, Y. K.,** Accumulation and chemical forms of cadmium and its effect on essential metals in rat spleen and pancreas, *J. Toxicol. Environ. Health*, 11, 727—737, 1983b.
247. **Suzuki, S. and Taguchi, T.,** Retention organ distribution, and excretory pattern of cadmium orally administered in a single dose to two monkeys, *J. Toxicol. Environ. Health*, 6, 783—796, 1980.
248. **Suzuki, S., Taguchi, T., and Yokohashi, G.,** Dietary factors influencing upon the retention rate of orally administered $^{115m}CdCl_2$ in mice with special reference to calcium and protein concentrations in diet, *Ind. Health*, 7, 155—162, 1969.
249. **Suzuki, S., Taguchi, T., and Yokohashi, G.,** Retention rate and organ distribution of cadmium following a single oral dose in a monkey (in Japanese), *Jpn. J. Ind. Health*, 14, 130—131, 1972.
250. **Suzuki, S., Taguchi, T., and Yokohashi, G.,** Gastrointestinal absorption and tissue distribution of cadmium after single oral exposure of monkeys (in Japanese), in Proc. 47th Annu. Meet. Jpn. Assoc. Ind. Health, Nagoya, March 29, 1974, 120—121.
251. **Suzuki, Y.,** The amount of cadmium bound to metallothionein in liver and kidney after long-term cadmium exposure (in Japanese), in Proc. 47th Annu. Meet. Jpn. Assoc. Ind. Health, Nagoya, March 29, 1974, 124—125.
252. **Suzuki, Y.,** Cadmium metabolism and toxicity in rats after long term subcutaneous administration, *J. Toxicol. Environ. Health*, 6, 469—482, 1980.
253. **Suzuki, Y. and Yoshikawa, H.,** Cadmium excretion in urine and feces of rats at different levels of cadmium toxicity, *Ind. Health*, 21, 43—50, 1983.
254. **Svartengren, M., Elinder, C.-G., Friberg, L., and Lind, B.,** Distribution and concentration of cadmium in human kidney, *Environ. Res.*, in press.
255. **Taguchi, T. and Suzuki, S.,** Influence of sex and age on the biological half-life of cadmium in mice, *J. Toxicol. Environ. Health*, 87, 239—249, 1981.
256. **Tanaka, K., Sueda, K., Onosaka, S., and Okahara, K.,** Fate of ^{109}Cd-labelled metallothionein in rats, *Toxicol. Appl. Pharmacol.*, 33, 258—266, 1975.
257. **Tanaka, M., Matsusaka, N., Yuyama, A., and Kobayashi, H.,** Transfer of cadmium through placenta and milk in the mouse, *Radioisotopes*, 21, 50—52, 1972.
258. Task Group on Lung Dynamics, Deposition and retention models for internal dosimetry of the human respiratory tract, *Health Phys.*, 12, 173—208, 1966.
259. Task Group on Metal Accumulation, Accumulation of toxic metals with special reference to their absorption, excretion and biological half-times, *Environ. Physiol. Biochem.*, 3, 65—107, 1973.
260. Task Group on Metal Interaction, Factors influencing metabolism and toxicity of metals: a consensus report, *Environ. Health Perspect.*, 25, 3—41, 1978.
261. Task Group on Metal Toxicity, in *Effects and Dose-Response Relationships of Toxic Metals*, Nordberg, G. F., Ed., Elsevier, Amsterdam, 1976.
262. **Tepperman, H. M.,** The effect of BAL and BAL-glucoside therapy on the excretion and tissue distribution of injected cadmium, *J. Pharmacol.*, 89, 343—349, 1947.
263. **Thompson, T. N. and Klaassen, C. D.,** Disposition of metals after portal and systemic administration to rats, *Toxicol. Appl. Pharmacol.*, 68, 442—450, 1983.
264. **Tobias, J. M., Lushbaugh, C. C., Patt, H. M., Postel, S., Swift, M. N., and Gerard, R. W.,** The pathology and therapy with 2,3-dimercaptopropanol (BAL) of experimental Cd poisoning, *J. Pharmacol.*, 87(Suppl. 102), 1946.
265. **Tohyama, C., Shaikh, Z. A., Nogawa, K., Kobayashi, E., and Honda, R.,** Elevated urinary excretion of metallothionein due to environmental cadmium exposure, *Toxicology*, 22, 289—297, 1981.

266. **Tomita, K.,** Sex differences of the time of biological half-life of cadmium injected subcutaneously to mice, *Jpn. J. Ind. Health*, 13, 46—47, 1971.
267. **Truhaut, R. and Boudene, C.,** Research on the fate of cadmium in the organs during intoxication in occupational medicine (in French), *Arh. Hig. Rada Toksikol.*, 5, 19—48, 1954.
268. **Tsuchiya, K., Seki, Y., and Sugita, M.,** Cadmium concentrations in the organs and tissues of cadavers from accidential deaths, *Keio J. Med.*, 25, 83—90, 1976a.
269. **Ullberg, S.,** Studies on the distribution and fate of ^{35}S labelled benzyl penicillin in the body, *Acta Radiol. Suppl.*, 118, 1—110, 1954.
270. **Uthe, J. F. and Chou, C. L.,** Cadmium levels in selected organs of rats fed three dietary forms of cadmium, *J. Environ. Sci. Health*, A14(2), 117—134, 1979.
271. **Vahter, M., Ed.,** Assessment of human exposure to lead and cadmium through biological monitoring, Report prepared for United Nations Environment Programme and World Health Organization by National Swedish Institute of Environmental Medicine and Department of Environmental Hygiene, Karolinska Institute, Stockholm, 1982.
272. **Valberg, L. S., Sorbie, J., and Hamilton, D. L.,** Gastrointestinal metabolism of cadmium in experimental iron deficiency, *Am. J. Physiol.*, 231, 462—467, 1976.
273. **Vander Mallie, R. J. and Garvey, J. S.,** Production and study of antibody produced against rat cadmium thionein, *Immunochemistry*, 15, 857—868, 1978.
274. **Vander Mallie, R. J. and Garvey, J. S.,** Radioimmunoassay of metallothioneins, *J. Biol. Chem.*, 254, 8416—8421, 1979.
275. **Waku, K.,** The chemical form of cadmium in subcellular fractions following cadmium exposure, *Environ. Health Perspect.*, 54, 37—44, 1984.
276. **Walsh, J. J. and Burch, G. E.,** The rate of disappearance from plasma and subsequent distribution of radiocadmium (^{115m}Cd) in normal dogs, *J. Clin. Med.*, 54, 59—65, 1959.
277. **Washko, P. W. and Cousins, R. J.,** Metabolism of ^{109}Cd in rats fed normal and low-calcium diets, *J. Toxicol. Environ. Health*, 1, 1055—1066, 1976.
278. **Waters, M. D., Nordberg, G. F., Cox, C. C., Vaughan, T. O., and Coffin, D. L.,** Cellular toxic effects of cadmium and other metals, in Int. Conf. Heavy Metals in the Environment, Toronto, Canada, 1975, IES, Toronto B54-B56.
279. **Webb, M.,** Biochemical effects of Cd^{2+}-injury in the rat and mouse testis, *J. Reprod. Fertil.*, 30, 83—98, 1972.
280. **Webb, M. and Etienne, A. T.,** Studies on the toxicity and metabolism of cadmium-thionein, *Biochem. Pharmacol.*, 26, 25—30, 1977.
281. **Welinder, H., Skerfving, S., and Henriksen, O.,** Cadmium metabolism in man, *Br. J. Ind. Med.*, 343, 221—228, 1977.
282. **Wong, K. L., Cochia, R., and Klaassen, C. D.,** Comparison of the toxicity and tissue distribution of cadmium in newborn, *Toxicol. Appl. Pharmacol.*, 56, 317—325, 1980.
283. **Worker, N. A. and Migicovsky, B. B.,** Effect of vitamin D on the utilization of zinc, cadmium and mercury in the chick, *J. Nutr.*, 75, 222—224, 1961.
284. **Yamagata, N., Iwashima, K., and Nagai, T.,** Gastrointestinal absorption of cadmium in normal humans (Japanese), in *Kankyo Hoken Rep. No. 31*, Japan Public Health Assoc., Tokyo, 1974, 84—85.
285. **Yoshikawa, H.,** Preventive effects of pretreatment with cadmium on acute cadmium poisoning in rats, *Ind. Health*, 11, 113—119, 1973.

Chapter 7

KINETIC MODEL OF CADMIUM METABOLISM

Tord Kjellström and Gunnar F. Nordberg

TABLE OF CONTENTS

I. INTRODUCTION

In the chapter on metabolism (Chapter 6) it was shown how complicated the metabolism of cadmium is. The degree of accumulation and retention varies greatly between organs and within organs and in certain organs the accumulation rates vary with the exposure intensity and dose.

A kinetic model of cadmium metabolism describes the process of absorption, distribution, biotransformation, and excretion in qualitative and quantitative terms in relation to time. Such a model is useful for several purposes. Quantitative knowledge of the kinetics of a metallic compound is essential for the diagnosis of poisoning. Sometimes, when metabolic data are combined with data on tissue sensitivity to damage, predictive estimates of risk of adverse effects may be possible. Whereas this has now also been recognized for metallic compounds other than cadmium,[4,30,32,44] the development of a metabolic model for cadmium has been one of the first applications of this general philosophy to make interpretations of empirical findings of adverse health effects in humans. Thus, it has been considered of interest to describe briefly, in a separate chapter, the development of our knowledge concerning the kinetic model of cadmium.

Biological indexes of exposure intensity or dose, such as urine, blood, or feces, can only be used if the kinetics are known. Knowledge of the half-time of cadmium in different tissues is important, in order to predict the accumulation that will occur after a certain type of exposure. Of particular interest is the kidney cortex which is the critical tissue for cadmium poisoning (Volume II, Chapter 9). The cadmium concentration in this tissue at ''normal'' low level exposures cannot, without great difficulty, be measured in vivo (Chapter 2).

A kinetic model can be more or less sophisticated describing the kinetics for only one tissue or several tissues. Mathematical expressions for the kinetics are used to calculate the half-time in each tissue as well as the quantitative relations between intake, accumulation, and excretion.

In Friberg et al.,[13] a one-compartment exponential model of the kidney kinetics was used to calculate half-time and dose-retention relationships. At that time the knowledge about the metabolism of cadmium was incomplete and it was not considered justified to set up a more elaborate model for cadmium accumulation in various organs or in the body as a whole. More experimental and epidemiological data are now on hand and it is possible to design a more elaborate model. For the calculations in this chapter, we will use the eight-compartment model developed by Kjellström and Nordberg[20] with some modifications. In another report by Nordberg and Kjellström,[31] this model was discussed further.

II. DIFFERENT APPROACHES USED IN KINETIC MODELS

The simplest model of cadmium metabolism, the one-compartment exponential model, is used each time the half-time in an organ is calculated from simple exponential decrease or accumulation curves. The decrease curve assumes a situation in which cadmium is only excreted from the organ and no new cadmium enters the organ. The curve follows the equation:[45]

$$C = C_o \times e^{-bt} \qquad (1)$$

where C = concentration in organ at time t; C_o = concentration in organ at time 0; b = elimination constant; and t = time since exposure ended. The (biological) half-time $t_{1/2}$ = ln2/b.

This type of situation usually occurs only in experiments where an animal or a person has been given a single dose with the purpose of studying the kinetics. A person with a high cadmium exposure, whose exposure ceases, also develops a tissue cadmium decrease curve similar to this one (Chapter 6, Section V.D.2), but the human decrease curve reflects several compartments and the equation above applies only approximately for parts of the curve.

When exposure is continuous, an accumulation in the tissue occurs, and the one-compartment model can be described with the equation:

$$A = a/b\,(1 - e^{-bt}) \tag{2}$$

where A = accumulated amount; a = fraction of daily intake taken up by the organ; b = elimination constant; and t = time since exposure started. This simple model for accumulation in a tissue has been extensively used by the International Commission for Radiation Protection (ICRP) for calculations regarding accumulation of radionuclides.[14] This has facilitated the prediction of risks due to such exposures.

A two-compartment model assuming the flow of cadmium first through one compartment and then through the next compartment can also be described mathematically in simple terms.[24] A two-compartment model for cadmium in blood was used by Järup et al.[15] and a three-compartment model for whole body cadmium was used by Shaikh and Smith.[38] An approximate fitting of the one-compartment decrease curve to separate parts of the more complex curve (a slow component and a fast component) has also been used.[35]

Several important tissues accumulate cadmium after long-term (decades) exposure (see Chapter 6, Section IV) and a multicompartment model is needed to include each tissue in the calculations. All transfers between compartments cannot be assumed to be linear and changes in exposure level or tissue weights with age make it difficult to express the model as a mathematical equation. A mathematical solution based on iterative calculation procedures and certain approximations will be used in a comprehensive model described further on (Section III.).

A. Animals

By using simple exponential model assumptions, half-times for cadmium in the whole body of mice, rats, rabbits, and monkeys (Chapter 6, Table 7) were calculated to be from several months up to years. The half-times in the slowest phase were 20 to 50% of the maximum life span of the animal (Chapter 6, Section V.D.1).

It is of interest that in studies by Nordberg[29] and Engström and Nordberg,[11] the longest half-times of ^{109}Cd in mice (about 2 years) were seen during continuous exposure. The mechanisms for these variations in kinetics are discussed in Chapter 6 (Section V.D.1). It is essential to mimic the exposure situation of humans as closely as possible if the results of animal experiments are to be used as a basis for a human model.

There are some reports indicating a variation of the whole body half-time with dose. Nomiyama and Nomiyama[27] and Nomiyama et al.[28] gave three groups of four to eight rabbits, for 3 to 5 weeks, daily subcutaneous injections of 0.5 or 1.5 mg Cd per kilogram body weight as $CdCl_2$. Then the tissue cadmium levels were followed up for 2 years during which time no further cadmium exposure occurred. The total injected dose was 17.5, 31.5, and 52.5 mg/kg body weight in the three groups, respectively. One-compartment exponential decrease curve assumptions were used to calculate half-times in liver and kidneys for separate parts of the decrease curve. Nomiyama and Nomiyama[27] reported half-times including the first more rapid phases of decrease. These half-times appeared to decrease with increasing dose, which is in agreement with the findings of an increased excretion in urine and feces of mice,[29] when the exposure level increased from 0.25 mg Cd per kilogram body weight to 0.5 mg/kg (Chapter 6, Section V.A.1).

Engström and Nordberg[10] exposed four groups of mice to single doses of radioactive-labeled $CdCl_2$ through gastric tube. The doses were 1, 25, 750, and 3750 μg Cd per kilogram body weight, respectively. After exposure the whole body retention was measured by gammaspectrometer for 140 days. The retention of cadmium as a percentage of given dose increased with increasing dose and the half-time in the slowest phase of the decrease curve was longest in the group with the highest dose, but there was no difference between the other three groups. An explanation for the difference in the results of this study[10] and the earlier studies may be that the earlier studies gave cadmium doses close to toxic doses, urinary excretion is then relatively higher than at low doses (Chapter 6, Section V.A.2).

Taguchi and Suzuki[42] also studied half-time for cadmium in the whole body of mice, in relation to sex and age. A carrier-free small dose of ^{109}Cd was injected intraperitoneally once and all feces were collected in order to measure whole body retention. Surprisingly, no cadmium was found to be excreted in urine, so this was disregarded. Individual tissues were measured at sacrifice 2 to 128 days postexposure. Three phases of whole body retention were found, each approximated by a one-compartment model and half-times were calculated. In the slowest phase the whole body half-times were between 200 and 600 days. The longest half-time was found in mice exposed at 1 year of age, with shorter half-times in mice exposed at younger age (Chapter 6, Table 7).

Half-times in individual tissues were established using one- or two-compartment models.[42] Each tissue was considered in isolation from the other tissues. The kidneys had the longest half-time in the slow phase, about 2400 and 2900 days (or 7 to 9 years) for male and female mice, respectively. No age dependence of the whole body half-time in mice after oral exposure was found by Engström and Nordberg,[11] but their study measured only the fast phase component with a half-time of about 70 days (Chapter 6, Section V.D.1).

The experiment by Taguchi and Suzuki[42] was very similar to the one by Matsubara-Khan,[24] who found a slow phase half-time in mouse kidney of 990 days. Her study of single exposure kinetics was followed by one on sequential exposure kinetics.[25] The accumulation curves of individual tissues, during 130 days of twice weekly carrier-free $^{109}CdCl_2$ intraperitoneal injections, were analyzed with exponential two-compartment models. Comparing the rate constants for the kidney model in this study — 0.001 and 0.30[25] — with those of an earlier study — 0.0007 and 0.30[24] — it appears that when exposing animals to small doses of carrier-free ^{109}Cd, kinetics are not influenced by exposure conditions.

The only more elaborate model for cadmium metabolism in animals that has been published[40] includes blood, kidney, liver, spleen, pancreas, testes, gastrointestinal tract, and residual carcass. The transport between compartments and blood is determined by rate constants assuming that a constant proportion of the amount in any tissue is transferred per unit time. Each transfer would thus be expressed by the same mathematical equation as that used in the one-compartment exponential model. Simultaneous transfer in either direction between several compartments can take place. The only excretion route included (in the model) was feces.

The rate constants were established by comparing the retention data from an acute exposure experiment on mice with calculated retention data from the model. A simulation computer program was used to fit the two sets of data. Three groups of mice received one, two, or three intravenous injections of ^{109}Cd at 48 hr intervals. The model was fitted to the kinetics during 6 days from the first injection. The rate constants from tissues to blood varied between 0.0014 and 0.0042 min^{-1}. The lowest value, 0.0014, was seen in kidney and this rate constant corresponds to a half-time of 8.25 hr.

The model designed by Shank et al.[40] is of very limited use for the study of human cadmium accumulation because it only displays the short-term situation after acute exposure. The distribution of cadmium from blood to various tissues as calculated using this model may, however, correspond to the distribution in humans. The rate constants were 0.66 min^{-1} for blood to liver, 0.067 for blood to kidney, 0.24 for blood to gastrointestinal tract, and

0.38 for blood to residual carcass. This means that about half of the cadmium in blood goes to liver and about 5% goes to the kidney. Further discussion of this short-term model is given by Shank and Vetter.[39]

B. Humans

The very long half-time of cadmium in the whole body of humans makes it very difficult to calculate the half-time from single exposure experiments accurately since very long observation times are required.

Four studies have been carried out (Chapter 6, Section V.D.2), but only in one[38] was the follow-up time long enough (2 years) to make a valid estimate of the slowest phase half-time (in the one case studied, it was estimated to be approximately 26 years). In another study,[35] a confidence interval for the slow half-time was estimated (130 days to infinity days). The age-related accumulation of cadmium in different tissues (Chapter 5) indicates half-times in kidney, liver, and muscles of years to decades (Chapter 6, Section IV.). Taking all data together it can be concluded that for humans as well as for animals (Chapter 6, Table 7), the whole body half-time is at least 20% of the average maximum life span (80 years for humans), which gives a half-time of at least 16 years.

The first attempts to use long-term accumulation curves for calculation of half-time were made by Tsuchiya and Sugita[47] and Kjellström.[16] Both these studies assumed one-compartment exponential models as a basis for calculations. Data from Tokyo indicated that the daily uptake of cadmium from food was 4 μg and at 50 years of age the "steady-state" body burden of cadmium was 30 mg, giving a whole body half-time of 16 years.[47] The authors noted that in Tokyo, urinary cadmium excretion increased with age up to age 30 to 40 years and then decreased. When they included an age-dependent variation of daily cadmium uptake from food in their model, the calculated whole body accumulation curve took on a similar age-related pattern to the urinary cadmium concentrations.

The model designed by Kjellström[16] was applied to the critical organ, the kidney,[21] and the calculated concentrations were compared with the age-related kidney cortex concentrations reported by Schroeder and Balassa.[37] An age-dependent variation in daily cadmium uptake from food parallel to the variation in energy intake was included in the model (Figure 1). It can be seen that the recommended energy intakes and the measured cadmium intakes agreed well.

The variation in kidney weight with age was also taken into consideration when calculating kidney cortex cadmium concentrations. In order to obtain a reasonable agreement between empirical and calculated accumulation curves, it was concluded that in the kidney the excretion rate would have to be less than 0.005%/day, corresponding to a half-time of greater than 38 years.[21] The model designed by Kjellström[16] was also described in detail in *Cadmium in the Environment*,[13] and was used for dose-response calculations.

The simple one-compartment model was improved in these studies[16,47] by incorporating age-dependent variations in cadmium intake and kidney weight. This improved the fit between the calculated and empirical accumulation curves, but it could not fully "recreate" the decrease in kidney cortex cadmium concentrations or urinary cadmium concentrations after age 40 to 50 years. Kjellström and Friberg[19] suggested that this decrease could be explained by three principally different factors related to cadmium intake: age-dependent variables such as dietary habits; time-dependent variables such as overall cadmium concentrations in foodstuffs; and cohort-dependent variables such as smoking habits. It was pointed out by Friberg et al.[13] that the empirical data, on which the comparisons were based, could have had age-dependent variations in smoker-nonsmoker ratios or male-female ratios, which could also cause an apparent decrease of kidney cortex cadmium concentrations at older age. Age-dependent renal morphological and functional changes were also suggested as a likely contributing factor.[13] Such changes were incorporated into the calculation procedure of the more elaborate model.[20]

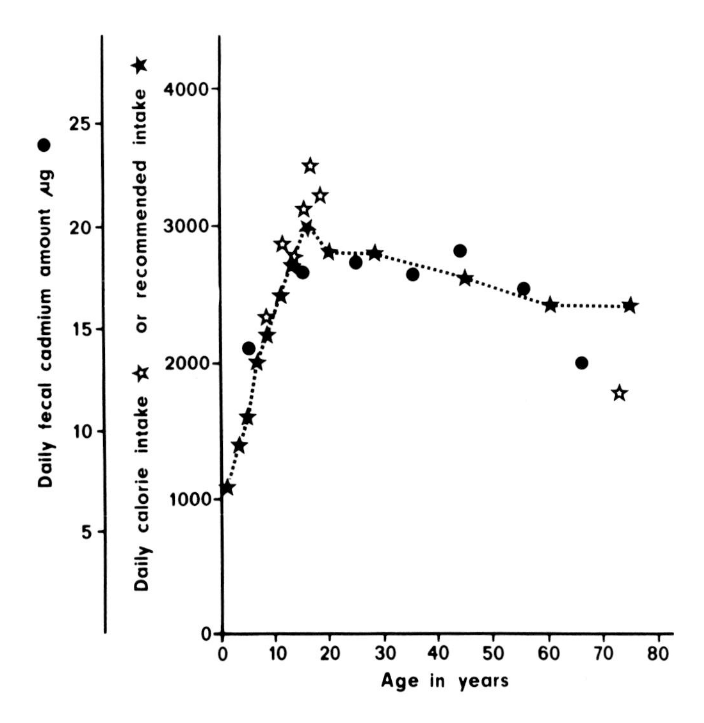

FIGURE 1. Variation of recommended and observed energy intake and measured cadmium intake in relation to age. (From Kjellström, T., Borg, K., and Lind, B., *Environ. Res.*, 15, 242—251, 1978. With permission.)

Travis and Haddock[46] further developed the one-compartment model to include age-dependent renal changes, as well as time- and age-dependent changes in cadmium exposure. The basic assumptions about the proportion of body burden in the kidney (1/3), the concentration ratio kidney cortex to whole kidney (1.5), and the gastrointestinal absorption rate (5%) were the same as in the model by Kjellström.[16]

The age variation of cadmium intake via food (assumed parallel to energy intake) was described by a series of linear approximations within age brackets. Travis and Haddock included cadmium exposure from tobacco smoke in their calculation, based on age-specific average smoking habit data for the U.S. in 1966. One cigarette was assumed to contribute 0.2 µg Cd and the pulmonary absorption rate was set at 25%.

The calculated curves were very similar to those produced by Kjellström.[16] A time-dependent increase of cadmium concentration in food in accordance with the increase found by Kjellström et al.[22] was next included in the calculation by Travis and Haddock.[46] The calculated kidney cortex cadmium concentrations had now decreased more clearly after 50 to 60 years of age, but the authors concluded that the time-related increase of food cadmium could not in itself explain the empirical data. Travis and Haddock[46] then added an age-dependent variation in renal excretion of cadmium to the calculation. Until age 30 years the renal excretion coefficient was constant and after that it increased linearly, which agrees with the assumption made by Kjellström and Nordberg[20] for their eight-compartment model (Section III.). A basis for this assumption was found in pathological studies showing the gradual loss of nephrons with age.[46]

The best match of the calculated data to the empirical data was achieved when the renal excretion coefficient was 0.02/year (or 0.0055%/day) at age 30 and it increased to 0.0575/year at age 80. The new calculated accumulation curve, which incorporated all the age-dependent variables mentioned above, fits in well between the two empirical data sets used

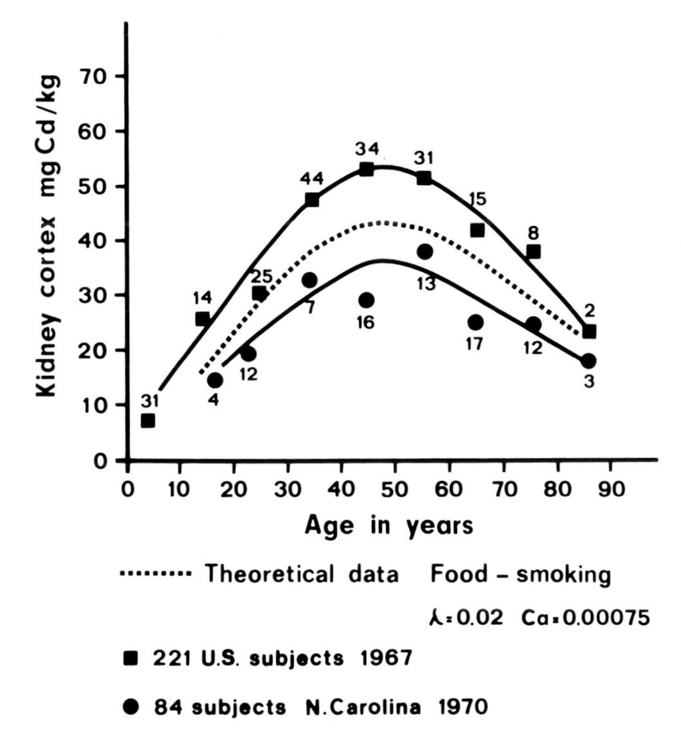

FIGURE 2. Comparison of calculated cadmium concentration in renal cortex (under the assumption of an age-dependent excretion rate constant) with actual autopsy data. (From Travis, C. C. and Haddock, A. G., *Environ. Res.*, 22, 46—60, 1980. With permission.)

as a comparison (Figure 2). The half-time for cadmium in kidney cortex corresponding to these excretion rates was estimated to be 34 years at birth and young age, decreasing to 11 years at age 80.[46]

III. COMPARTMENTS AND PATHWAYS IN A COMPREHENSIVE MODEL

Kjellström and Nordberg[20] and Nordberg and Kjellström[31] described an eight-compartment kinetic model of cadmium metabolism, which has the advantage of being able to calculate not only accumulation in kidney, but in other tissues as well. It should be pointed out that the model is based on a number of approximate assumptions, but it appears to be able to calculate the long-term accumulation of tissue levels under a number of different exposure situations with reasonable accuracy.[20] As it describes qualitatively the interrelationship between several body compartments and as it is the most detailed model of cadmium metabolism published, we will present it in some detail in this chapter. It should be pointed out that new data on the kinetics of cadmium in blood (Chapter 6, Section III.) and on the uptake of cadmium from blood into the kidney (Chapter 6, Section IV.) are not incorporated in the described model. The calculations of cadmium in blood should therefore be considered tentative. The general framework of the model is shown in Figure 3. The coefficients C1-C19 determine the transfer between compartments. In most cases, the daily transfer is assumed to be a fixed proportion of the accumulated amount in the compartment. The model can be divided into four parts with different functions: absorption and uptake; transport and distribution; excretion; and retention and accumulation. The calculations in the model utilize an iterative procedure with the time unit equal to 1 day.[20]

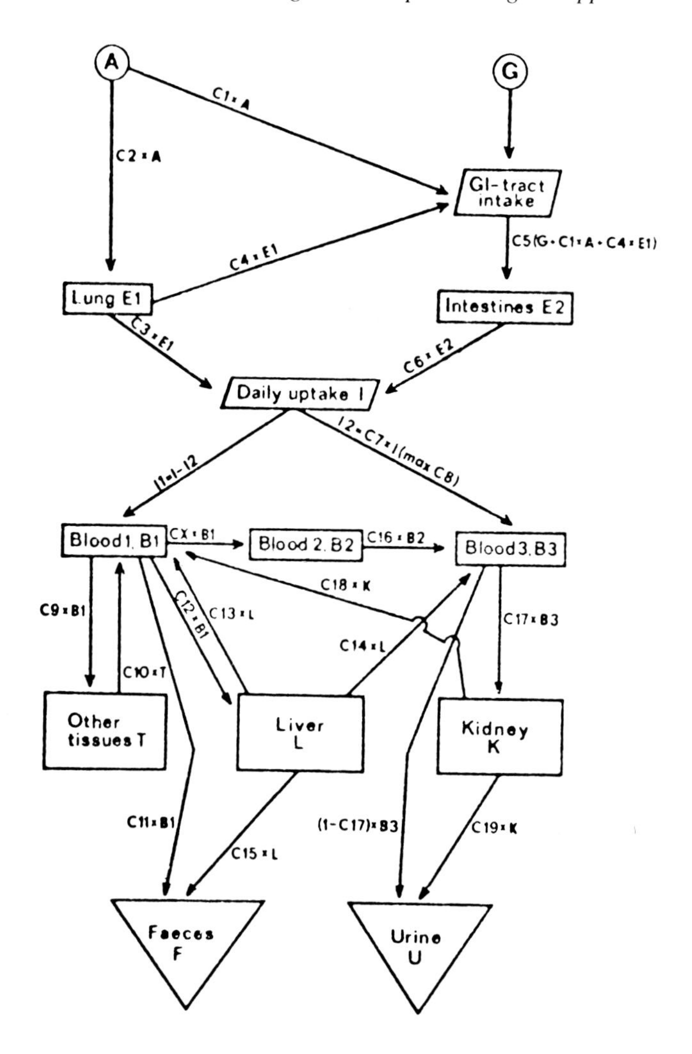

FIGURE 3. Flow scheme of the kinetic model of cadmium metabolism. (From Kjellström, T. and Nordberg, G. F., *Environ. Res.*, 16, 248—269, 1978. With permission.)

A. Absorption and Uptake

Cadmium absorption occurs almost exclusively by the pulmonary or the gastrointestinal route (Chapter 6, Section II.). Cadmium compounds are inhaled as particulate matter, either as fumes with very small particle size or as dust. The general principles for deposition and absorption of particulate matter described by the Task Group on Lung Dynamics[43] and by the Task Group on Metal Accumulation[45] were taken to be valid for cadmium (Chapter 6, Section II.A) and were used in this model.

The respiratory cadmium intake (A) (Figure 3) can be diverted to the gastrointestinal tract (C1 × A) due to clearance of cadmium deposited on the mucosa of nasopharynx, trachea, or bronchi. It can also be deposited in the alveoli (C2 × A) and from there be absorbed into the blood (C3 × E1). The remainder of the respiratory intake is exhaled.

Some of the cadmium in the alveoli is transported via alveolar clearance back to the bronchi (C4 × E1) and eventually to the gastrointestinal tract after swallowing. Based on data given by the Task Group on Lung Dynamics,[43] C1 was estimated at 0.1 to 0.2 for cadmium fumes (e.g., cigarette smoke) and at 0.4 to 0.9 for cadmium dust. Calculations

Table 1
ASSUMED AND MODELED VALUES OF COEFFICIENTS

Coefficients	Initially assumed ranges[a]		Values fitting to empirical data
C1	0.1—0.2 (cigarette smoke)		0.1
	0.4—0.9 (factory dust)		0.7
C2	0.4—0.6 (cigarette smoke)		0.4
	0.1—0.3 (factory dust)		0.13
C3	0.01—1	day^{-1}	0.05
C4	0.1 × C3 = 0.001—0.1	day^{-1}	0.005
C5	0.03—0.1		0.048
C6	0.05	day^{-1}	0.05
C7	0.2—0.4		0.25
C8	0.5—5	μg	1
C9	0.4—0.8		0.44
C10	0.00004—0.0002	day^{-1}	0.00014
C11	0.05—0.5		0.27
C12	0.1—0.4		0.25
C13	0—0.0001	day^{-1}	0.00003
C14	0.0001—0.0003	day^{-1}	0.00016
C15	0—0.0001	day^{-1}	0.00005
C16	0.004—0.015	day^{-1}	0.012
C17[b]	0.8—0.98		0.95
C18	0—0.0001	day^{-1}	0.00001
C19[c]	0.00002—0.0002	day^{-1}	0.00014
CX[d]	0.01—0.05		0.04
C20	0.05—0.5		0.1
C21	0—0.000002	day^{-1}	0.0000011

[a] If no unit is given, this means that the coefficient is a unitless proportion.
[b] C17 decreases from age 30 to age 80 years by 33%.
[c] C19 increases from age 30 years with C21 each year.
[d] Cx = 1–C9–C11–C12.

with different values were carried out (Section IV.B) and a best fit between calculated and empirical values was found for C1 = 0.1 (fume) and 0.7 (dust). In accordance with the difference in distribution of small (fume) and large (dust) particles, C2 was estimated to be 0.4 to 0.6 for fume and 0.1 to 0.3 for dust. The best fit values for all coefficients are listed in Table 1. The alveolar clearance is likely to be small in comparison with the rest of the lung clearance[4,43] and we assumed C4 to be 0.1 × C3.

Cadmium intake via the gastrointestinal tract consists of food cadmium (G) and cadmium cleared from alveoli (C4 × E1) and respiratory tract (C1 × A). Most of the cadmium in the intestinal lumen will pass unabsorbed and the retention C5 was assumed to be in the range 0.03 to 0.1 (see Chapter 6, Section II.B.). The cadmium retained in the intestinal walls will accumulate to a certain extent[41] before being absorbed into blood. C6 was assumed to be 0.05 (day^{-1}), but available data are insufficient to estimate this coefficient with accuracy. In any case, C6 does not influence the shape of the long-term accumulation curves, but only the short-term (days to weeks) curves. The total amount of cadmium absorbed into blood each day (C3 × E1 + C6 × E2) is called daily uptake (= I μg/day).

B. Transport and Distribution

Blood is the major distribution medium for cadmium within the body (Chapter 6, Section III.). Cadmium in blood is mainly bound to albumin, metallothionein, or red blood cells

(Chapter 6, Section III.). Within the blood cells, cadmium is also bound to metallothionein (Chapter 4), but this cadmium is not readily exchangeable for plasma cadmium.[29,33]

In the model the blood was divided into three compartments: the albumin-bound cadmium (B1), the cell-bound cadmium (B2), and the metallothionein-bound cadmium (B3). The turnover of cadmium in B1 and B3 is very rapid and all cadmium input into these compartments is assumed to have continued to other compartments within less than a day. Thus, the contribution of B1 and B3 to whole blood cadmium concentration is less than the calculated amounts in these compartments. We assumed this fraction (C20) to be in the range 0.05 to 0.5.

The part of cadmium uptake (C7 × I) which is bound to metallothionein (B3) will continue mainly to kidney and urine. As about a third of body burden after long-term exposure is in the kidneys (Chapter 6, Section IV.A.2), we assumed C7 to be 0.2 to 0.4. The B3 compartment has a limited number of binding sites (Chapter 4) and therefore, the daily flow from I to B3 was maximized by C8 (0.5 to 5μg/day).

Accumulation in B2 is determined by the turnover rate of red blood cells. The mean life of erythrocytes is 120 days[2] which implies a half-time of 83 days and C16 would be 0.008 (day^{-1}). For the modeling, we assumed C16 would be in the range 0.004 to 0.015 (day^{-1}).

From B1, cadmium is transferred to red blood cells (B2), liver (L), and other tissues (T), and via intestinal wall cells to feces (F). The proportions of B1 distributed to L and T were assumed to agree approximately with their proportion of whole body burden of cadmium (16% for L, 50% for T) (Chapter 6, Section IV.A.2). Thus, C12 was assumed to be 0.1 to 0.4 and C9 was set at 0.4 to 0.8 (Figure 3). The liver is a main organ for metallothionein production (Chapter 4) and it was assumed that most of the cadmium in B3 came from the liver (C14 × L). From B2, metallothionein-bound cadmium will add to the B3 compartment and the B3-cadmium is cleared through kidney glomeruli. Some cadmium is reabsorbed in the proximal tubuli (C17 × B3) and adds to kidney accumulation (K) and the rest is excreted via urine (U).

About 95% of the glomerular filtrate of Cd-metallothionein is reabsorbed in the renal tubuli of mice.[34] We assumed C17 to be in the range 0.8 to 0.98. Tubular reabsorptive capacity decreases with age. Between age 30 and 80 years it decreased 33%.[26] In the model we assumed that a similar decrease would occur in C17.

Cadmium is transported back from liver, kidney, and other tissues to the blood (Figure 3). This is assumed to occur mainly to the B1 (albumin) compartment (C10 × T, C13 × L, and C18 × K), but the liver also contributes to B3 (metallothionein) (C14 × L).

C. Excretion

Almost all cadmium in the body is excreted via feces and urine (Chapter 6, Section V.). Fecal cadmium consists mainly of the nonabsorbed part of ingested cadmium. "True" fecal excretion originates from blood via the intestinal wall (C11 × B1) and from bile (C15 × L) (Chapter 6, Section V.B.). Recent studies (Chapter 6, Section V.B.) have shown that the main part of biliary cadmium is correlated with the amount of cadmium in liver. We assumed C15 to be in the range 0 to 0.0001 (day^{-1}). With long-term low level exposure, fecal and urinary excretion are about the same (Chapter 6, Section V.). Urinary excretion is mainly a function of body burden (Chapter 6, Section V.A), but a part of this excretion is directly dependent on blood cadmium. We have taken this into consideration in the model by splitting urinary excretion into two parts: (1 − C17) × B3 coming from blood and C19 × K coming from kidney (Figure 3). At steady state, the total daily excretion would be the same as total daily uptake. In Sweden, the average adult daily cadmium intake via food is about 16 μg[23] and the average body burden of nonsmokers is about 5 mg at age 50.[6] With a gastrointestinal absorption rate of 5%, the daily uptake (0.8 μg) would be 0.016% of body burden.

Average adult (30 to 60 years) urinary excretion is approximately 0.35 μg/day.[7] Thus, the daily excretion rate for urine would be 0.007% of body burden and, by subtraction from the estimated total excretion, the fecal excretion would be 0.009% of body burden. As discussed in Chapter 6 (Section V.), animal data also favor a somewhat higher excretion via feces than via urine. The coefficients C11, C15, C17, and C19 in the model were fitted so that the excretion rates were similar to these values, taking into consideration other parameters that had to be fitted to empirical data.

D. Retention and Accumulation

The main part of body burden will be found in the liver (L), the kidneys (K), and in other tissues (T) (particularly muscles, skin, and bone) (Chapter 6, Section IV). Using a one-compartment model, the half-time in liver was estimated to be 7 years.[48] We set C13 at 0 to 0.0001 (day^{-1}), and C14 at 0.001 to 0.003 (day^{-1}) which in combination with C15 gave a half-time in liver between 4 and 19 years.

The half-time in kidney has been estimated at between 14[48] and 38 years.[16] It was estimated that about 0.007% of whole body burden was excreted daily in urine. In order to fit with these values, C19 was assumed to be in the range 0.00002 to 0.0002 (day^{-1}), and C18 in the range 0 to 0.0001 (day^{-1}). The corresponding range of kidney half-times would be 6 to 38 years.

The decrease in tubular reabsorptive capacity mentioned above may be an effect of a decrease in the total number of tubular cells.[5] It is likely that such a cell loss would lead to an increased cadmium loss from the kidney. This could go to the blood (C18 × K) or to the urine (C19 × K). The latter is more likely because the morphological damage in affected tubular cells usually involves damage to the brushborder cell membrane.[3] We assumed that C19 increased linearly after age 30 with C21 each year. Initially, C21 was set at 0 to 0.000002 (day^{-1}).

Very little data are available regarding half-times in other tissues. It was found that the age-dependent accumulation curves for cadmium in muscle indicate an even longer half-time than that for kidney.[17,18] With long-term low level exposure about half of the body burden is in "other tissues", indicating that major accumulation occurs there as well as in liver and kidney. We assumed that C10 was in the range 0.00004 to 0.00020 (day^{-1}), corresponding to half-times between 9 and 47 years.

IV. MATHEMATICAL EXPRESSION OF THE COMPREHENSIVE MODEL

A. Equations and Calculation Algorithm

The transfer of cadmium from one compartment to the next is calculated with an iterative procedure, in which all the changes are assumed to take place at one point in time once a day (iteration step = 1 time unit = 1 day). In long-term calculations, the maximum error is the change during 1 day, which for compartments with great accumulation constitutes a very small proportion of the accumulated amount. Calculations with longer and shorter time units than 1 day were tested,[20] but it was found that the shorter times gave no increase in calculation accuracy and longer times resulted in increasing errors. The transfers can be expressed with a series of equations for cadmium amounts in each compartment.[20]

The initially assumed ranges for the coefficients as well as the final values fitted with the model are listed in Table 1.

B. Fitting Transfer Coefficients to Empirical Data

The coefficient values were determined by comparing calculated age-specific cadmium concentrations in kidney cortex, liver, and urine with empirical data for Sweden.[20] Tissue weight data used for calculations of concentrations were derived from handbooks containing

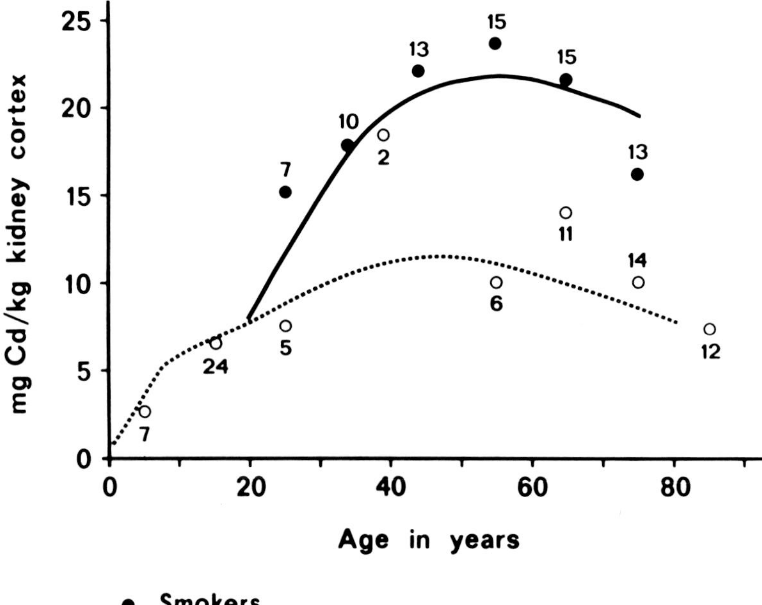

FIGURE 4. Calculated and empirical data on cadmium concentrations in kidney cortex in Sweden. Dotted line: calculated concentrations, nonsmokers. Unfilled circles: empirical concentrations, number of observed persons indicated. Filled circles and solid line: corresponding data for smokers (20 cigarettes per day). (From Kjellström, T. and Nordberg, G. F., *Environ. Res.*, 16, 248—269, 1978. With permission.)

basic biological data. In Kjellström and Nordberg,[20] a ratio of 1.50 between cadmium concentrations in renal cortex and whole kidney was used. As recent data (Chapter 6, Section IV.B.1) indicate that the ratio is more likely to be 1.25 after long-term low level exposure, there is a need to reassess the coefficients with this new ratio in mind.

The type of agreement achieved is shown in Figure 4 (kidney cortex) and Figure 5 (urine). The data for kidney, urine, and liver of nonsmokers were used to calculate the square deviations between individual values at different ages and the calculated data for the same ages. The coefficients were varied within their given ranges until an approximate minimum point for square deviation was reached experimentally. No attempt was made to define an exact minimum point, because of the large number of coefficients involved and the uncertainties of some of the assumptions underlying the model. In order to ascertain that the coefficient values selected gave a reasonable fit, the calculated cadmium concentrations in muscle and blood of adults were also checked to determine agreement with published data from Sweden.

The comparisons mentioned above were mainly affected by those coefficients that had a large influence on long-term accumulation in the main storage organs. In order to find values for the coefficients that influenced short-term changes to a large extent (e.g., C7, C8, C16), a comparison was made between short-term calculated data and measured cadmium concentrations in blood and urine of newly employed factory workers with high cadmium exposure.[20] Great individual variation made the comparison difficult, but the general trend of the calculated data (fast component half-time in blood about 1 month; initial 40% increase of urine cadmium; slow increase of urine cadmium during 1 year) was fitted to resemble the observed data. The coefficient values eventually selected for the model are listed in Table 1.

FIGURE 5. Calculated and empirical data on cadmium concentrations in urine in Sweden. Solid line: calculated concentrations, nonsmokers. Unfilled circles: empirical concentrations, number of observed persons indicated. (From Kjellström, T. and Nordberg, G. F., *Environ. Res.*, 16, 248—269, 1978. With permission.)

C. Testing the Model

Long-term cigarette smoking more or less doubles cadmium concentrations in tissue (Chapter 6, Section IV). For the model calculations we estimated that smoking 20 cigarettes per day gives rise to a cadmium fume intake of 3 µg/day (Chapter 3, Section IV.). With respiratory absorption rate (C2) at 40%, the kidney cortex accumulation curve agrees well with empirical data (Figure 4). Urine, blood, and liver values agreed reasonably well.[20]

Further comparison of both calculated and empirical values was carried out for Japanese people with "background" exposure, Japanese people with high exposure, and Swedish factory workers with high exposure.[20] There was good agreement between observed and calculated values with the exception of blood cadmium and urine cadmium of workers with very high acute exposure (Chapter 6, Section V. A). It is likely that the intake and distribution parts of the model do not incorporate all of the relevant pathways taken by cadmium after very high intensity exposures.

The model calculations have been tested in a study of cadmium in blood, liver, and bile.[8] Specimens were collected during routine gallstone operations of 23 patients in the age range 26 to 66 years. There was a significant correlation between liver cadmium and blood cadmium (preoperation) concentrations (Figure 6) and the relationship curve calculated with the model was situated in the middle of the two regression lines. The results of bile analysis were questioned because the cadmium concentrations found corresponded to a 10 times higher bile cadmium transport than predicted from the model. If all this bile cadmium is excreted, the half-time of cadmium in liver would be much shorter than that indicated by other data. The explanations given by Elinder et al.[8] were erroneous analysis of bile (because no quality control cross-check could be done); high bile cadmium because of gallstones or the operation itself; and an enterohepatic circulation of cadmium.

Further opportunities to test the model have become available due to the use of in vivo neutron activation analysis to study cadmium concentrations in liver and kidney.[9,36] For mixed groups of active and retired cadmium workers without cadmium-induced renal damage, these authors calculated the regression line for liver cadmium and kidney cortex cadmium (Figure 7). The model calculated different curves for a worker whose exposure started at age 30 and who worked for 10 years, depending on whether he was active or had retired from cadmium work (Figure 7). The reason for this is that in the model calculations the liver cadmium level starts to decrease as soon as exposure ceases, whereas the kidney cadmium continues to increase for 5 years and then slowly decreases. This agrees with

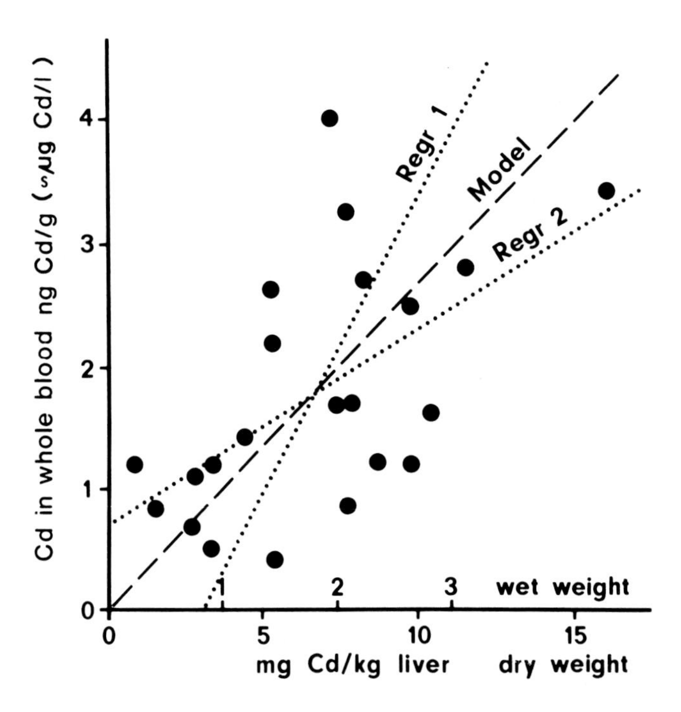

FIGURE 6. Comparison between cadmium in liver biopsy material and cadmium in whole blood of 22 Swedes who underwent gall bladder operations. Regression lines and the relationship calculated with the metabolic model by Kjellström and Nordberg[20] are included. (From Elinder, C.-G., Kjellström, T., Lind, B., Molander, M. L., and Silander, T., *Environ. Res.*, 17, 236—241, 1978b. With permission.)

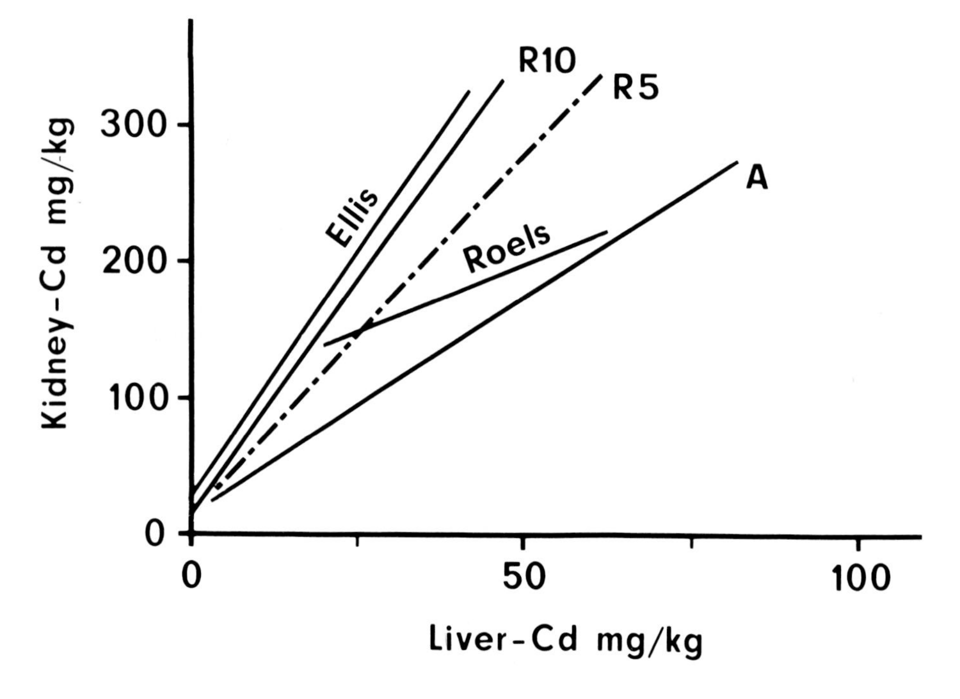

FIGURE 7. Relationships between cadmium concentrations in liver and kidney cortex for active (A) and retired workers (5 or 10 years retired, R5 and R10) according to the model and as observed in two in vivo neutron activation studies.[9,36]

animal and human data (Chapter 6). There is good agreement between the observed and calculated curves (Figure 7).

V. USE OF THE METABOLIC MODEL

A. Interpolation and Extrapolation of Data

Once established, the model can be used to compare and add information to new data sets. The original autopsy studies, which were the basis for the model, included all age groups. New studies can be limited to selected age groups and the level of the whole curve can be adjusted to fit the selected age groups. Any change in their level (if in the same direction in each group) could be caused by a general change in cadmium exposure. Intermediate age group cadmium levels can then be predicted from the adjusted curve.

Time relationships can be studied. The model can be used to calculate the tissue cadmium levels each day. A group of workers may be sampled once a month. The results can be used to construct time curves for the cadmium concentration in each tissue. The concentrations at intermediate points and before or after the sample collection period can be calculated using the model.

One use of the model, presently being investigated at the Swedish battery factory extensively studied by Friberg[12] and others (Volume II, Chapter 9), is to use the individual blood cadmium and urine cadmium concentrations and the individual ages and exposure histories to calculate present and future kidney cortex levels.[40a] If this were successful it would be an effective way of preventing excessive kidney cortex cadmium concentrations on an individual basis.

B. Interorgan Relationships

As shown in some of the examples given above, interorgan relationships have been measured in cross-sectional studies. An understanding of the time or age dependency of some of the data may be essential to explain the findings. The model makes it easier to see how ''anomalies'' in the data emerge. As an example we calculated the relationship between liver cadmium and kidney cortex in workers who had been exposed to cadmium for 10 years from age 30 (Figure 7). As mentioned above, when high cadmium exposure ends abruptly, the liver cadmium concentration will immediately start decreasing with a half-time of about 9 years (Section IV.C). The kidney cadmium concentrations will go on increasing for about 5 years and then start to decrease slowly. A shift of cadmium from liver to kidney takes place.

The model can help us analyze the data in such a way that the time relationships discussed in Section IV.C become obvious. If the liver-kidney cadmium relationship is used to establish individual or group kidney cortex cadmium concentrations based on measured liver cadmium concentrations, the location of the relationship curve becomes very important. At the same liver value, the retired workers' curves will give falsely high kidney cortex values for a still-active worker and vice versa (Figure 7).

C. Indexes of Exposure and Retention in Organs

In the way the model was developed, it is evident that in the calculations blood cadmium will be strongly related to recent (months) exposure and urine cadmium will be strongly related to body burden. As discussed in Chapter 6 (Section VI), a part of urine cadmium is related to recent exposure and a part of blood cadmium related to body burden. These parts can be quantified by model calculations and taken into consideration when analyzing empirical data.

Calculated fecal cadmium excretion is another indicator which is of interest. For adults with cadmium intake via food only, the fecal and urinary cadmium excretions are roughly equal. Normally this means excretions of 0.2 to 1 μg/day, which cannot be detected in feces containing 10 to 20 μg/day unabsorbed cadmium from food. Additional cadmium exposure via air will add to fecal cadmium excretion,[1] both in the form of unabsorbed cadmium and as "true" excretion. During the first year, a worker exposed to 50 μg/m^3 would increase his "true" fecal excretion from 0.33 to 13 μg/day and his total fecal output from 18.6 to 227 μg/day (according to the model). The difference (i.e., 196 μg) originates from gastrointestinal translocation of cadmium deposited on the mucociliar mucosa of the respiratory tract in connection with inhalation exposure.

Fecal cadmium can possibly be used as an indicator of recent cadmium dust exposure, but the difficulties in quantifying the relationship (because of variations in particle size, individual swallowing and spitting habits, etc.) make fecal cadmium a very approximate indicator (Chapter 6, Section VI.). Due to the relationship between cadmium in blood and feces, the "true" fecal excretion will be elevated, corresponding to the blood level, even after air exposure ceases. After 10 years of exposure and 1 year of nonexposure, the fecal excretion is calculated to be 4.8 μg/day, which may be possible to detect, particularly if a diet very low in cadmium is eaten for a week before the feces test.

D. Predictions of Unknown Kinetics

One advantage of a multicompartment model is that a number of interorgan relationships, that have not or cannot be studied "in vivo", can be depicted. The calculation result of a continuous increase in calculated kidney cortex concentrations, for more than 5 years after a 10-year cadmium dust exposure period, has important implications for the interpretation of tissue data from workers. This aspect of cadmium accumulation has not been studied and would be very difficult to study even with in vivo neutron activation analysis, which as yet is not accurate enough to detect the small changes taking place in the kidney (Chapter 2, Section V.).

If kidney damage is a function of not only the kidney cortex cadmium level, but also the duration of time over which a certain level is maintained, then prevention of damage could possibly be affected by stimulating the decrease in cadmium levels, without influencing kidney function. It should be remembered that kidney damage that occurs in industry hardly ever occurs at steady state, but at a time when both kidney and liver concentrations are increasing. If it takes some time for damage to occur, this means that dose-effect studies of workers give falsely high estimates for the individual critical concentrations[36] (Volume II, Chapter 9).

An interesting aspect of using blood cadmium as an indicator of recent cadmium intake is the relationship of blood cadmium to age. Due to the higher energy intake per kilogram body weight at young ages, and our assumptions of a correlation between energy intake and cadmium intake via food, the blood cadmium concentrations in a population with stable cadmium concentration in food are likely to reach a peak at age 5 to 6 years. This would be a useful age group to test if high cadmium intake via food were suspected in an area, but no detailed age-specific data on human blood cadmium are available (Chapter 5).

E. Calculations of Dose-Retention Relationships

Probably the most important use of metabolic models is the calculation of selected organ concentrations as a function of exposure level and exposure duration. This enables us to predict the concentrations in an organ after a certain dose has been achieved (dose-retention relationship), and with the use of estimates of critical concentration for different severity of effects, a dose-effect relationship can be established.

One of the more straightforward dose-retention relationships was calculated by Kjellström and Nordberg.[20] This is the relationship between average daily cadmium intake via food (at age 50) and the average cadmium concentration in kidney cortex at age 45 (Figure 8).

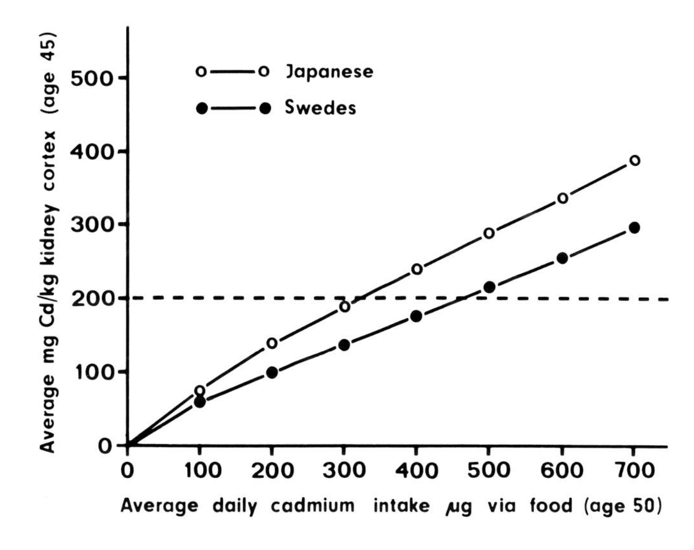

FIGURE 8. Cadmium concentrations in kidney cortex of Swedes and Japanese at age 45 as a function of daily cadmium intake via food. The maximum concentrations are reached at age 45. Daily cadmium intake assumed to vary with age like calorie intake. (From Kjellström, T. and Nordberg, G. F., *Environ. Res.*, 16, 248—269, 1978. With permission.)

Different curves were found for Japanese and Swedes because of different body weights. Note that these are average kidney cortex concentrations. Half of the exposed population will have higher concentrations and half will have lower. Due to the relatively large individual scatter in age-specific cadmium concentrations in tissues[6] and the finding that this coincides with the scatter in individual exposure levels,[17] it becomes essential to take into consideration the frequency distribution of kidney cortex cadmium concentrations around these averages when using them for risk assessment (Volume II, Chapter 13).

VI. SUMMARY AND CONCLUSIONS

Simple one-compartment kinetic models of cadmium metabolism in humans have been used to calculate the half-time of cadmium in renal cortex and the dose-retention relationship. Such calculations are of great value in risk assessments of different types of cadmium exposure, because the half-time is so long that data from, e.g., animal experiments, cannot be directly applied to human exposure situations.

A more elaborate eight-compartment kinetic model has been proposed. This model has been modified somewhat and is presented in detail here. A number of assumptions have to be made regarding certain aspects of the kinetics of cadmium, but the proposed model appears to be able to calculate the cadmium accumulation in several tissues with reasonable accuracy.

The advantage of the more elaborate model is that it can calculate the long-term expected cadmium concentrations in several tissues, including indicator tissues such as blood and urine. This opens a number of possibilities for use of the model to improve our understanding of cadmium kinetics. It also makes possible dose-effect and dose-response calculations based on tissue data as well as on cadmium intake data.

REFERENCES

1. **Adamsson, E., Piscator, M., and Nogawa, K.,** Pulmonary and gastrointestinal exposure to cadmium oxide dust in battery workers, *Environ. Health Perspect.*, 28, 219—222, 1979.
2. **Berlin, M.,** Determination of red blood cell life span, *JAMA*, 188, 375—378, 1964.
3. **Black, D. and Jones, N. F.,** *Renal Disease*, 4th ed., Blackwell Scientific, Oxford, 1979.
4. **Camner, P., Clarkson, T. W., and Nordberg, G. F.,** Routes of exposure, dose and metabolism of metals, in *Handbook on the Toxicology of Metals*, Friberg, L., Nordberg, G. F., and Vouk, V., Eds., Elsevier, Amsterdam, 1979, 355—381.
5. **Darmady, E. M., Offer, J., and Woodhouse, M. A.,** The parameters of the ageing kidney, *J. Pathol.*, 109, 195—207, 1973.
6. **Elinder, C.-G., Kjellström, T., Friberg, L., Lind, B., and Linnman, L.,** Cadmium in kidney cortex, liver, and pancreas from Swedish autopsies, *Arch. Environ. Health*, 31, 292—302, 1976.
7. **Elinder, C.-G., Kjellström, T., Linnman, L., and Pershagen, G.,** Urinary excretion of cadmium and zinc among persons from Sweden, *Environ. Res.*, 15, 473—484, 1978a.
8. **Elinder, C.-G., Kjellström, T., Lind, B., Molander, M. L., and Silander, T.,** Cadmium concentrations in human liver, blood and bile: comparison with a metabolic model, *Environ. Res.*, 17, 236—241, 1978b.
9. **Ellis, K. J., Morgan, W. D., Zanzi, I., Yasumura, S., Vartsky, D., and Cohn, S. H.,** Critical concentrations of cadmium in human renal cortex: dose-effect studies in cadmium smelter workers, *J. Toxicol. Environ. Health*, 7, 691—703, 1981.
10. **Engström, B. and Nordberg, G. F.,** Dose dependence of gastrointestinal absorption and biological half-time of cadmium in mice, *Toxicology*, 13, 215—222, 1979a.
11. **Engström, B. and Nordberg, G. F.,** Factors influencing absorption and retention of oral ^{109}Cd in mice: age, pretreatment and subsequent treatment with non-radioactive cadmium, *Acta Pharmacol. Toxicol.*, 45, 315—324, 1979b.
12. **Friberg, L.,** Health hazards in the manufacture of alkaline accumulators with special reference to chronic cadmium poisoning (Doctoral thesis), *Acta Med. Scand.*, 138 (Suppl. 240), 1—124, 1950.
13. **Friberg, L., Piscator, M., Nordberg, G. F., and Kjellström, T., Eds.,** *Cadmium in the Environment*, 2nd ed., CRC Press, Boca Raton, Fla., 1974.
14. International Commission on Radiological Protection, Report of the Task Group on Reference Man, Rep. No. 23, Pergamon Press, Oxford, 1975.
15. **Järup, L., Rogenfelt, A., Elinder, C.-G., Nogawa, K., and Kjellström, T.,** Biological half-time of cadmium in blood of workers after cessation of exposure, *Scand. J. Work Environ. Health*, 9, 327—331, 1983.
16. **Kjellström, T.,** A mathematical model for the accumulation of cadmium in human kidney cortex, *Nord. Hyg. Tidskr.*, 53, 111—119, 1971.
17. **Kjellström, T.,** Accumulation and Renal Effects of Cadmium in Man. A Dose-Response Study, Doctoral thesis, Karolinska Institute, Stockholm, 1977.
18. **Kjellström, T.,** Exposure and accumulation of cadmium in populations from Japan, the United States and Sweden, *Environ. Health Perspect.*, 28, 169—197, 1979.
19. **Kjellström, T. and Friberg, L.,** Interpretation of empirically documented body burdens by age of metals with long biological half-times with special reference to past changes in exposure, in *Proc. 17th Int. Congr. Occup. Health*, available through the Secretariat, Av. Rogue Saenz, Pena, 110-2 piso-Oficio 8, Buenos Aires, Argentina, 1972.
20. **Kjellström, T. and Nordberg, G. F.,** A kinetic model of cadmium metabolism in the human being, *Environ. Res.*, 16, 248—269, 1978.
21. **Kjellström, T., Friberg, L., Nordberg, G., and Piscator, M.,** Further consideration on uptake and retention of cadmium in human kidney cortex, in *Cadmium in the Environment*, Friberg, L., Piscator, M., and Nordberg, G., Eds., CRC Press, Boca Raton, Fla., 1971, 140—155.
22. **Kjellström, T., Lind, B., Linnman, L., and Elinder, C.-G.,** Variation of cadmium concentration in Swedish wheat and barley. An indicator of changes in daily cadmium intake during the 20th century, *Arch. Environ. Health*, 30, 321—328, 1975.
23. **Kjellström, T., Borg, K., and Lind, B.,** Cadmium in feces as an estimator of daily cadmium intake in Sweden, *Environ. Res.*, 15, 242—251, 1978.
24. **Matsubara-Khan, J.,** Compartmental analysis for the evaluation of biological half-lives of cadmium and mercury in mouse organs, *Environ. Res.*, 7, 54—67, 1974.
25. **Matsubara-Khan, J. and Machida, K.,** Cadmium accumulation in mouse organs during the sequential injections of cadmium-109, *Environ. Res.*, 10, 29—38, 1975.
26. **Miller, J. H., McDonald, R. K., and Shock, N. W.,** Age changes in the maximum rate of renal tubular reabsorption of glucose in males, *J. Gerontol.*, 7, 196—200, 1952.
27. **Nomiyama, K. and Nomiyama, H.,** Biological half-time of cadmium in rabbits (in Japanese), in *Kankyo Hoken Rep. No. 38*, Japan Public Health Association, Tokyo, 1976, 156—158.

28. **Nomiyama, K., Nomiyama, H., Yotoriyama, M., and Taguchi, T.,** Some recent studies on the renal effect of cadmium, in *Cadmium 77, Proc. 1st Int. Cadmium Conf., San Francisco, 1977*, Metal Bulletin Ltd., London, 1978, 186—194.
29. **Nordberg, G. F.,** Cadmium metabolism and toxicity. Experimental studies on mice with special reference to the use of biological materials as indices of retention and the possible role of metallothionein in transport and detoxification of cadmium, *Environ. Physiol. Biochem.*, 2, 7—36, 1972a.
30. **Nordberg, G. F.,** Models used for calculation of accumulation of toxic metals, in *Proc. 17th Int. Congr. Occup. Health 1972*, available through the Secretariat, Av. Rogue Saenz, Pena, 110-2 piso, Oficio 8, Buenos Aires, 1972b.
31. **Nordberg, G. F. and Kjellström, T.,** A metabolic model for cadmium in man, *Environ. Health Perspect.*, 28, 211—217, 1979.
32. **Nordberg, G. F. and Strangert, P.,** Estimations of a dose-response curve for long-term exposure to methylmercuric compounds in human beings taking into account variability of critical organ concentration and biological half-time: a preliminary communication, in *Effects and Dose-Response Relationships of Toxic Metals*, Nordberg, G. F., Ed., Elsevier, Amsterdam, 1976, 273—282.
33. **Nordberg, G. F., Piscator, M., and Nordberg, M.,** On the distribution of cadmium in blood, *Acta Pharmacol. Toxicol.*, 30, 289—295, 1971.
34. **Nordberg, M. and Nordberg, G. F.,** Distribution of metallothionein-bound cadmium and cadmium chloride in mice: preliminary studies, *Environ. Health Perspect.*, 12, 103—108, 1975.
35. **Rahola, T., Aaran, R. K., and Miettinen, J. K.,** Half-time studies of mercury and cadmium by whole-body counting, in *Assessment of Radioactive Contamination in Man*, IAEA-SM-150/13, International Atomic Energy Agency, Unipublisher, New York, 1972, 553—562.
36. **Roels, H. A., Lauwerys, R. R., Buchet, J. P., Bernard, A., Chettle, D. R., Harvey, T. C., and Al-Haddad, I. K.,** *In vivo* measurement of liver and kidney cadmium in workers exposed to this metal: its significance with respect to cadmium in blood and urine, *Environ. Res.*, 26, 217—240, 1981.
37. **Schroeder, H. A., and Balassa, J. J.,** Abnormal trace metals in man: cadmium, *J. Chronic Dis.*, 14, 236—258, 1961.
38. **Shaikh, Z. A. and Smith, J. C.,** Metabolism of orally ingested cadmium in humans, in *Mechanism of Toxicity and Hazard Evaluation*, Holmstedt, B., Lauwerys, R., Mercier, M., and Roberfroid, M., Eds., Elsevier, Amsterdam, 1980, 569—574.
39. **Shank, K. E. and Vetter, R. J.,** Model description of cadmium in transport in a mammalian system, in *Cadmium in the Environment, Part 2, Health Effects*, Nriagu, J. O., Ed., John Wiley & Sons, New York, 1981, 583—590.
40. **Shank, K. E., Vetter, R. J., and Ziemer, P. L.,** A mathematical model of cadmium transport in a biological system, *Environ. Res.*, 13, 209—214, 1977.

40a. **Spang,** personal communication.

41. **Suzuki, S., Taguchi, T., and Yokohashi, G.,** Dietary factors influencing upon the retention rate of orally administered $^{115m}CdCl_2$ in mice with special reference to calcium and protein concentrations in diet, *Ind. Health*, 7, 155—162, 1969.
42. **Taguchi, T. and Suzuki, S.,** Influence of sex and age on the biological half-life of cadmium mice, *J. Toxicol. Environ. Health*, 87, 239—249, 1981.
43. Task Group on Lung Dynamics, Deposition and retention models for internal dosimetry of the human respiratory tract, *Health Phys.*, 12, 173—208, 1966.
44. Task Group on Metal Toxicity, in *Effects and Dose-Response Relationships of Toxic Metals*, Nordberg, G. F., Ed., Elsevier, Amsterdam, 1976.
45. Task Group on Metal Accumulation, Accumulation of toxic metals with special reference to their absorption, excretion and biological half-times, *Environ. Physiol. Biochem.*, 3, 65—107, 1973.
46. **Travis, C. C. and Haddock, A. G.,** Interpretation of the observed age-dependency of cadmium body burden in man, *Environ. Res.*, 22, 46—60, 1980.
47. **Tsuchiya, K. and Sugita, M.,** A mathematical model for deriving the biological half-life of a chemical, *Nord. Hyg. Tidskr.*, 53, 105—110, 1971.
48. **Tsuchiya, K., Seki, Y., and Sugita, M.,** Organ and tissue cadmium concentration of cadavers from accidental deaths, in *Proc. 17th Int. Congr. Occup. Health*, available through the Secretariat, Av. Rogue Saenz, Pena, 1100-2 piso, Oficio 8, Buenos Aires, 1972.

INDEX

A

B

C

D

E

F

G

H

I

J

K

L

M

N

O

P

Q

R

T

U

V

W

X

Y

Z